FUTURE MECHATRONICS AND AUTOMATION

Studies in Materials Science and Mechanical Engineering

eISSN: 2333-6560

Volume 1

PROCEEDINGS OF THE 2014 IMSS INTERNATIONAL CONFERENCE ON FUTURE MECHATRONICS AND AUTOMATION (ICMA 2014), BEIJING, 7–8 JULY 2014

Future Mechatronics and Automation

Editor

Guohui Yang

International Materials Science Society, Hong Kong, Kowloon, Hong Kong

CRC Press
Taylor & Francis Group
Boca Raton London New York Leiden

CRC Press is an imprint of the
Taylor & Francis Group, an **informa** business

A BALKEMA BOOK

Published by:
CRC Press/Balkema
Schipholweg 107C, 2316 XC Leiden, The Netherlands

CRC Press/Balkema is an imprint of the Taylor & Francis Group, an informa business

Visit the Taylor & Francis Web site at
http://www.taylorandfrancis.com

and the CRC Press Web site at
http://www.crcpress.com

Typeset by MPS Limited, Chennai, India

ISBN: 978-1-138-02648-3 (Hardback)
ISBN: 978-1-315-76218-0 (Ebook PDF)

Future Mechatronics and Automation – Yang (Ed.)
© 2015 Taylor & Francis Group, London, ISBN 978-1-138-02648-3

Table of contents

Future Mechatronics and Automation – Yang (Ed.)
© 2015 Taylor & Francis Group, London, ISBN 978-1-138-02648-3

Preface

2014 IMSS International Conference on Future Mechatronics and Automation (ICMA 2014) was held on July 7–8, 2014 in Beijing, China. The conference was an international forum for the presentation of technological advances and research results in the fields of Intelligent Robotics, Mechatronics, and Mechanical Engineering. The conference brought together leading researchers, engineers and scientists in the domains of interest from around the world. We warmly welcomed previous and prospective authors to submit their new research papers to ICMA 2014, and share valuable experiences with scientists and scholars from around the world.

In the past twenty years, Intelligent Robotics, Mechatronics, and Mechanical Engineering have become involved in many varied applications throughout the world, with multiple products and rapid market services. They have has not only provided industries with new methods, new tools and new products, but also changed the manner, philosophy and working environments of people in the manufacturing field.

The ICMA 2014 program consisted of invited sessions, technical workshops and discussions with eminent speakers covering a wide range of topics. This rich program provided all attendees with the opportunity to meet and interact with one another.

All the papers in the conference proceedings have undergone an intensive review process performed by the international technical committee, and only accepted papers are included. This volume comprises the selected papers from the subject areas of Intelligent Robotics, Mechatronics, and Mechanical Engineering.

We hope that the contents of this volume will prove useful for researchers and practitioners in developing and applying new theories and technologies in Intelligent Robotics, Mechatronics, and Mechanical Engineering.

Finally we would like to acknowledge and give special appreciation to our keynote speakers for their valuable contributions, our delegates for being with us and sharing their experiences, and our invitees for participating in ICMA 2014. We would also like to extend our appreciation to the Steering Committee and the International Conference Committee for the devotion of their precious time, advice and hard work to prepare for this conference.

Future Mechatronics and Automation – Yang (Ed.)
© 2015 Taylor & Francis Group, London, ISBN 978-1-138-02648-3

Organizing Committee

HONORARY CHAIR

Tianharry Chang, *IEEE SYS Brunei Darussalam Chapter Past Chair, Brunei Darussalam*

GENERAL CHAIRS

Enke Wang, *International Materials Science Society, Hong Kong*
Mark Zhou, *Hong Kong Education Society, Hong Kong*

PUBLICATION CHAIR

Guohui Yang, *International Materials Science Society, Hong Kong*

ORGANIZING CHAIRS

Khine Soe Thaung, *Society on Social Implications of Technology and Engineering, Maldives*
Tamal Dasas, *Society on Social Implications of Technology and Engineering, Maldives*

PROGRAM CHAIR

Barry Tan, *Wuhan University, China*

INTERNATIONAL COMMITTEE

S. Sugiyama, *Ritsumeikan University, Japan*
Lijing Dong, *North China Electric Power University, China*
Hong Dong, *Naval University of Engineering, China*
J.M. Liu, *Forestry College of Beihua University, China*
Wangyang Yu, *Jilin University, China*
Duo Qi, *Air Force Engineering University, China*
Tiantian Liu, *Cloud Computing Center, Chinese Academy of Sciences, China*
Yi Jiang, *Wuhan Polytechnic University, China*
Binghua Jiang, *China Three Gorges University, China*
Yongjun Wang, *Guilin University of Aerospace Technology, China*
Zhengqing Chu, *Anhui Xinhua University, China*
Yanming Zhou, *Lushan College of Guangxi University of Science and Technology, China*
Hua Guo, *Shandong University of Science and Technology, China*
Xianqian Hu, *National University of Defense Technology, China*

Section 1: Mechanical engineering

Future Mechatronics and Automation – Yang (Ed.)
© 2015 Taylor & Francis Group, London, ISBN 978-1-138-02648-3

Analysis of measurement uncertainty for aircraft docking and assembly

YuCheng He, GuoXi Li, BaoZhong Wu & JingZhao Yang
College of Mechatronic Engineering and Automation, National University of Defense Technology, Changsha, China

ABSTRACT: For the difficulty of estimating measurement uncertainty during aircraft docking, the paper proposes an evaluation method for measurement uncertainty based on a three-dimensional model; The author uses VC++ and Visual Basic to do secondary development on Spatial Analyzer which is a measuring software to develop a simulation software for measurement; A procedure was designed for measuring the position and orientation of the aircraft components, and for simulating measurement procedures based on the Monte Carlo Theory. This paper evaluates the results of the simulation, which indicates that the measuring procedure is feasible and can provide guidance for the rapid deployment of measurement and docking for aircraft parts.

Keywords: Monte Carlo Theory, uncertainty, aircraft docking, measurement, position and orientation

1 INTRODUCTION

In the field of aircraft assembly, assembly work used to be completed manually using rigid tooling. It has been transformed into relying on a digital flexible assembly system [1]. An important prerequisite for completing a digital assembly is that the measurement system can accurately obtain the position difference between two separate aircraft parts, The position difference is then fed back to the motion control system. Laser Tracker is widely used in the field of aircraft measurement and assembly. Take "FARO Laser Tracker X" for example: its measuring range can reach seventy metres and its accuracy can reach 0.001. However, a good measuring result includes not only the measuring value, but also includes its confidence intervals [2]. After considering uncertainty, the result is still able to meet the precision of aircraft docking and such a measurement result is credible. Currently, there are four main methods of evaluating measurement uncertainty: a statistical method, an analytical method, an expert empirical method and a computer simulation method [3]. In statistical methods, the workers repeat measuring the workpiece many times and this method can provide reliable assessment. However, aircraft production is too complex and large and the measuring environment too complicated for the statistical method to be suitable. The analytical method needs to solve the sensitivity coefficient from error sources to results, synthesizing the impacts of error sources. In the process of aircraft measurement, error sources are numerous and their transitive relationships are complex. Therefore, the analytical method is not appropriate. Expert empirical method relies on the experience of empirical staff excessively, its standardization is so low that it is not suitable for universal application.

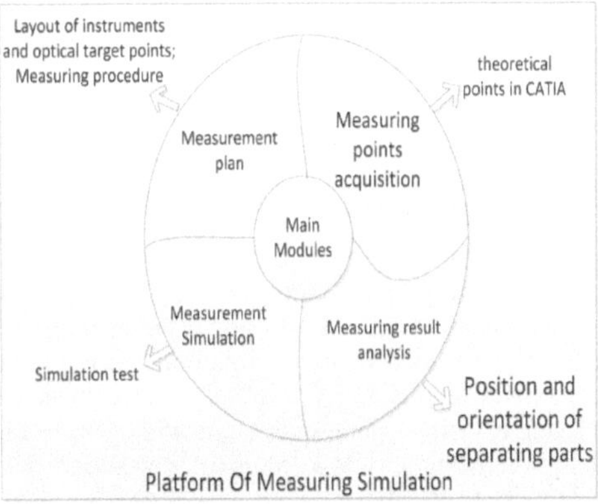

Figure 1. Main modules of measuring simulation platform.

The main idea of the computer simulation method is that a constructing measurement system model, based on the transitive relationship between error sources and the measurement results, reproduces the important sources of error in the measurement system's model. Finally, it is necessary to calculate the uncertainty through simulation.

In the process of measuring aircrafts' separate components there are many factors affecting measurement accuracy, these factors include not only errors of the laser tracker itself, but also include the temperature of the workshop, vibration, and deformation of the workpiece. Since errors are distributed randomly, the preferred choice of measuring an aircraft's position and orientation is the computer simulation method.

Based on the aforementioned information, the author developes a computer simulation measuring platform and its main modules are shown in Figure 1, At the same time, the author has designed a

Figure 2. The model of aircraft docking and assembly.

measurement plan for an aircraft docking model. This model is shown in Figure 2 and the plan is tested with Monte Carlo method, which indicate the uncertainty of the results which highlight the significance of the measurement results.

2 ANALYSIS OF MEASURING UNCERTAINTY

2.1 *Collecting measuring points*

Before the simulation of measuring aircraft components' position and orientation, it is necessary to obtain the theoretical coordinate of optical target points. If each point of coordination is obtained through the basic operation of CATIA, a lot of time will be spent and what is worse, operational faults will occur easily. Therefore, the author did a secondary development on CATIA based on VBA.

Through these commands:

```
ReDim InputObjectType(5)
InputObjectType(0) = "Face,Point"
```

We can set the feature being selected be face or point and so on. Through these commands:

```
sel.SelectElement2(InputType, "User Select", True)
sel.Item(1).GetCoordinates Coo
```

We can obtain coordinates of mouse click [4]. Then save the points' coordinates data to a text file by the VB program. The operations in CATIA will be simplified greatly if we depend on these methods.

2.2 *Measuring process*

The point groups extracted from CATIA were loaded into spatial analyzer: a measuring software. The point groups consist of ERS points, theoretical ERS points, initial-position points, common points and target-position points.

2.2.1 *ERS points*

The function of ERS (Enhanced Reference Points) is to establish an assembly benchmark. The first choice of ERS points are terrestrial reference points or fixed points in an assembly plant. During the measuring process, the first step is measuring the "ERS points" and obtaining their coordinates. Secondly, acquire the coordinates of corresponding "theoretical ERS points" in CATIA. Thirdly, fit "theoretical ERS points" to "ERS points" and obtain the transform matrix. Through these three steps, the relationship between actual environment and virtual environment is constructed.

2.2.2 *Common points*

An aircraft's shape is complex and large, all the measuring points cannot be measured by a single instrument. So two or more instrument are needed to accomplish the measuring task. However, the coordinate system of each instrument is works independently. In order to unify the laser trackers in the same network, it is necessary to measure the common points, constructing a USMN (Unified Spatial Metrology Network).

2.2.3 *Initial-position points and target-position points*

"The initial-position points" refer to the measuring points on the moving part before docking. "The target-position points" refers to the same points, but the difference is that separating parts have been docked precisely.

Figure 3 shows the measuring procedure.

2.3 *Simulation*

According to the Monte Carlo Theory, when simulation times tends to infinity, the uncertainty is closest to real value, but SA basic operations support only one time simulation. Therefore, the author did a secondary development work on SA. Based on the interface provided by SA, every step of the SA operations was converted into C++ code. The loop was exerted to realize simulate for 10,000 times or more. After simulation, the expectation or standard deviation was calculated, together with other relevant mathematical characteristics. Before simulation in the SA, optical interference checking is essential to guarantee that the optical target point would not be blocked. Thereafter, the points which could be measured by a laser tracker were stored and these remaining points were loaded into the SA, ignoring the CAD model. This saves a lot of running time, especially if the simulation frequency rises to 10,000.

2.4 *Analysis of measuring result*

Simulate the measuring procedure 60,000 times, collect the data (including three moving parameters and three rotating parameters) generated by the simulation,

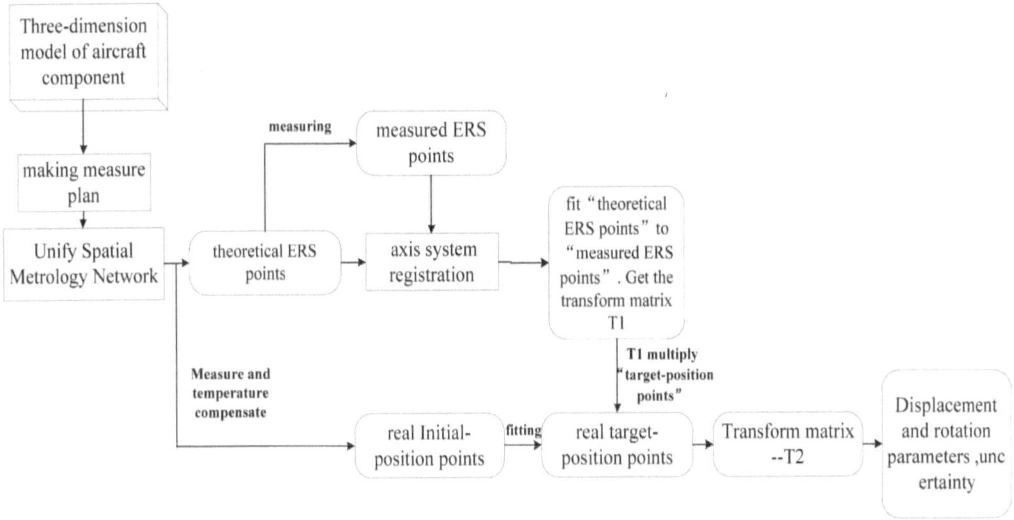

Figure 3. Measuring Procedure.

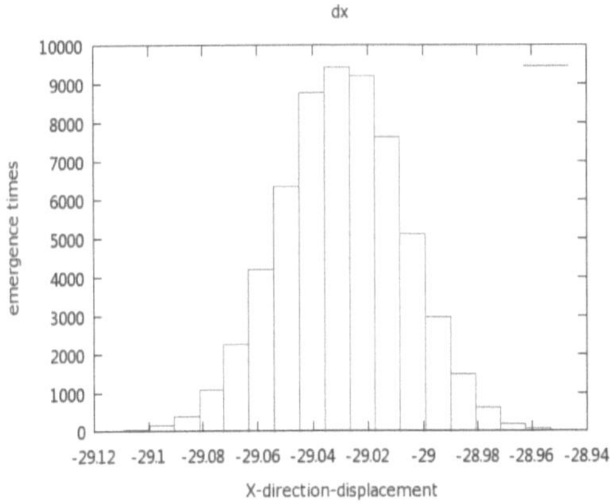

Figure 4. Distribution of moving parameter in the X direction.

and draw a histogram. Figure 4 shows the distribution of a moving parameter in the X direction and obviously it is an approximate normal distribution, calculating the averages and the standard deviations of these six parameters. Table 1 shows the results of position and orientation.

According to the definition of standard deviation, about 99.7% of the data falls in the scope which deviates from the mean within three standard deviations. For instance, the moving value in X-direction is likely to be located in [−29.069, −28.962], and its possibility will be 99.7%. In order to verify the credibility of the uncertainty. Figure 5 was drawn to analyse its convergence. If the evaluation result is not stable and the wave fluctuates significantly, the sample size can be increased and the simulation test can be repeated. After

Table 1. Measuring results when simulating 60,000 times.

	Adjustment of position and orientation	Uncertainty (1σ)	Uncertainty (3σ)
Dx (mm)	−29.029	0.022	0.067
Dy (mm)	−200.537	0.005	0.014
Dz (mm)	−0.226	0.068	0.205
Rx (′)	−285.103	0.038	0.113
Ry (′)	−314.188	0.013	0.038
Rz (′)	−153.849	0.013	0.040

these steps, if uncertainty still remains and the trend of convergence does not emerge, the measurement procedure or algorithm is examined as necessary. Figure 5 shows that the uncertainty tends to be stable when the simulation times rise to 30,000 and therefore, the evaluation result is credible.

3 CONCLUSION

There are many influential factors of uncertainty during the aircraft measurement process, so it is difficult to analyse the uncertainty. The problem is that, an aircraft is so large, complex, and expensive that repeating measurements and statistical methods are not feasible.

Evaluation techniques of aircraft measurement are very rare in domestic aviation. This paper presents a method of evaluating measurement uncertainty based on a 3D model and simulation test. The author studied and simulated aircraft measurement and ultimately provided a measured result and uncertainty, filling a

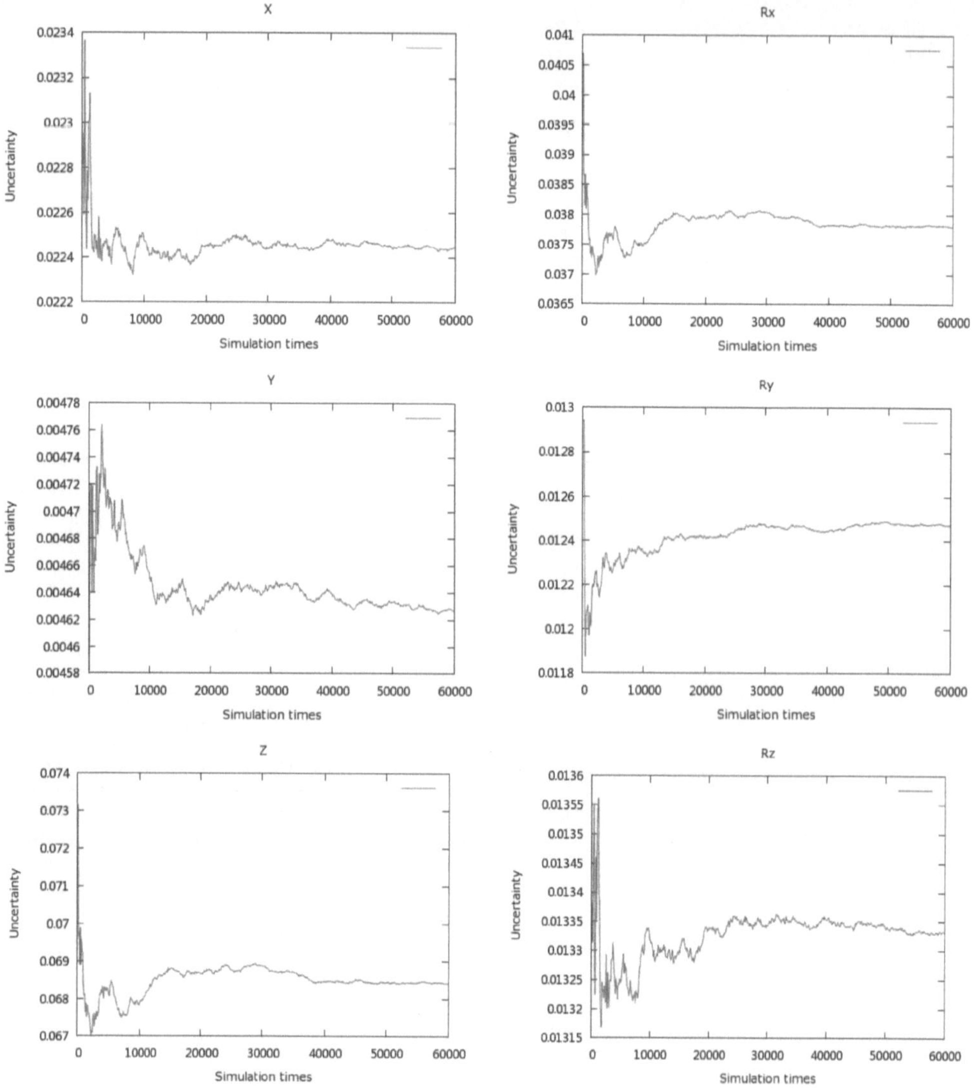

Figure 5. Convergence tendency of uncertainty.

blank in evaluating aircraft measurement. The method can be extended to measuring cars, ships, and other large size objects. Future research should focus on how to arrange instruments, how to distribute optical target points properly, and how the quantity of optical target points impact on the resultant measurement.

REFERENCES

[1] HeShengqiang. Digital assembly technology system of aircraft [J]. Aeronautical Manufacturing Technology 2010(23): 32–37.

[2] WangJie. Research on laser tracker measuring technology based on uncertainty [D]. National University of Defense Technology. Changsha. 2011.

[3] ShiZhaoyao, ZhangYu. Uncertainty evaluation of measuring gear contour using three coordinate measuring machine [J]. Precision Engineering, 2012(4):23–26.

[4] Hu Ting, Wu Lijun. Basic secondary development for CATIA [M]. Beijing: Press of electronics industry, 2006.

Future Mechatronics and Automation – Yang (Ed.)

The application of the China EDPF-NT + DCS in a power plant for an FGD project

LiJing Dong, WeiPing Liang & YuPing Wang
Control and Computer Engineering, North China Electric Power University, NCEPU, Baoding, China

ABSTRACT: FGD projects have been expanded rapidly with environmental protection issued by the government and domestic DCSs have been widely used in the major desulfurization projects. With the domestic desulfurization project being a booming development, the levels of domestic DCSs are maturing and the monopoly of the foreign DCS market has been already broken. This document describes the application of the GuoDianZhiShen Company's EDPF-NT + Distributed Control System (DCS) based on a power plant Flue Gas Desulfurization (FGD) project.

Keywords: DCS, desulfurization technology, EDPF-NT +

1 INTRODUCTION

The 2 × 300 MW Thermal Power Co., a direct air-cooling heating unit, has simultaneously constructed two flue gas desulfurization units, configured with one furnace and one tower. Flue gas desulfurization uses an ammonia method process, from the Jiangsu Province Chemical Industry Research Institute Co., Ltd. Design and Management. Unit #1 has been in intermittent operation since the end of 2010 but, after 168 trial runs over 153 days, the ammonia desulfurization device has not run properly. After frequent failures, desulfurization efficiency standards are unable to meet the environmental requirements and the unit cannot run continuously. At the same time, as the coal market supply is unreliable, the actual burning of coal conditions and design coal there is a certain bias. According to the actual operation of the power plant, desulfurization devices are currently operating. Considering the future uncertainty in the coal market, the existing two ammonia transformation desulfurization devices were converted to a limestone-gypsum wet flue gas desulfurization process.

The project was conducted in September 2013 by the electrical system, enter debug. Each subsystem debugging was completed on September 20, 2013. On 24 September 2013, the trial operation was set to start. FGD systems were introduced at 18:00 hrs on 7 October 2013 beginning with a 168 hour test run. At 18:00 on 14 October 2013, the 168 hour test run was successfully concluded. All the performance parameters met the design requirements.

System Configuration, The DCS configuration in the power plant flue gas desulfurization system consists of one engineer station, one operator station which has also been used as a previous station, and one SIS interface station. The system's configuration is shown in Figure 1. The project uses the EDPF-NT + 1.3 version of the system hardware and software developed by the Beijing-based Guo Dian Zhi Shen Company. The control system on this platform was designed to achieve centralized monitoring and distributed control of the 2 × 300 MW FGD project.

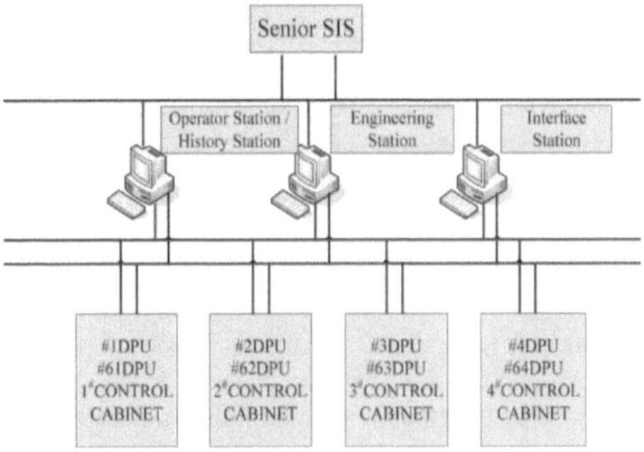

Figure 1. FGD DCS configuration.

After completing the power installation of the Industrial Process Computer (IPC), VISO2003, ADOBE READER, and other essential software was first installed. Then the network card was configured and installed into the system's integration configuration, but the choices made by the A, B card are random and in order to make the card number correspond to the A, B network, the card needs to be checked for correct configuration, in principle, to define the left side of the card for the A network and the right side for the B network [1]. Finally, the purpose of the system's configuration is to ensure the stable operation of the XP system and to ensure a minimization of the system's resources in order to maintain a long-term operation. The main operations include the following aspects: to redefine the IPC in the computer name of the computer system's properties (e.g. eng191), to remove the system's resources from the project, to cancel system restore, to disable system updates, and to turn off the auto-play U disk and the removable hard disk functions.

Upon completion of the industrial control configuration, the software installation, including the installation of the engineer station and the operator station running EDPF-NT + files on the selected engineer and operator site installation, the installation prompts a check on the corresponding components in order to complete the installation. It is worth noting that, on the engineering station, in order to save the configuration logic, is important that all source files and converted files out of the SAMA and the entire logic configuration files are downloaded to the server of each DPU throughout the project. At the same time, the DPU in line parameter can also be a project engineer uploaded to the server station. Therefore, in order to configure and to have easy maintenance, under normal circumstances, engineers and the engineering station should coexist on the same server computer.

After the installation is completed, the software installation DPU and the EDPF-NT + system software should be installed. The configuration software comes with DPU: DIS. Before configuring, the PC network connection must first be tested. The DPU initial IP DPU module is 172.101.1.254. The IP address is different from the IP network configuration rules NT + system, and, therefore, segment PC IP (e.g. 172.101.1.1) needs to be added to the same network. After editing the initial ping IP, normal communications can be observed in the test packets and are not lost. Instead, the network status must be checked. After testing the connection, the DPU is automatically installed by the DIS software.

2 CONTROL DESIGN

2.1 *General design*

The desulfurization DCS should meet the following requirements:

- The desulfurization control using distributed control systems, respectively, is used as an auxiliary network subsystem integrated into the existing systems.
- The Data Acquisition Control System (DAS), analog control (MCS), sequence control (SCS), and other functions to meet a variety of operating conditions of the desulfurization system requirements [2]. A DAS system MTBF of not less than 8600 hours, an average of SCS, and a MCS system MTBF of not less than 24,000 hours.
- The DCS should be easy to configure, easy to use, and easy to expand [3]. The monitoring, the alarm and diagnostic functions of the system are highly concentrated on the LCD display and can be printed on the printer.
- The design of the DCS should have appropriate redundancy configuration and a diagnosis module level since the diagnostic function is highly reliable. A failure in any component parts of the system should not affect any work within the system.
- The DCS can communicate a failure or any running fault in the operational stations and in the LCD, in order to ensure the safety of the desulfurization system's outage.

2.2 *Requirements for working conditions*

Instrumentation and control systems should be designed to meet the following conditions:

- The desulfurization system in the boiler is 30%~100% BMCR in normal operation conditions.
- When the desulfurization system output with the boiler load and flue gas volume changes, it must ensure that the desulfurization efficiency is greater than 95%.
- A rapid removal of flue gas desulfurization in system C does not affect the normal operation of the boiler.
- When MFT is triggered, the system should avoid further deterioration of the boiler's parameters. The desulfurization FGD-DCS and the power plant MIS, together with the SIS systems' networked communications, can monitor the status of the operation, through the communication interface, which is implemented on the MIS and SIS main plant desulfurization equipment to monitor the operational status.

2.3 *Control system reliability measures*

This process system includes the flue gas system and the absorber system which belong to the unit system, the first level gypsum dewatering system, the secondary level gypsum dewatering system, the discharge systems, the lime grinding system, the desulfurization waste water treatment system, and the process water systems which belong to the desulfurization common system [2]. Also included are a 6 kV electrical power system, a 380 V power system, desulfurization transformers, security power systems, and DC systems, while a UPS adds a desulfurization island monitoring system. There is a total of 1909 points in the process; an I/O list and cards layout (e.g. #1DPU) is as follows in Table 1 & 2.

To ensure the reliability of the control system the following spare headroom for the future expansion of the system is needed:

- Within each type of each cabinet the I/O channel has at least 10% spare capacity, including a hardwired alternate point of contact points; the remaining points of the I/O assignment and control systems are produced inside.

Table 1. I/O list.

Signal type	#1FGD	#2FGD	Public	Slurry Preparation
AI (4~20 mA)	26	26	40	16
RTD	66	66	0	24
AO (4~20 mA)	2	2	6	3
DO	146	146	171	164
DI	213	213	301	234
PI	6	6	8	4
SOE	6	6	6	2
Total	465	465	53	447

Table 2. The layout of # 1 DPU cards.

Front					
A	Module	Base	B	Module	Base
A1(01H)	AO8	DZ-32	B1(07H)	AI16(mA)	DZ-32
A2(02H)	DIO32	DZ-DIO	B2(08H)	AI16(mA)	DZ-32
A3(03H)	DIO32	DZ-DIO	B3(09H)	PI8	DZ-32
A4(04H)	RTD16	DZ-64	B4(0AH)	RTD16	DZ-64
A5(05H)	RTD16	DZ-64	B5(0BH)	RTD16	DZ-64
A6(06H)	DI32	DZ-64	B6(0CH)	DI32	DZ-64
Back					
C	Module	Base	D	Module	Base
C1(0DH)	DIO32	DZ-DIO	D1(13H)	DIO32	DZ-DIO
C2(0EH)	DIO32	DZ-DIO	D2(14H)	DIO32	DZ-DIO
C3(0FH)	DIO32	DZ-DIO	D3(15H)	DIO32	DZ-DIO
C4(10H)	DIO32	DZ-DIO	D4(16H)	DIO32	DZ-DIO
C5(11H)	RTD16	DZ-64	D5(17H)	DIO32	DZ-DIO
C6(12H)	DI16	DZ-32	D6(18H)		

- When they are most busy, the processing power controller and the operator station processing power must operate at a 60% margin.
- The capacity of the internal memory is not using more than 50% of its capacity and the capacity of the external memory is not using more than 40%.
- The Ethernet communication bus loading rate is not greater than 20%, with a token ring communication bus load of not more than 40%.
- The operator station server allows a maximum capacity of 50,000 tags.
- Relay in case of relay should not only satisfy the corresponding number of DO channel numbers, but also keep some spare location (including relay installed base and terminal block) in order to expand.

2.4 *Control system design*

2.4.1 *HMI screen*

The HMI screen display includes analog data, a digital alarm and status display, a pump and valve operation panel, and sequence control panels etc.

The slurry circulating pump and valve opening current feedback signals are analog. The analog input module AI16 collected a 4~20 mA signal to the DCS by setting the range and conversion factor. The real-time analog value is displayed on the screen and the digital input module DI32 collected a 0, 1 signal to the DCS, by determining the amount of 0.1 discoloration alarm signal settings and start-stop switch status display to slurry circulating pump for example, when the circulation pump is running red, green when the outage, other fault indications yellow; the control panel display, opening the pump or valve operation panel, there are start, stop, suspend, hang and fail to confirm the solution five basic operations command, according to different circumstances add cast/cut interlocking, cast/cut backup and other buttons on the screen operator actions, DCS DO module via remote boot field devices. DCS then collected DI valve position feedback signal to determine whether the site equipment starts/action or whether the valve is open/close.

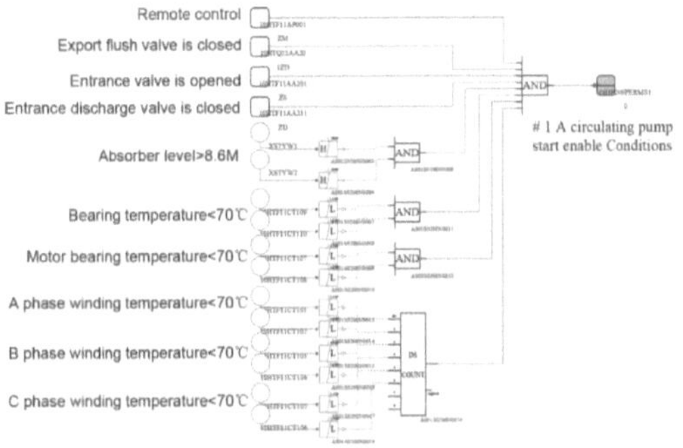

Figure 2. Start enable conditions in SAMA.

2.4.2 *Configuration design*

Slurry circulation pumps, fans, and oxide mills comprise 6 kV important equipment for starting enabling conditions and interlocking protection conditions which become extremely important in the desulfurization system design process. An explanation of #1 absorber slurry circulation pumps serves as a simple example; start enable conditions in SAMA, such as shown in Figure 2.

To facilitate the operation personnel to operate, the SCS sequence control system design of the absorber slurry circulation automatic start/stop step sequence was used, which can achieve automatic tracking step sequence start-up, stop valve, and its related equipment functions. What needs to be stressed is that an experiment of steps to stop was made. This included feedback signals from the traditional state feedback points, changes for the slurry pump current points, (AI), and state feedback (DI) two feedback signals. The state feedback point of failure was simulated and then the sequence control was started, before it was evident that the first improvement of sequence control could not be normal. The improved sequence control is not affected by the bad points of the feedback signal and can be carried out in accordance with the normal step sequence and the effect is obvious.

3 CONCLUSIONS

In this paper, a true power plant desulfurization project was used to explain how to use the EDPF-NT + system to achieve automatic control of the whole desulfurization DCS. Taking the example of the absorber slurry circulation pump, details were shown about how to design and improve the application of DAS and the SCS automation level of the system. It is predicted that the system's design is very efficient and a change from the traditional single signal feedback to the multi-signal feedback system. It improves the safety and reliability of the system as new technologies are added for its use in industrial plants. The DCS will continue to improve and its application will be a big advance in development in the future.

ACKNOWLEDGMENT

I would like to show my deepest gratitude to my advisor, Dr. Liang Weiping, a respectable, responsible, and resourceful scholar, who has provided me with valuable guidance in this writing, and helped me conquer many difficulties in both my study and my life during my graduate study. To my other committee member, Wang Yuping, I am very grateful for her probing questions and validation of the worthiness of my research. Thank you very much for your help and support.

REFERENCES

[1] Beijing Guo Dian Zhi Shen Control Technology Co., Ltd. "Guo Dian Zhi Shen EDPF-NT + system user manual," 2010.

[2] Dong Yuqiang and Bai Yan, "DCS application in power plant desulfurization system," Control and Instruments in Chemical Industry, vol. 39, pp. 1561–1566, 2012.

[3] Xu Peng, "Application of DCS in Desulphuration System of Power Station," Industrial Control Computer 2011, 24(4): 23–24, 26.

Future Mechatronics and Automation – Yang (Ed.)
© 2015 Taylor & Francis Group, London, ISBN 978-1-138-02648-3

Research on the integrated test system of dynamical balance and the correction of optical axis of the coordinator

Peitian Cong & Hui Han
School of Mechanical Engineering, Shenyang Ligong University, Shenyang, China

ABSTRACT: Coordinator is a single pivot frame type gyro rotor with three degrees freedom, and the machining errors of the coordinator rotor will lead to the unbalance vibration and the nutation of the spherical mirror optical axis on the coordinator. In this paper, a micro-acceleration vibration sensor was used to detect the vibration of the coordinator pivot in the test system, and the laser and PSD sensor were adapted to measure the nutation of the coordinator optical axis. Then, the auto-tracing filter was also used to improve signal quality, and the correcting masses of unbalance vibration and the nutation of the spherical mirrors optical axis can be calculated by the microprocessor. After 2–3 times correction, the coordinator will achieve a qualified status.

Keywords: Coordinator; Dynamical balance; nutation

1 INTRODUCTION

Coordinator is an important component to achieve the target tracing for the guided weapon [1]. The object and structure schematic of one kind of coordinator rotor are shown in Fig. 1 and Fig. 2. From Fig. 2, it can be seen that the coordinator is a single pivot frame type three-degree-of-freedom gyro rotor, and D is the supporting point. Point F is the fixed part of the supporting shaft, and C is the spherical mirror which used to collect and reflect the target light. The speed of the coordinator rotor is 6000 r/min in the working process.

For the ideal coordinator rotor, the centrifugal force caused by all of the rotating quality is zero at point D, and the inertia axis of coordinator overlaps with the optical axis of spherical mirror C while rotating at a high speed [2]. However, during the manufacturing process, the machining errors will lead to the unbalance vibration and the nutation of the spherical mirrors optical axis on the coordinator. Therefore, at the position A (see in Fig. 1a), twenty screw holes were designed evenly on the circumference of coordinator to add the screw mass, and at the position B, the circle trapezoidal slots were designed as shown in Fig. 1a, which were used to add the lead wire mass. Accordingly, if applying the proper masses at the position A and B, the vibration of the coordinator and the nutation of the optical axis can be eliminated effectively.

At the present, the mass (m_1) and angle of the adding lead wire on correct surface A were determined by vibration testing of the supporting bench, which can eliminate the force at point D, so all the processes above is called dynamic balance. Observing the nutation of the spherical mirrors by optical method can determine the mass and location of adjusted screw (m_2) on the correct surface B, which can used to

Figure 1. Structure schematic figure.

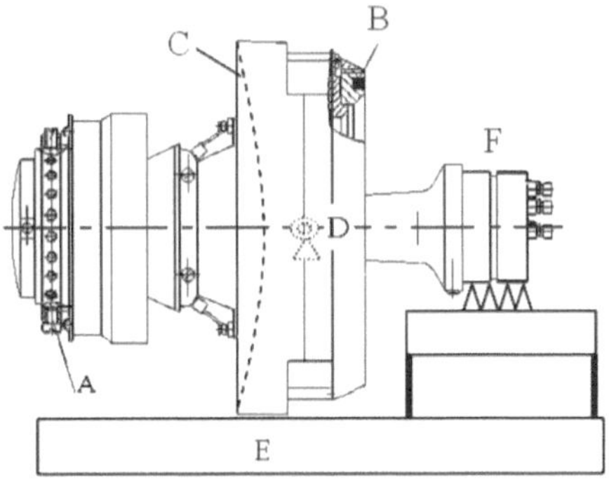

Figure 2. Object figure.

eliminate the nutation of the spherical mirrors [4]. In order to avoid repeated adjustment and reduce the complexity of the operation and the frequency of corrective actions for the missile coordinator dynamical balance and the optical axis nutation in the manufacturing process, an integrated test system of the missile coordinator dynamical balance and the correction for optical axis were studied. In the integrated test system, the two stations, dynamical balance adjustment and rotating adjustment were merged into one station, and the mass and angle of A, B are detected only once. At the same time, the accuracy and efficiency of operations were increased.

2 MECHANICAL MODEL AND ANALYSIS

The coordinator rotor was designed symmetrically, and the material mass deviation produced in the manufacturing process can be equivalent to the unbalance mass m_1 on the correct surface A and m_2 on the correct surface B. When the rotor rotates at angular velocity of ω, the centrifugal forces generated by the two unbalance mass, m_1 and m_2, were expressed as $F_1 = m_1 R_1 \omega_2$ and $F_2 = m_2 R_2 \omega_2$, where R_1, R_2 are the radius of the m_1 and m_2 circumference, respectively. F_1 and F_2 were shifted to the rotor pivot D (the origin of coordinates point O) and can be equivalent to the force F and torque G. Here, F and G are the vectors, and m_1 and m_2 are expressed as vectors $\vec{m_1}$ and $\vec{m_2}$, which can be estimated as followed:

$$F = C_{11}\vec{m}_1 + C_{12}\vec{m}_2 \tag{1}$$

$$G = C_{21}\vec{m}_1 + C_{22}\vec{m}_2 \tag{2}$$

2.1 *Motion analysis of the coordinator centroid*

The total quality of coordinator is M, and the vibration $x(t)$ produced under the action of the vector force F can be expressed as:

$$\frac{d}{dt^2} x(t) = F(t)/M \tag{3}$$

Vibration displacement vector X is expressed as:

$$\vec{X} = -\frac{1}{M\omega^2} F \tag{4}$$

Here, the relationship between the vibration displacement X and force F is linear, and according to the Eq. 1, the linear relationship of m_1, m_2 and F, There is

$$\vec{X} = A_{11}\vec{m}_1 + A_{12}\vec{m}_2 \tag{5}$$

2.2 *Analysis of the rotation of optical axis*

Fig. 3 shows that nutation was generated because the optical axis OO_1 deviated from inertia axis OZ under the equivalent moment G, which were shown as followed:

$$J_X\ddot{\alpha} - H\dot{\beta} = G_Y = G\sin(\omega t + \phi_0) \tag{6}$$

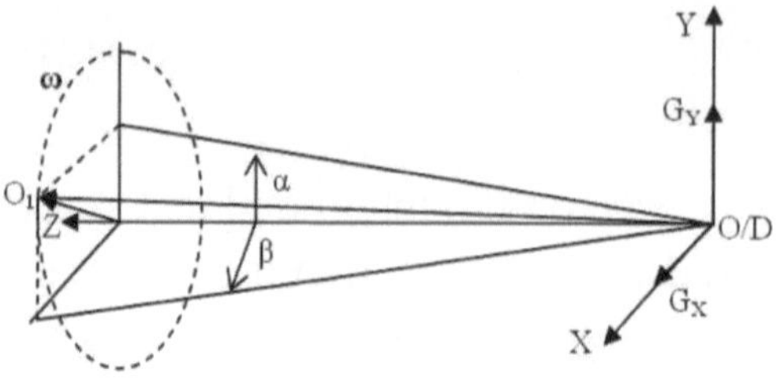

Figure 3. Mechanical model of the optical axis rotation.

$$-J_y\ddot{\beta} + H\dot{\alpha} = G_X = G\cos(\omega t + \phi_0) \tag{7}$$

Here, α and β are the nutation angles of coordinator rotor around the x-axis and y-axis, respectively, J_x and J_y are the inertia moments around the x-axis and y-axis, $J_x = J_y$. ω is the angular velocity of the coordinator rotor itself, and $H = J_z\omega$ is the angular momentum of coordinator rotor. J_z is the inertia moment of the rotor to polar axis (z-axis). α and β were solved as:[3]

$$\alpha = -KG\sin(\omega t + \phi_0) \tag{8}$$

$$\beta = KG\cos(\omega t + \phi_0) \tag{9}$$

Rotation angle α is expressed as vector $\vec{\alpha}$, and the relationship between $\vec{\alpha}$ and G is also linear.

$$\vec{\alpha} = K\vec{G} \tag{10}$$

According to the Eq. 2, $\vec{\alpha}$ and m_1, m_2 are the following linear relationship

$$\vec{\alpha} = A_{21}\vec{m}_1 + A_{22}\vec{m}_2 \tag{11}$$

By detecting the vibration displacement X of coordinator and vector $\vec{\alpha}$ of optical axis rotation, the magnitude and angle of $\vec{m_1}$ and $\vec{m_2}$ can be obtained.

3 INTEGRATED TEST SYSTEM OF THE COORDINATOR DYNAMICAL BALANCE AND THE CORRECTION OF OPTICAL AXIS

The test system is shown in Fig. 4, it can be observed that the components of coordinator were fixed onto the balancing test bench, and the coordinator rotor can rotate at speed of 6000 r/min. Datum signal sensor is a coil with 200 turns and diameter Φ 0.05, which is usually used to detect the position of rotor magnetic pole. After translating a datum signal shaping circuit into pulse signal V_f, its frequency will be the rotor rotation frequency f_0. The vibration sensor is a kind of accelerometer sensor, and the voltage signal V_2 was outputted through the charge amplifier and V_2 is the sinusoidal signal with frequency of f_0. Additionally, V_2 and the vibration displacement X are in linear relationship.

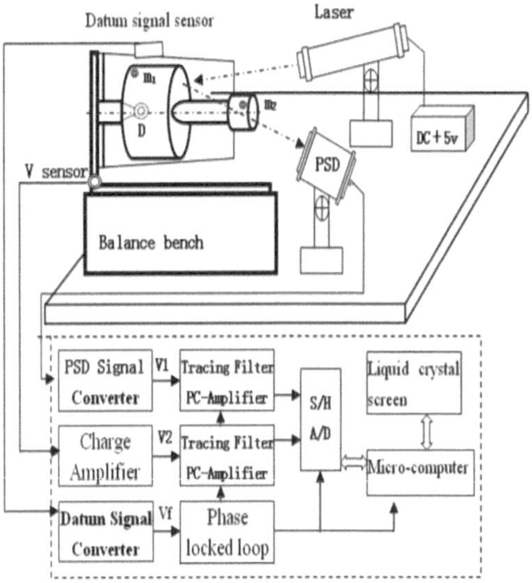

Figure 4. Schematic of test system.

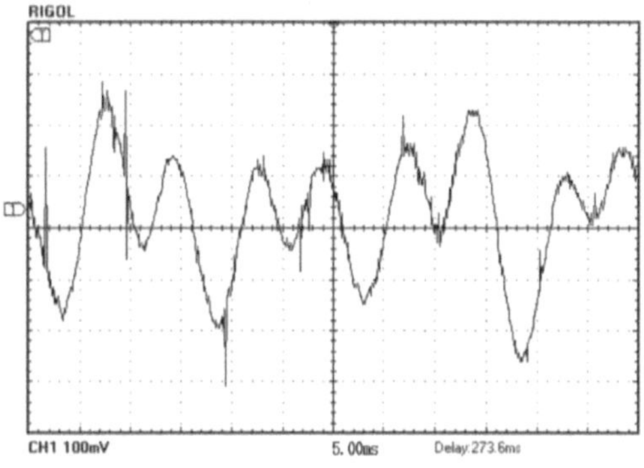

Figure 5. Waveform before filtering.

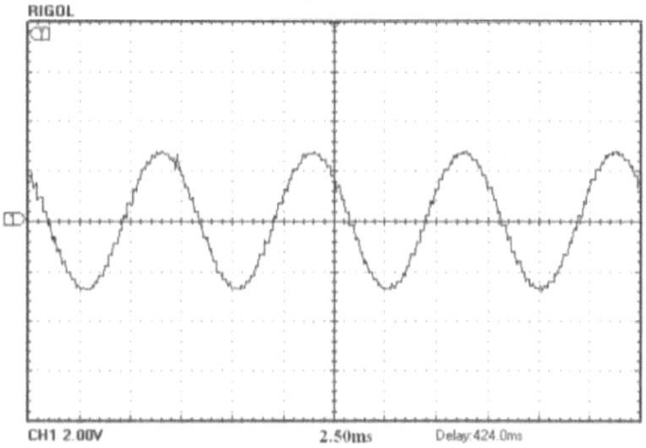

Figure 6. Waveform after filtering.

The laser produces a laser beam with power of 5 mw and wavelength of 650 nm. The laser beam were shoot onto the coordinator spherical mirror, and the reflected beam falls on the PSD detector of 8 × 8 mm and produces a signal V_1 through the PSD signal converter. The moving of reflection spot on the PSD responses the rotation angle displacement $\vec{\alpha}$ of coordinator optical axis. The voltage signal V_1 is sinusoidal signal with its frequency f_0.

Phase locked loop (PLL) frequency multiplier is composed of phase lock loop chip CD4046 and 12-bit binary counter CD4040. Pulse signal with frequency f_0 was input, and square-wave signal with frequency from $256f_0$, $128f_0$, $64f_0$ … to $2f_0$ and f_0 was output. PLL frequency multiplier is also used to control the tracking filter, start A/D and count for speed. Tracking filter is composed of switched-resistance filter and N-path filter (N = 32). Center frequency f_C equal to the rotating frequency of rotor f_0, at the same time, it is used to filter out the unwanted noise and improve the Signal to Noise Ratio (SNR) of V_1 and V_2. The waveforms before and after filtering of the optical axis rotation signal are shown in Fig. 4 and Fig. 5, respectively.

Microcomputer system is the controlling core of the test system, and the vector signals $\vec{V}_1$ and $\vec{V}_2$ were obtained based on the programmable amplifier controls, A/D data acquisition, and calculations of the amplitude and phase angle of V_1 and V_2 by using the related analysis principle. Then the system calibration and the calculations of magnitude and angle of unbalanced amount m_1, m_2 were completed. The color liquid crystal screen displays the input datas and the results of the test system.

Therefore, it can be obtained that $\vec{V}_1$ responses the optical axis rotation $\vec{\alpha}$ and $\vec{V}_2$ responses the vibration displacement X of coordinator, respectively.

According to Eq. 5 and Eq. 10, the formula can be gained,

$$\vec{m}_1 = E_{11}\vec{V}_1 + E_{12}\vec{V}_2 \tag{12}$$

$$\vec{m}_2 = E_{21}\vec{V}_1 + E_{22}\vec{V}_2 \tag{13}$$

Here E_{11}, E_{12}, E_{21} and E_{22} are correlation factors in plural form. The process of obtaining E_{11}, E_{12}, E_{21} and E_{22} is system calibration. Here it is assumed that the initial masses are $\vec{m}_1$ and $\vec{m}_2$, and the vibration and rotation signals of $\vec{V}_1$, $\vec{V}_2$, according to Eq. 12 and Eq. 13, the vibration and rotation signals $\vec{V}_{11}$ and $\vec{V}_{12}$ were measured when the known mass $\vec{m}_{10}$ is added on the zero degree of correction plane 1 where $\vec{m}_1$ existed; While the vibration and rotation signals $\vec{V}_{21}$ and $\vec{V}_{22}$ were measured when the known mass $\vec{m}_{20}$ is added on the zero degree of correction plane 2 where $\vec{m}_2$ existed. The relationship is shown as follows:

$$\vec{m}_1 + \vec{m}_{10} = E_{11}\vec{V}_{11} + E_{12}\vec{V}_{12} \tag{14}$$

$$\vec{m}_2 + 0 = E_{21}\vec{V}_{11} + E_{22}\vec{V}_{12} \tag{15}$$

$$\vec{m}_1 + 0 = E_{11}\vec{V}_{21} + E_{12}\vec{V}_{22} \tag{16}$$

$$\vec{m}_2 + \vec{m}_{20} = E_{21}\vec{V}_{21} + E_{22}\vec{V}_{22} \tag{17}$$

Table 1. Balancing test data of coordinator rotor.

	m_1	Φ_1	m_2	Φ_2
Initial	185.2 mg	23.4°	98.4 mg	165.7°
After NO. 1 cor.	17.4 mg	45.8°	9.2 mg	154.3°
After NO. 2 cor.	6.3 mg	5.2°	4.4 mg	176.8°
After NO. 3 cor.	4.3 mg	32.5°	3.2 mg	155.4°

Calibration parameters E_{11}, E_{12}, E_{21} and E_{22} can be calculated according to Eq. 12 to Eq. 17 and then these parameters will be stored in E^2PROM by the computer. Since then, as long as the vibration and nutation signals were measured, the unbalance masses $\vec{m}_1$ and $\vec{m}_2$ can be calculated according to Eq. 12 and Eq. 13. Thus, adding the masses $\vec{m}_1$ and $\vec{m}_2$ directly, the unbalanced vibration of coordinator and the optical axis can be eliminated easily.

4 EXPERIMENTAL RESULTS AND CONCLUSIONS

Experiment was carried out in our laboratory after the integrated test system of the coordinator dynamical balance and the correction for optical axis was completed. The experimental conditions are those the mass M of coordinator rotor is 445 g, the radius R1 at B-side is 20 mm, the radius at A-side is 11 mm, the calibration test mass m_{10} is 200 mg, the calibration test mass m_{20} is 80 mg, and the experimental speed n is 5800 r/min.

After finishing calibration, the m_1, m_2 were measured four times, and the masses have been corrected three times. Experimental data are in Table 1.

Rotor eccentricity e is expressed as followed according to the final state ($m_1 = 4.3$ mg, $m_2 = 3.2$ mg):

$$e < (m_1 R_1 + m_2 R_2) / M = 0.35 \mu m$$

The final state of rotor was measured using the conventional optical methods, and the nutation angle of optical axis is less than 0.002°. Based on the proper calibration and test, the double-sided unbalance of coordinator rotor were measured accurately at one time. Balancing vibration X of coordinator rotor and the nutation angle α of optical axis were reduced to the required range after adjusting.

CORRESPONDING AUTHOR

Hui Han, 13898199051, zlhanhui@163.com

REFERENCES

Sheping Tian, Dan Hou, Xiaoying Sheng. Digital Method for Measuring Dynamical Balance of the Coordinate Gyrorotor. Chinese Journal of Scientific Instrument, Vol. 24 (2003), pp. 8–9.

Hongquan Zhou, Ping Cai, Xiaorong Chen. Integrated Detection System about Coordinator Gyroscope. Foreign Electronic Measurement Technology, No. 6 (2002), pp. 22–25.

Zhihui Yu. Dynamic Balance of Coordinator Gyro. Aero Weaponry, No. 5 (1994), pp. 24–26.

Yue Chen, Haiqing Chen, Zhaoshu Liao. Research of Automatic Measuring Instrument for Dynamical Balance of a Coordinator Gyrorotor. Modern Electronics Technique, No. 14 (2005), pp. 87–88.

Future Mechatronics and Automation – Yang (Ed.)

Comparison between different Sub-Grid Scale models for the nonreacting flow in a Lean Direct Injection combustor

Hong Dong
College of Power Engineering, Naval University of Engineering, Wuhan, China

Xue-you Wen
The 703 Research Institute of CSIC, Harbin, China

ABSTRACT: Large Eddy Simulation (LES) is a credible and feasible tool for researching a combustor, but appropriate Sub-Grid Scale (SGS) models must be adopted to gain the best performance from an LES. The Smagorinsky-Lilly (S-L) model, the Dynamic Smagorinsky-Lilly (Dynamic S-L) model, the Wall-Modelled LES (WMLES) model, and the Dynamic Kinetic Energy SGS model are used for large eddy simulation of the nonreacting flow in a Lean Direct Injection (LDI) combustor. The LES results were compared with the RANS results and with the experimental data, and they showed that all the LES results were more reasonable than the RANS results. Among these LES results, the WMLES model and the Dynamic Kinetic Energy SGS model showed the best performance. The Dynamic S-L model can also provide reasonable results, while the S-L model is not good enough.

Keywords: Large Eddy Simulation; Sub-Grid Scale model; LDI combustor

1 INTRODUCTION

With the increasing concern for environmental protection, low pollutant emissions are an important component in modern gas turbines[1]. An LDI combustor is a design concept for low pollutant emissions. In the LDI system, the fuel is injected directly into the flame without being premixed or prevaporized and is burnt under fuel-lean conditions for making the lowest possible flame temperature. Therefore, it is important to achieve a fine atomization and mixing of the fuel and air quickly and uniformly. An LDI is a concept that depends heavily on the swirler designs. Researchers at the NASA Glenn Research Center did lots of work on the LDI combustor by experimental and numerical methods[1–3]. Yongqiang Fu et al. (2005) did experimental research on the nonreacting flow of an LDI combustor in detail; Farhad et al. (2006) used a RANS method to simulate the flow field of an LDI combustor; H. El-Asrag et al. (2007) studied the LDI combustor by LES methods with a Dynamic S-L model. These results show that the LESs give a better performance than RANS. However, all these result do not match the experimental data well enough, especially the flow structure which cannot represent in intense turbulent regions downstream, the exit of the convergent-divergent venture. Therefore, further work needs to be done.

Large eddy simulation is a powerful tool for the study of a combustor with reasonable precision and with an acceptable workload. LES methods filter the Navier-Stokes (N-S) equation by introducing a filter. Large eddies are then resolved directly by an N-S equation. Small eddies are modelled and the relationship of small eddies and large eddies is established by SGS models. Therefore, the SGS model is a key factor in the performance of the LES. Different SGS models treat this relationship diversely, and affect the simulative results directly[4].

In this paper, we focus on the LES simulations of nonreacting flows in a single element LDI combustor. Four SGS models are used in the LES calculation. All results will be analysed, compared, and validated with the experiment's data, and the applicability of these SGS models for LDI combustor simulation will then be discussed.

2 NUMERICAL METHODOLOGY

2.1 *Governing equations and SGS models*

For the incompressible flow, after being LES filtered, the continuity equation and the Navier-Stokes equation are given by:

$$\frac{\partial \overline{u_i}}{\partial x_i} = 0 \tag{1}$$

Figure 1. LDI single element geometry

$$\frac{\partial \overline{u_i}}{\partial t}+\frac{\partial \overline{u_i u_j}}{\partial x_j}=-\frac{1}{\rho}\frac{\partial \overline{p}}{\partial x_i}+\nu\frac{\partial^2 \overline{u_i}}{\partial x_j \partial x_j}+\frac{\partial\left(\overline{u_i u_j}-\overline{u_i}\,\overline{u_j}\right)}{\partial x_j} \quad (2)$$

where $\overline{u_i}$ and $\overline{u_j}$ are mean velocity components, ρ is density, $\overline{p}$ is mean pressure. The unclosed source at the right part of Equation 2 is defined as the sub-grid stress tensor:

$$\overline{\tau_{ij}}=\left(\overline{u_i u_j}-\overline{u_i}\,\overline{u_j}\right) \quad (3)$$

$\overline{\tau_{ij}}$ is the sub-grid stress tensor, which is the momentum transport between large-scale turbulences and unresolved small eddies which are filtered out.

The closed models of sub-grid stress tensor are important to realize the LES. Until now, scholars have established many sub-grid scale models to resolve this problem. In the paper, the Smagorinsky- Lily model[5], the Dynamic Smagorinsky-Lily model[6], the Algebraic Wall-Modeled LES model (WMLES)[7], and the Dynamic Kinetic Energy Sub-grid Scale model[8] are used to make an LES of an LDI combustor. The first three models are essentially algebraic models, while last one is a k-equation model.

2.2 *The geometry of the single element combustor and the mesh*

The research object is a single element LDI combustor, which is the same as the combustor in Figures 2 and 3. The single element LDI combustor (Fig. 1) consists of a cylindrical air passage with air swirlers and a converging-diverging venturi section, extending to a confined 50.8-mm square flame tube. The air swirlers have helical, axial vanes with vane angles of 60 degrees. The air is swirled swiftly as it passes through the 60 degree swirlers and enters the flame tube.

The mesh uses 2,243,859 hexahedron elements by using ICEM software, which is twice over the mesh used by H. El-Asrag et al. (2007). The fine mesh is useful for obtaining the most accurate simulated result for the LES method. The unstructured mesh is shown in Figure 2.

2.3 *Boundary condition*

The boundary condition used is the same as in Figure 3. An inflow bulk velocity of 20.14 m/sec is provided through a tube upstream from the swirl injector. For the non-reactive case, the inflow air is at temperature, $T_0 = 294$ K, and at a pressure of 1 atm. All walls are treated as no-slip and adiabatic.

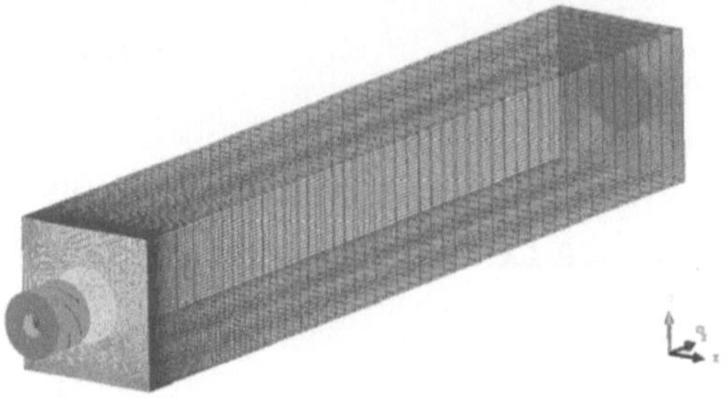

Figure 2. The unstructured mesh of LDI combustor.

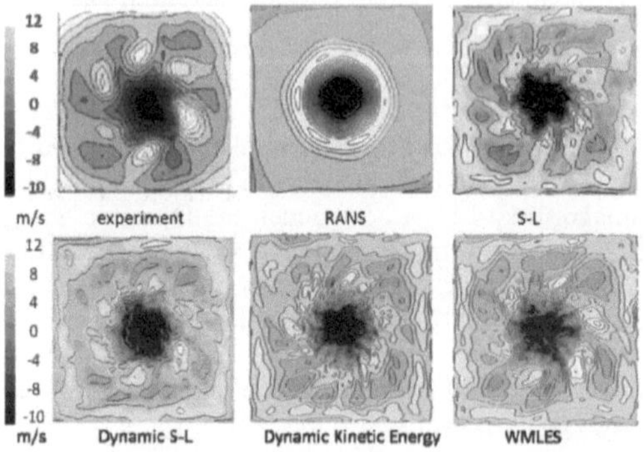

Figure 3. Comparison of the mean axial velocity (*Ux*) iso-contours at X = 3 mm.

2.4 *Calculation method*

The Mach number is blow 0.3; the air is set to be impressible. The pressure-based coupled solver is adopted. The time step is 5e−6 s. All residual sums must decrease to less than 1e−4 in every time step after iterations.

3 RESULT AND ANALYSIS

In this section, four LES results and a RANS result, which is based on a realizable k-ε model, are provided and compared with the experimental data obtained. The comparisons between numerical results and the experimental data are shown in Figures 3 to 11. The experimental data is obtained from Figures 2 and 3.

Figure 3 illustrates the comparison of the mean axial velocity iso-contours of LES, the axial velocity iso-contours of RANS, and the experimental data at the x = 3 mm section. The section is close to the exit of the venture, where turbulent intensity is very high, flow structure is complicated, and both of which will promote the mixing process of air and fuel. The four LES results compare well with the experimental data. Six high-speed pockets are shown. When they are merged together in a RANS simulation, the result is the same as

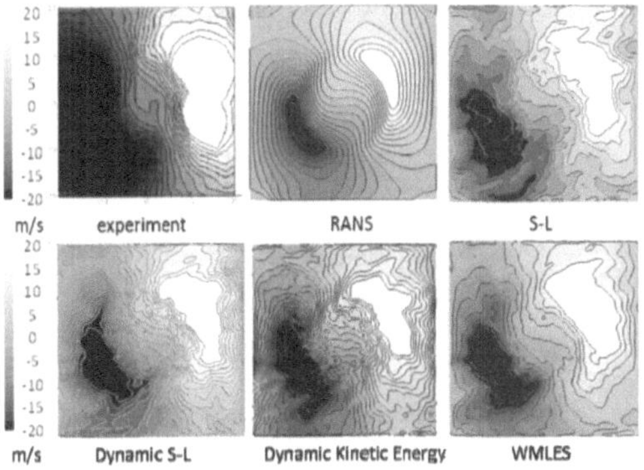

Figure 4. Comparison of the mean radial velocity (Uz) iso-contours at X = 3 mm.

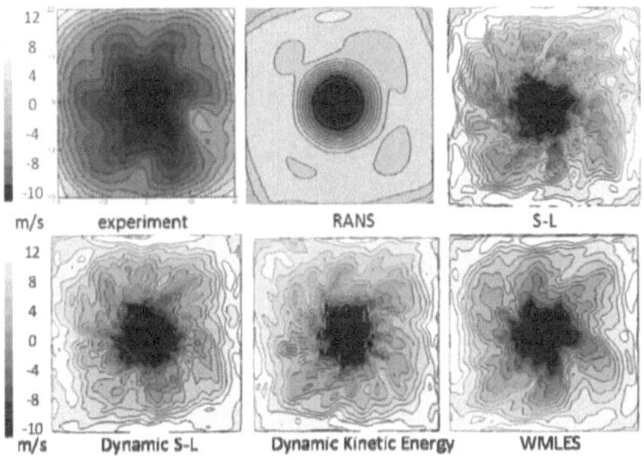

Figure 5. Comparison of the mean axial velocity (Ux) iso-contours at X = 5 mm.

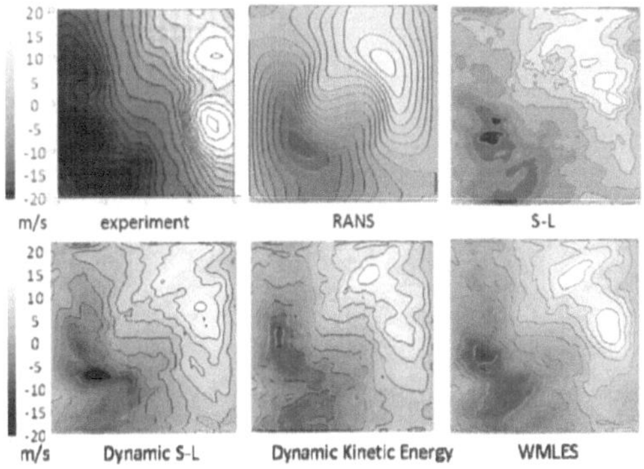

Figure 6. Comparison of the mean radial velocity (Uz) iso-contours at X = 5 mm.

the LES results in Figure 3. The reason maybe that the mesh used by H. El-Asrag et al. was not fine enough.

The performance of the LES depends partly on the mesh. Figure 4 compares the iso-contours of the mean radial velocity at x = 3 mm section. Four LES results also compare well with experimental data.

Figure 5 compares the mean axial velocity iso-contours, and Figure 6 compares the mean radial velocity iso-contours at x = 5 mm section. We find

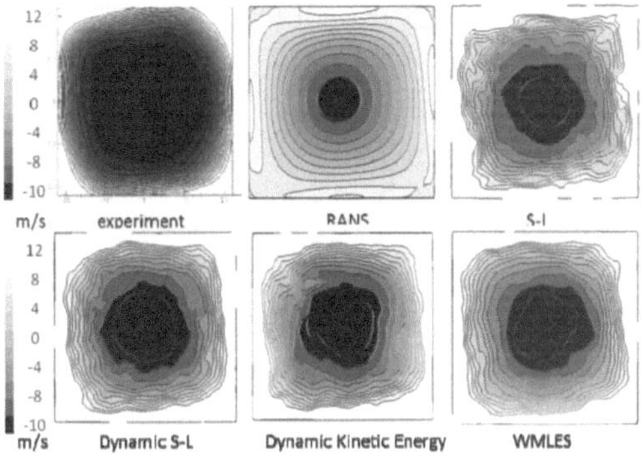

Figure 7. Comparison of the mean axial velocity (Ux) iso-contours at X = 12 mm.

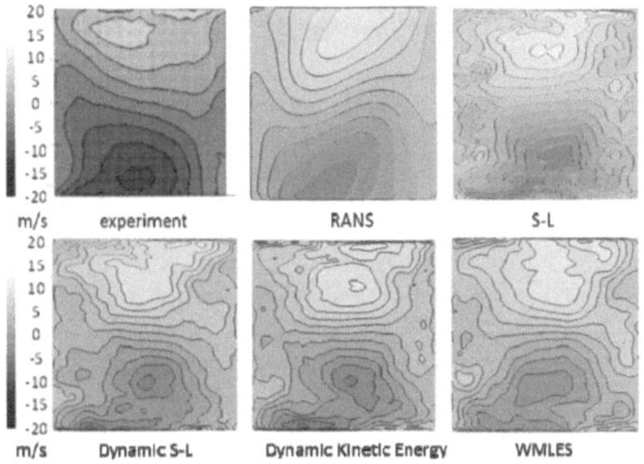

Figure 8. Comparison of the mean radial velocity (Uz) iso-contours at X = 12 mm.

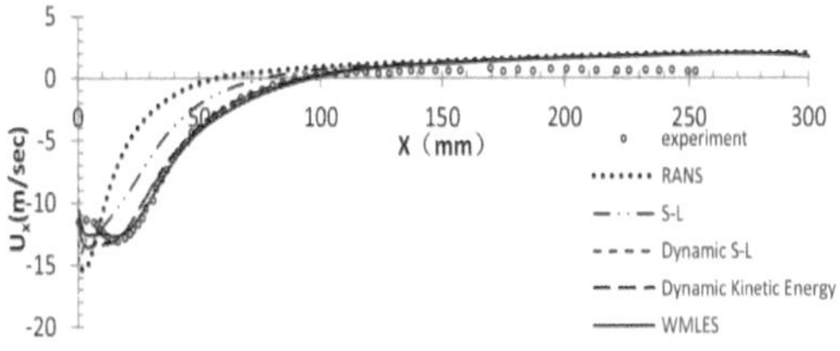

Figure 9. Comparison of the mean axial velocity (Ux) at the centreline.

that the Dynamic Kinetic Energy model and WMLES model results compare better with the experimental data than the others.

Figure 7 compares the iso-contours of the mean axial velocity, and figures 8 compares the iso-contours of the mean radial velocity at x = 12 mm section. There is a distance between the section and the venture exit. Turbulent fluctuations decrease quickly and the flow structures become more uniform. Figure 7 shows that the four LES results compare well with the experimental data. However, they slightly underestimate the width of the recirculation zone. Dynamic S-L, Dynamic Kinetic Energy and the WMLES model compares well with the experimental data in Figure 8.

Figure 9 compares the mean axial velocity at the centreline. Dynamic S-L, Dynamic Kinetic Energy,

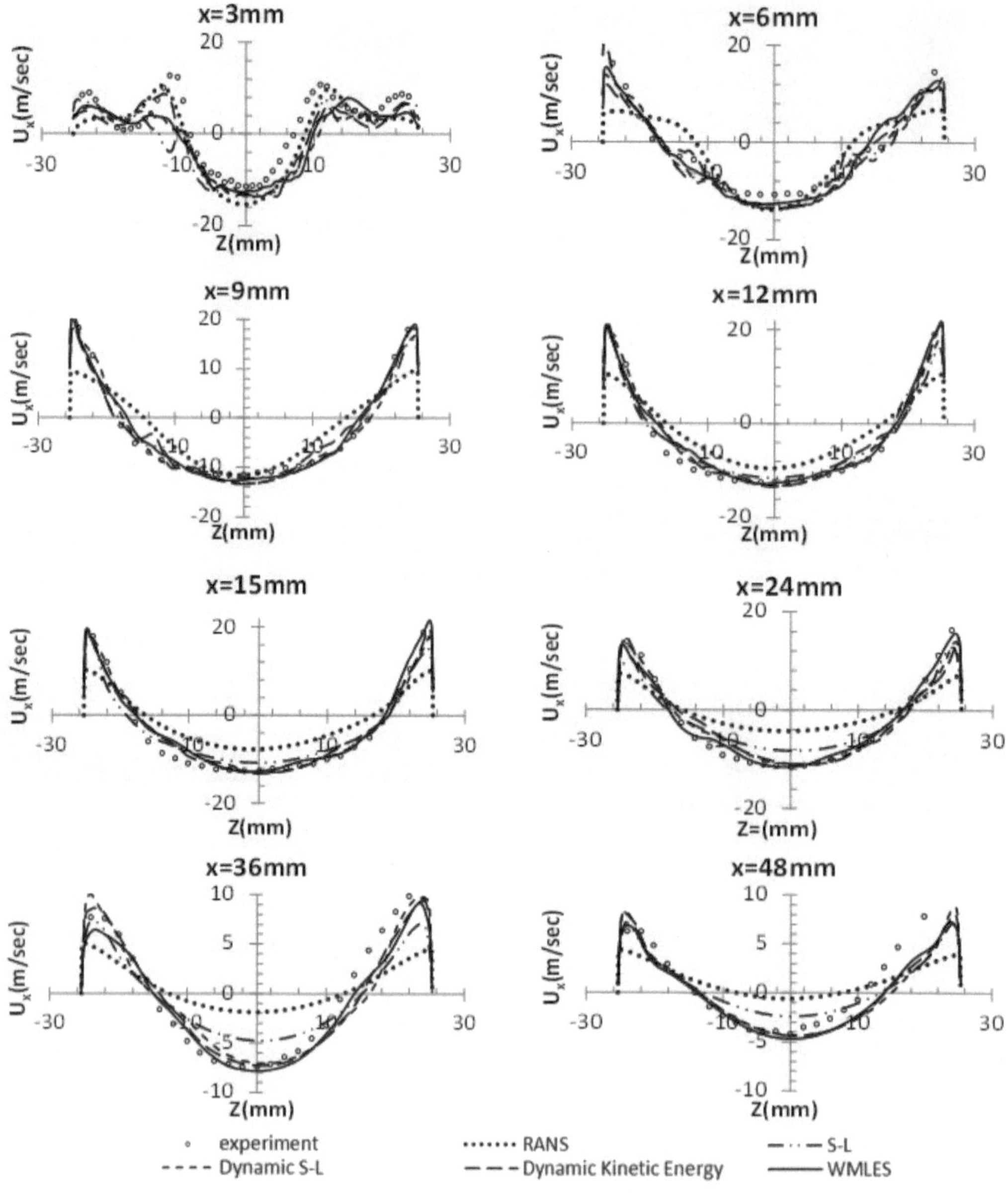

Figure 10. Comparison of the mean axial velocity (Ux) at different axial station.

and the WMLES model LES results predict the length of the recirculation zone very well, while the S-L model LES result and the RANS result have errors.

Figure 10 compares the mean axial velocity at different axial stations between simulative results and experimental data. Dynamic S-L, Dynamic kinetic energy, and WMLES models LES results compare well with the experimental data; they all predict the width of the recirculation zone and the value of back flow velocity, while the S-L model LES and RANS underestimate these. In addition, at x = 3 mm, the measured axial velocity shows emergency of peak velocity near the wall which was not shown either by the RANS results or by the LES results in Figures 2 and 3, while the four SGS models and LES in this paper capture the peak velocity very well. This may be owing to the fine treatment of mesh near the wall.

Figure 11 compares the mean cross-stream velocity at different axial stations. At x = 12 mm and x = 15 mm, all simulation results' error are slightly large. At other locations, Dynamic S-L, Dynamic kinetic energy and WMLES model LES results compare well with the experimental data. However, they underestimate this velocity component slightly.

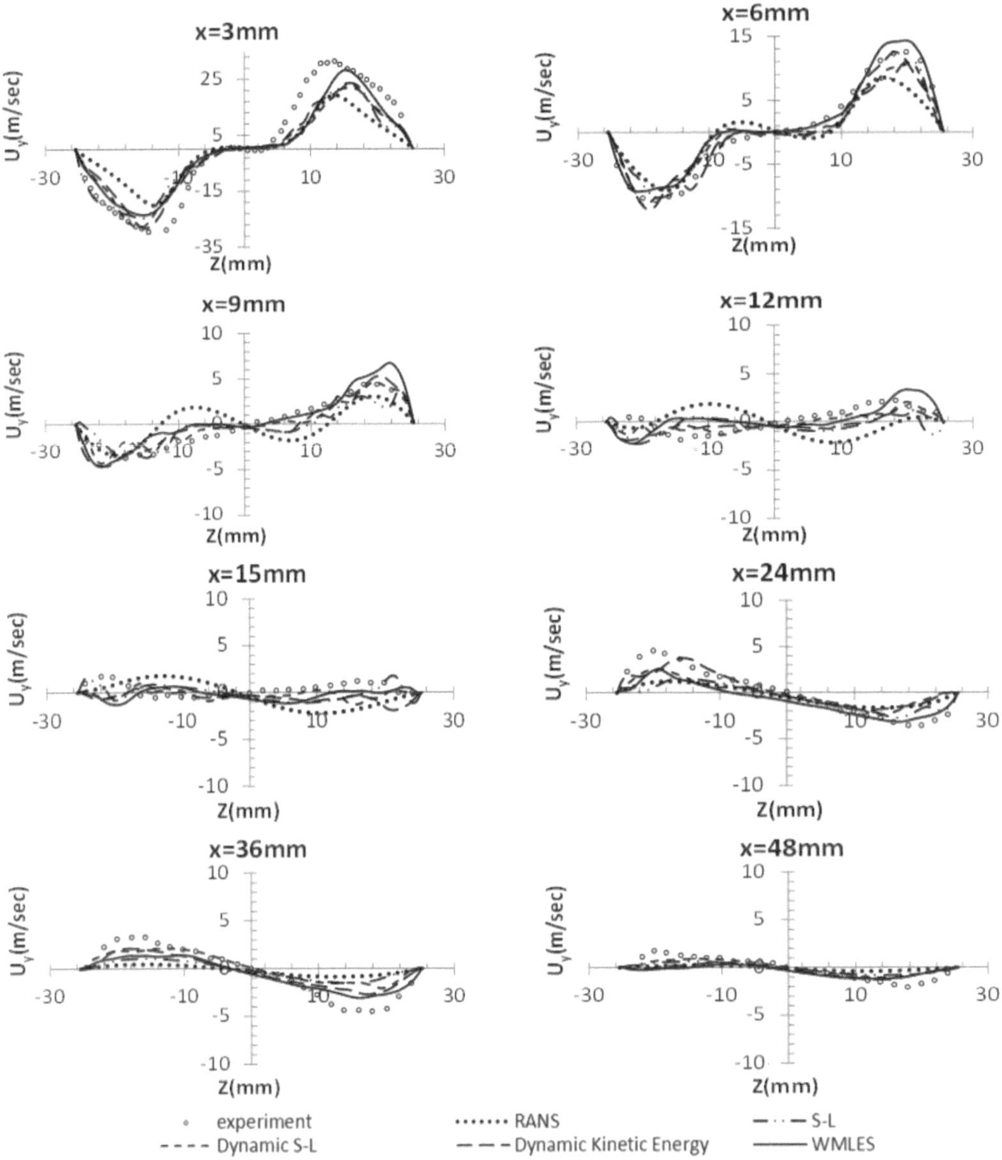

Figure 11. Comparison of the mean cross-stream velocity (Uy) at different axial station.

4 SUMMARY AND CONCLUSION

The fine hexahedral mesh is used for discretization of the LDI combustor control volume. Four kinds of SGS model are used for an LES of the nonreacting flow in a LDI combustor. When the results are compared with a RANS result and the experimental data, we can obtain the following conclusions:

(1) The LES technique has been shown to give a better prediction of turbulent flows in LDI combustor than the RANS, especially the shape of the recirculation zone and the distribution of velocity in the recirculation zone is represented well.

(2) In comparison with the LES results in Figure 3, we find that the mesh is important in the performance of an LES, especially, a fine mesh with reasonable treatment of the near wall mesh is propitious to obtain accurate simulation results.

(3) The LES methods with Dynamic S-L, Dynamic kinetic energy and WMLES SGS models all can give reasonable results compared with the experimental data, and the results from the Dynamic kinetic energy and WMLES SGS models are more accurate than other results. In addition, the calculated quantity of LES with Dynamic kinetic energy SGS model is the biggest among the four SGS models.

REFERENCES

[1] Fu, Y., Jeng, S.M. & Tacina, R. 2005. Characteristics of the Swirling Flow Generated by an Axial Swirler, Proceedings of GT2005 ASME Turbo Expo 2005: Power for Land, Sea and Air June 6–9, 2005, Reno-Tab, Nevada, USA.

[2] Farhad, D., Nan-Suey, L. & Jeffrey, P. M. 2006, Investigation of swirling air flows generated by axial swirlers in a flame tube. NASA/TM-2006-214252, GT2006-91300

[3] El-Asrag, H., Ham, F. & Pitsch, H. 2007. Simulation of lean direct injection combustor for the next high speed civil transport (HSCT) vehicle combustion systems. In Annual Research Briefs, Stanford Calif.: Center for Turbulence Research, pp. 241–253.

[4] Yuelong, H., Yuanhao, D., Yingwen, Y. et al. 2012. Large-eddy simulation of two-phase reacting flows and combustion performance in model combustor [J]. Journal of Aerospace Power, 2012, 9(27):1939–1942.

[5] Smagorinsky, J. 1963. General Circulation Experiments with the Primitive Equations. I. The Basic Experiment. Month. Wea. Rev... 91. 99–164. 1963.

[6] Lilly, D. K. 1922. A Proposed Modification of the Germano Subgrid-Scale Closure Model. Physics of Fluids. 4. 633–635. 1992.

[7] Shur, M.L., Spalart, P.R., Strelets, M.K. & Travin, A.K. 2008. A Hybrid RANS-LES Approach with Delayed-DES and Wall-Modelled LES Capabilities. International Journal of Heat and Fluid Flow. 29: 6. December 2008. 1638–1649.

[8] Kim, W.W. & Menon, S. 1997. Application of the localized dynamic subgrid-scale model to turbulent wall-bounded flows. Technical Report AIAA-97-0210. 35th Aerospace Sciences Meeting, Reno, NV American Institute of Aeronautics and Astronautics. January 1997.

Future Mechatronics and Automation – Yang (Ed.)

In-vehicle information system embedded software developing approach based on QNX RTOS

Hao Cheng & Zhiyuan Liu
Department of Control Science and Engineering Harbin Institute of Technology, Harbin, China

ABSTRACT: This paper gives a software architecture of In-vehicle information (IVI) system with real-time microkernel operating system, lists basic functions and bus technology needed by IVI. With an example, this paper describes the realization of highly reliable audio and video transmission technology in vehicle.

Keywords: In-vehicle Information System, QNX, RTOS, MOST

1 INTRODUCTION

With the progress of the electronics industry, IVI become a complex system with navigation, audio and video entertainment, mobile communication sand intelligent vehicle management. The complexity of IVI terminal does not only increase the difficulty of software development, but also makes the vehicle bus technology faces more challenges.

Now Linux has been used in IVI system, such as Cadillac CUE system [1]. But Linux is not a real-time OS. There liability can be assured. For highly safety require field, such as Rearview camera and auxiliary reversing, Linux is difficult to meet the application requirements. QNX is a microkernel RTOS (real-time operating system), there is a number of applications in security field, such as rail transport, aviation, medical and auto motive. It has a reliable HMI and has a good prospect in IVI system. Paper [3] gives Object-oriented software architecture of IVI. In the third chapter, he makes a detailed analysis on the requirements, described the complexity of in-vehicle information systems. However, due to restrictions of object-oriented programming that only used in display programs in the security field the structure is difficult to realize.

Paper [4] gives a describe of AVB video transmission protocol, But AVB is just a transmission protocol, the OEMs prefer a complete solution to ensure the reliability, And MOST bus physic transmission with two wires UTP is a lower cost than Ethernet. MOST bus is a shared ring bus. Paper [5] gives an application of audio transmission, and discussed the hardware design. Paper [6] gives a full-duplex communication design with MOST. But these papers didn't have a design of software architecture, the advantages of MOST bus is a low-cost transmission of high-quality video. But there are no comments on it in these papers.

As IVI becomes a system which has a variety of functions such as Multi-sensor data acquisition and processing, information interaction and display, fault diagnosis and so on. Well-designed software architecture helps improving reusability of code. It is more conducive to test code automatically. The period of developing and testing will be shortened. More over, the traditional vehicle bus has been difficult to meet the high quality audio and video transmission needs. Over all soft ware architecture of IVI on QNX real-time operating system is presented in this paper. And it expounds the method using MOST bus to transmit high-quality video with low cost.

The first chapter of this paper discusses the basic functions used in QNX development; the second chapter describes the vehicle information system architecture design on QNX; the third chapter describes the development of embedded software vehicle mainly solves the problem of the system; the fourth chapter build a test environment, test results are given finally.

2 QNX OPERATING SYSTEM FEATURES

The micro kernel of QNX mainly contains the following parts: Process manager, file system manager, device manager, network manager. Micro kernel and drivers, user applications, third party components use message bus to communicate. The micro kernel system services include the process and thread, thread scheduling, synchronization service, clock and timer, interrupt handling.

Microkernel structure as shown in Figure 1, Programs use kernel call to call the microkernel provides.

A preemptive scheduling strategy is used in QNX. Programs have 256 Priorities 0~255. Priority 0 is idle thread, Priority 1~64 are Non privileged, Priority 65~255 are privileged. With the same priority thread

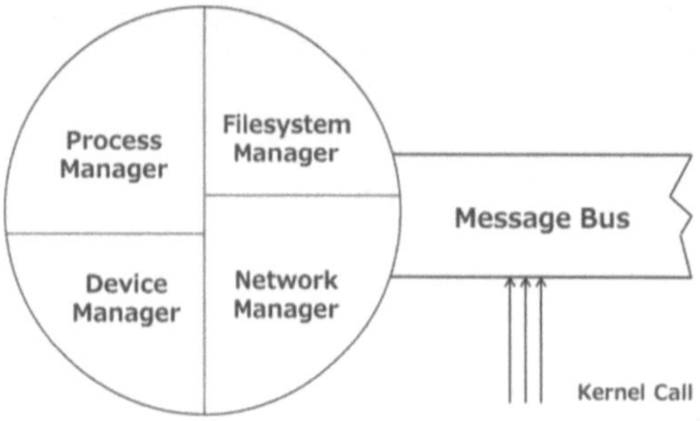

Figure 1. The micro kernel structure of QNX.

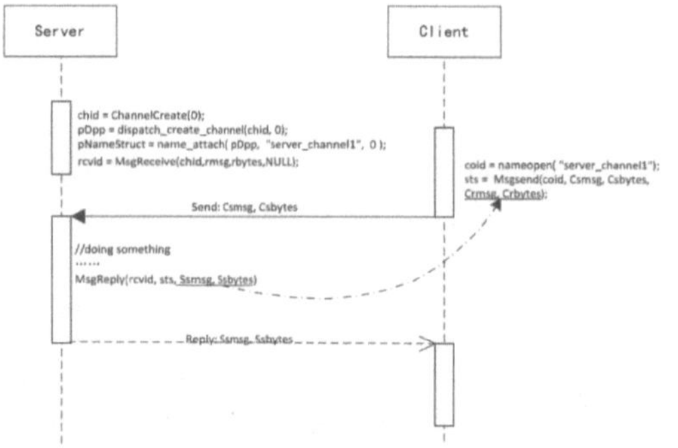

Figure 2. The QNX message mechanism.

scheduling policy, QNX provides scheduling policy: FIFO, Round-Robin, sporadic.

A FIFO scheduling thread runs continually until it relinquishes control voluntarily or it is preempted by a higher-priority thread.

A Round-robin scheduling thread runs continually until it relinquishes control voluntarily, or it is preempted by a higher-priority thread, or it consumes its times lice.

Sporadic scheduling is a scheduling strategy that guarantees a certain priority programs can get a certain time in a certain period. The scheduling of threads will have a front ground (high priority) state, and a background (low priority) state. Operating system ensures the execution time of front ground.

General programs are executed in the Round-Robin scheduling priority 11, Some real-time applications such as IMMO which send Manchester code with GPIO. Precision is required in 7~10 us. It is designed to run in FIFO scheduling priority 20.

Process communication is completed with IPC (inter process communication). When using IPC, firstly you must build a connection channel. Then message will be send through these channels.

QNX has two kinds of IPC: synchronous and asynchronous messaging passing

Synchronous message used Msg Send(), Msg Wait(), Msg Reply() to communicate. The calls is shown in Figure 2.

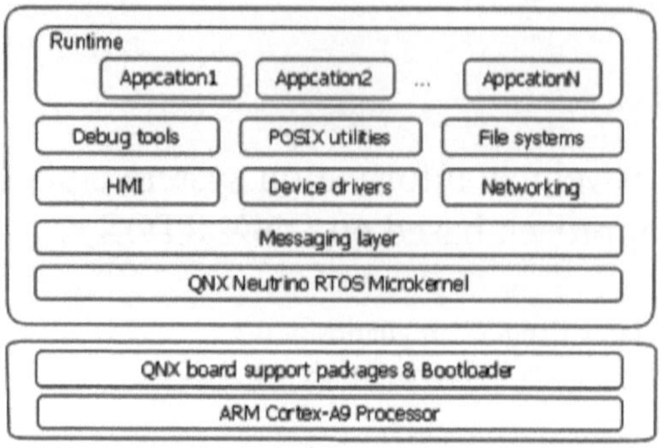

Figure 3. IVI system software architecture.

Asynchronous message (Pluse) does not require message response; the kernel uses message queues to buffer them.

3 VEHICLE INFORMATION SYSTEM SOFTWARE ARCHITECTURE DESIGN ON QNX

Vehicle information system from the bottom to the upper layer application software organization is very complex, a good framework construction of the whole software system can contribute to achieve faster and better.

3.1 *Software architecture of IVI system with QNX*

IVI system software architecture with QNX is shown in Figure 3.

Boot loader as the underlying software modules, will complete initialization of CPU and initialization of basic external devices such as external RAM, the system clock, serial port, network, complete the operating system image writing and preparation for the operating system running environment. The startup component of QNX provided by BSP provides all kinds of information needed by operating system. On the bottom of the upper layer, there are QNX Microkernel and message pass layer. In the middle, there are some of the components needed by IVI, like HMI, file system components, hardware drivers, net components, some POXIS interface components and debug related tools. On the top, there are applications according to IVI functions, like navigation, A/V play and so on.

3.2 *QNX message deadlock and message transfer structure design*

In the design of a real-time system, it is difficult to deal with deadlock and priority inversion of a complex system. The two problems are generally caused by using lock operation of message passing or critical resource accessing. Therefore in the message passing, when the two low priority clients both send messages

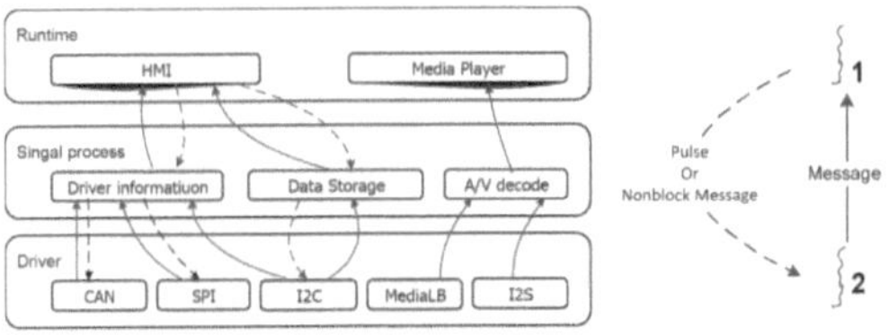

Figure 4. IVI message passing architecture.

to server, make server thread inherits priority of the higher priority thread. So in the beginning higher priority thread will first obtain the communication rights, ensure real-time better.

Deadlock is caused by existing of loop in message passing between two threads that wait for the other. It is very difficult to solve when the programs has been written. Because in a complex system, message passing is very complex. In the overall design of software two threads cannot send synchronous message to each other. A layered message passing proposal is necessary to prevent the message loops.

When designing the entire system, each thread is divided into some layers. Lower layer's threads can only send message to upper layers'. Upper layers' threads use Asynchronous message (Msg SendPulse()) or nonblack message to inform the lower layer's.

Here is a basic messaging model shown in Figure 4, Whole IVI system is divided into three basic levels: runtime, Signal process, Driver. Runtime layer is mainly responsible for the user input information collection and the information presented. Signal Process layer primarily responsible for collecting the information from underlying drivers and process them. Driver layer is theabstraction of hardware devices.

CAN bus driver is used to access a variety of vehicle information. SPI and I2C are buses which are used to communicate with extern devices. Such as power management using I2C interface, AD sampling using the SPI interface. Media LB is responsible for the transmission of audio and video and system online upgrades. Implementation will be described in Chapter 3.

4 EMBEDDED SOFTWARE DEVELOPMENT OFMOST BUS

4.1 *MOST bus network layer and link layer software development*

IVI system soft ware is complex, Due to space limitations, we just describe the development of MOST bus and Media LB driver.

MOST (Media Oriented Systems Transport) is a ring bus, using UTP, optical fiber, or coaxial as physic media. MOST network model contains all of the seven-layer structure of ISO/OSI model [6] There are three transmission protocol MOST25 MOST50 MOST150. MOST use frames to form data and sample frames on the bus in 48 kHz.

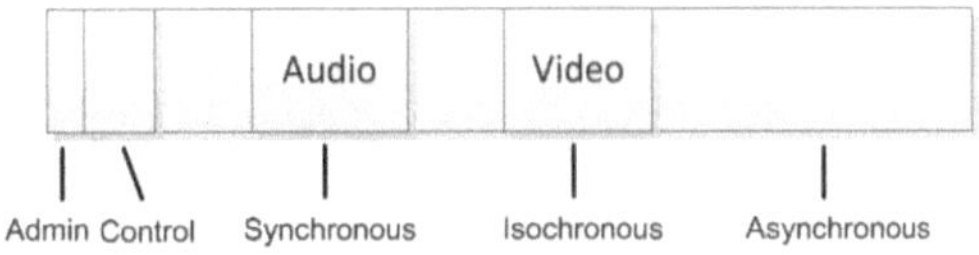

Figure 5. MOST frame structure.

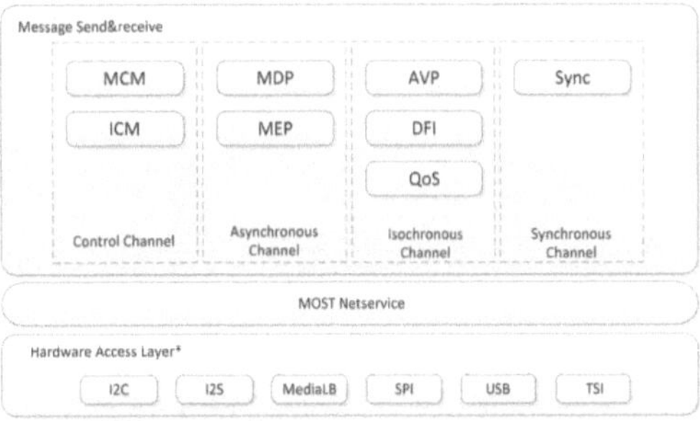

*AccessINIC Use one or muti

Figure 6. MOST software architecture.

MOST frame structure is shown in Figure 5.

In the actual software development, as some features have been packaged in INIC (Intelligent interface chip), and provided related Net Service in standard C code. Generally follows the software architecture to develop MOST bus.

Hardware access layer need to select the appropriate hardware interfaces based on the specific needs. However, not all hardware interfaces can support four frame types. This paper select Media LB who support all frame types. The middle layer is generally transplant Net Service codeas a management of INIC smart chip, MOST networks, MOST interfaces for data transmission. The following briefly describes the function of the four frame types and their applications.

Channel contains two kinds of packets: MCM (MOST Control message), ICM (INIC Control Message), MCM used to complete the MOST bus configuration, ICM is configured for mutual communication between the INIC. Control Message is in the head of frame, used to transmitting and receiving small data, shared by all the nodes. Arbitration mechanism and verification mechanism is provided.

Synchronous Channel Mainly for the transmission of audio signals can only be assigned once in MOST frames. No address is used. All synchronization domain of each frame must be filled. Each individual amplifier takes the same data at the same time. So there is no delay.

Asynchronous Channel is mainly used for transmission of data packets, Packet size is not fixed. Asynchronous domain is allowed to have a blank. Arbitration mechanism is provided Asynchronous is shared by all the nodes.

Isochronous Channel has no address and can only be assigned once in MOST frames. Isochronous domain is allowed to have a blank. This channel is used to transmit Ethernet data and compressed video.

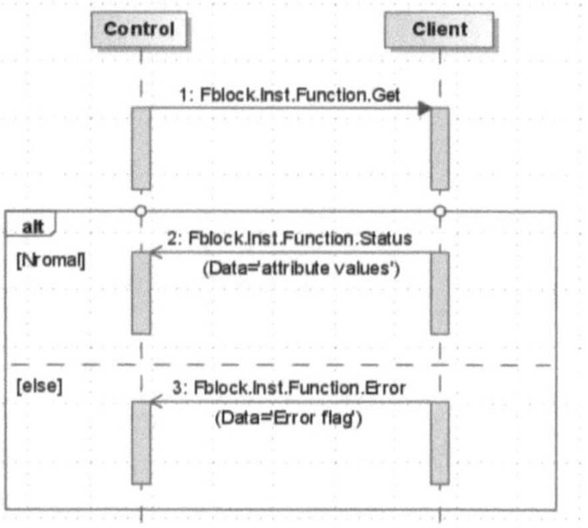

Figure 7. Get operation (Fblock ID. Inst ID. Fkt ID. OP Type. Data).

MOST bus use Fblock that achieves certain function. When accessed using object (Fbloc kID), examples (Inst ID), methods (Fkt ID), operation (OP Type) model, for example see Figure 7.

4.2 *MediaLB bus driver development*

MediaLB bus uses a corresponding agreement with the MOST bus structure, is also divided into four kinds of channels. All the MediaLB of MPU is IP core from Microchip.

Media LB module uses a modular design approach. The general control register and data transfer register are separated. Common control register use direct addressing. Data related to the control register (CTR) use indirect addressing. There is a built-in RAM for buffering data.

When the module is initialized, you need to initialize the pin configuration, initialize the system PLL, MediaLB clock is provided by INIC, and then the internal clock needs to be synchronized with the INIC clock.

Establish MediaLB data sockets, MediaLB the Channel Table Register defines three types of registers:

1. CDT (Channel Descriptor Table): 64, Describing the state of the logical channel FIFO, corresponding to each logical channel identifier CL (Channel Label).
2. CAT (Channel Allocation Table): Provides a logical channel and DMA FIFO links, There are 128 of which FIFO before 64 (CAT for MediaLB) represents a logical channel, after 64 (CAT for HBI)-one correspondence with the user allocated memory. Each CAT has a CL.
3. ADT (AHB Descriptor Table): 64, corresponding to HBI, is management of coping data from FIFO to memory through DMA.

A socket is containing one CAT for MediaLB, one CDT, one CAT for HBI. Their connection is CL. Since ADT have Correspondence with CAT for HBI, a Socket is done.

Sockets described in CTR are shown as Figure 9.

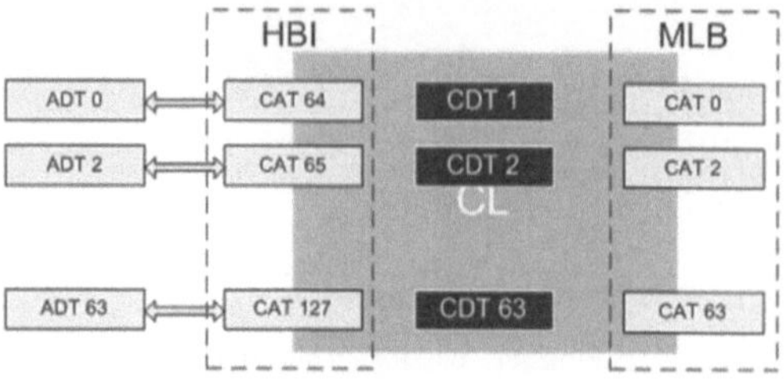

Figure 8. CTR's structure of MediaLB.

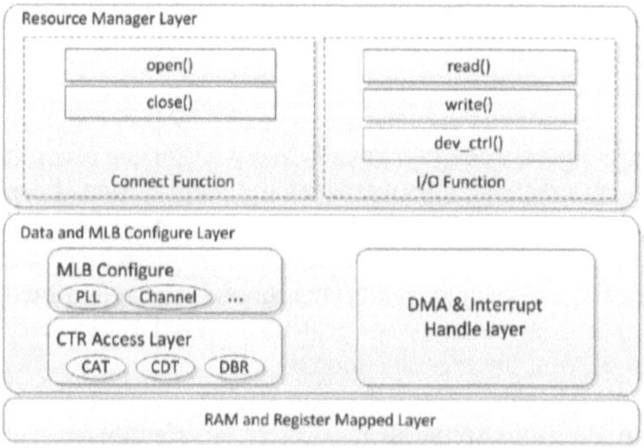

Figure 9. The software architecture of MediaLB driver.

Figure 10. Experimental system.

According to the structure of QNX resource manager write the drivers as in Figure 9.

The driver provides external interfaces are divided in two parts. One is response for link management. The other part is I/O functions, providing data interface and configure the interface. Devctr linter faces provides dynamic configuration interface which contains fps MLB module configuration and sockets configuration. DMA data layer is responsible for DMA and all register configuration. Address mapping layer is responsible for mapping the registers and physical memory to the program address area.

5 EXPERIMENT

Use frees cale I.mx6 Quad SABRE AI system as experimental platform.I.mx6 is ARM Cortex-A9 quad-core

Figure 11. Play online video.

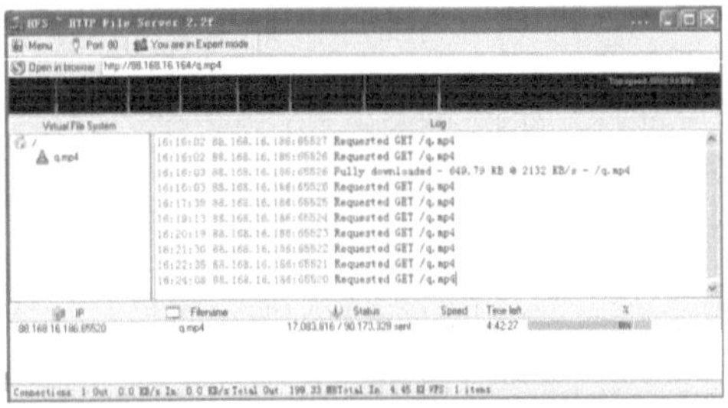

Figure 12. HTTP server's feedback.

processor who's Operating frequency of a single core is 1.5 GHz. Video output is connected to directly to the monitor with HDMI. Audio output is connected to an external speaker. The MAMAC protocol implemented by MOST50 is used for network packet transmission. Construction of the experimental system is shown in Figure 11.

Write a build file of QNX to generate a QNX image and download to the I.MX6 through tftp, Verify remote video playback results are as shown in Figure 11 & 12.

Player running on QNX read 1080p video files on the HTTP file server, it is played smoothly. Other components running normally, No crash is found in a long period.

6 CONCLUSIONS

This paper summarizes the soft architecture of in-Vehicle Infotainment system based on QNX real-time operating system, and gives a solution using MOST a low cost bus to transmit audio and video, describes the MOST bus developed software architecture. It provides a foundation to develop a complex IVI system. We can develop some other functions upon this architecture.

ACKNOWLEDGEMENT

Supported by National High Technology Research and Development Program of China (863 Program) (2012AA110701).

REFERENCES

CUE page, http://www.cadillac.com/cadillac-cue.html

Customer page, http://www.qnx.com/solutions/industries/defense.html#customers

Chen W, Huang Y P, Chen B, et al. Design and Implement of Multimedia Transmission Based on MOST [J]. Journal of Jilin University. Information Science Edition, 2010, 28(2): 141–146.

Lee S Y, Park S H, Choi H S, et al. MOST network system supporting full-duplexing communication [C]//Advanced Communication Technology (ICACT), 2012 14th International Conference on. IEEE, 2012: 1272–1275.

MOST: the automotive multimedia network; from MOST25 to MOST150[M]. Franzis, 2011.

Steinbach T, Lim H T, Korf F, et al. Tomorrow's in-car interconnect? A competitive evaluation of IEEE 802.1 AVB and Time-Triggered Ethernet (AS6802)[C]//Vehicular Technology Conference (VTC Fall), 2012 IEEE. IEEE, 2012: 1–5.

Tanenbaum A S. Modern operating systems [M]. Prentice Hall Press, 2007.

QNX Software Systems Limited, QNX® Neutrino® RTOS Programmer's Guide[EB/OL], http://www.qnx.com/download/feature.html?programid=26179, 2014.

Zhou H, Zhang G, He P, et al. Object-oriented framework design for in-vehicle information platform [C]//Electrical and Control Engineering (ICECE), 2011 International Conference on. IEEE, 2011: 5350–5353.

Future Mechatronics and Automation – Yang (Ed.)
© 2015 Taylor & Francis Group, London, ISBN 978-1-138-02648-3

Research on steering angle tracking control approach for Steer-By-Wire system

Mingyuan Zhang & Zhiyuan Liu
Department of Control Science and Engineering, Harbin Institute of Technology, Harbin, China

ABSTRACT: The steering angle tracking control system is the basis for Steer-by-Wire (SBW) system. In this paper, a steering angle tracking control approach using linear observer and H_∞ controller are presented for SBW system. The strong nonlinear of road reaction torque and the inaccuracies of steering system parameters are well solved by this approach. Firstly the observer is designed to compensate road reaction torque. Secondly the H_∞ controller is designed based on the system model with compensator. At last the observer and controller are verified by vehicle dynamics simulation software veDYNA under typical steering conditions.

Keywords: Steer-by-Wire, Disturbance Compensation, H_∞ Control, Robust Control

1 INTRODUCTION

While steer-by-wire (SBW) system eliminated the mechanical steering column, it brought a series of research questions such as the system control strategy, the road feel torque planning, steering angle Tracking control, fault diagnosing and treatment. In [1]–[4], the torque control strategy, the speed control strategy, the angle control-torque feedback strategy, the torque drive-angle feedback strategy had been proposed. Tracking control system of steering angle was the basis of these Strategies in SBW system. The main problem in steering angle control was the suppression of the road reaction torque suppression. In [5], author considered the impact of aligning torque caused by wheel lateral force, the sticky-coulomb friction torque and the motor disturbance torque, and designed the sliding mode controller for steering angle tracking control. In [6], on the basis of the sliding mode control, author designed a Lipschitz continuous incremental controller to solve the switch vibration problem.

Considering that the road reaction torque is large and difficult to calculate directly during vehicle steering. In this paper we give a control method using disturbances compensator and H_∞ controller to reduce the tracking error caused by the road reaction torque. The main contents are as follows: In section 2, the modeling the steer system and analyzing the mathematical description of road reaction torque. In section 3, designing the road reaction torque observer and the H_∞ controller. In section 4, simulating with veDYNA to verify the observer and controller.

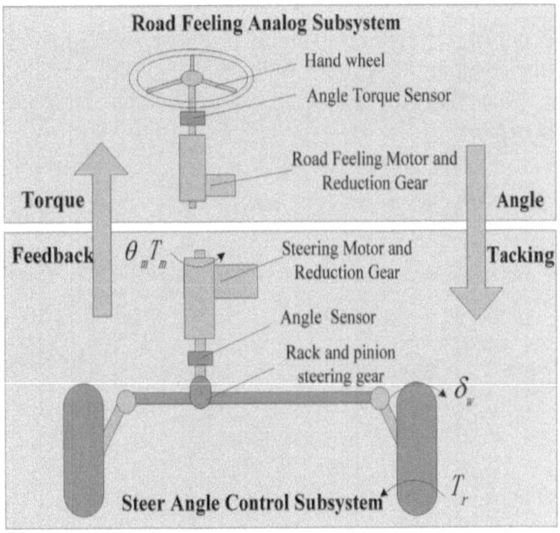

Figure 1. Steer by wire system architecture.

2 DYNAMICS MODEL

Figure 1 shows the SBW system architecture. In Figure 1, steer angle tracking subsystem is research object in this paper which is consist of the steering motor, reduction gear, steering gear and vehicle wheel.

To simplify the model, make the following assumptions:

(1) Ignore the elastic deformation of Transmission.
(2) Ignore the Steer Angle differences between two wheels caused by Ackermann steering mechanism.

(3) Friction torque in steering mechanism can be description by viscous friction and Coulomb friction.

According to Newton's mechanical laws, the dynamical equation of Steer angle tracking system is

$$\begin{cases} J_m\ddot{\theta}_m = T_m - T_l - B_m\dot{\theta}_m - T_{cm} \\ J_w\ddot{\delta}_w = kNT_l - B_w\dot{\delta}_w - T_{cw} - T_r \\ \theta_m = kN\delta_w \end{cases} \quad (1)$$

where θ_m, δ_w are the rotation angle of motor and vehicle wheel. J_m, J_w, B_m, B_w are the inertia and Viscous friction coefficient of motor and steering mechanism. T_m, T_l, T_{cm} are the motor torque, load torque and coulomb friction torque on motor shaft. T_{cw}, T_r are the coulomb friction torque and road reaction torque on steering kingpin. k, N are the ratio of reduction gear and steering gear.

Motor torque control subsystem is generally achieved through the current loop [7]. To simplify the model, we use a first-order inertia element to describe it as follows:

$$\frac{T_m(s)}{T_m^*(s)} = \frac{1}{\tau_m s + 1} \quad (2)$$

where T_m^* is the expected output torque, τ_m is the Inertia time constant of Motor torque control system.

According to (1) and (2), we have the state space description

$$\begin{cases} \dot{X}_p = A_p X_p + B_{1p} T_m^* + B_{2p} T_d \\ \delta_w = C_p X_p \\ X_p = \begin{bmatrix} T_m & \delta_w & \dot{\delta}_w \end{bmatrix}^T \end{cases} \quad (3)$$

where

$$A_p = \begin{bmatrix} -\frac{1}{\tau_m} & 0 & 0 \\ 0 & 0 & 1 \\ \frac{kN}{J_{eq}} & 0 & \frac{B_{eq}}{J_{eq}} \end{bmatrix} \quad B_{1p} = \begin{bmatrix} \frac{1}{\tau_m} & 0 & 0 \end{bmatrix}^T$$

$$B_{2p} = \begin{bmatrix} 0 & 0 & -\frac{1}{J_{eq}} \end{bmatrix}^T \quad C_p = \begin{bmatrix} 1 & 0 & 0 \end{bmatrix}$$

$$J_{eq} = k^2N^2J_m + J_w \qquad B_{eq} = k^2N^2B_m + B_w$$

T_m^* is the control input, T_d is the disturbance torque input, δ_w is output. T_d is consist of two part:

$$T_d = T_c + T_r \quad (4)$$

where

$$T_c = kNT_{cm} + T_{cw} = (kNF_{sm} + F_{sw})sign(\dot{\delta}_w)$$

F_{sm}, F_{sw} are the coulomb friction torque, T_r is the road reaction torque. As T_c is very small compared with T_r, so we consider T_r the major impact in this paper. T_r is consist of two part:

$$T_r = T_{fw} + T_a \quad (5)$$

where T_{fw} is the friction torque between tire and road, and T_a is the Aligning torque.

T_{fw} is the largest during spot turn, and decreases with the increase of the wheel turning radius. The largest friction torque can be approximated by the empirical formula

$$M_{Sch} = k\mu\sqrt{\frac{F_z^3}{p}} \quad (6)$$

T_a is mainly generated by the tire lateral force. In steady-state, aligning torque caused by lateral force has relationship with vehicle mass m, wheelbase l,the distance between center of mass and rear axle l_H, vehicle speed v, steering angle δ_w, sideslip angle of front and rear wheel α_V α_H and total drag distance of rear wheel n_V. Approximate expression is as follows:

$$T_a = \frac{ml_H v^2 n_v(\delta_w - \alpha_V + \alpha_H)}{l^2} \quad (7)$$

The above analysis shows that many variables are associated with T_r, and some of them are changing with the vehicle state and the driving environment. It is difficult to calculate T_r directly in real-time control system.

3 DESIGN OBSERVER AND CONTROLLER

3.1 *Road reaction torque observer design*

As the road reaction torque changes slowly during the sampling period, it can be considered a constant value in a short time. Thus it can be described by random walk model. Therefore (1) can be rewritten as state space description:

$$\begin{cases} \begin{bmatrix} T_d \\ \dot{\delta}_w \\ \hline \delta_w \end{bmatrix} = \left[\begin{array}{cc|c} 0 & 0 & 0 \\ -\frac{1}{J_{eq}} & -\frac{B_{eq}}{J_{eq}} & 0 \\ \hline 0 & 1 & 0 \end{array}\right] \begin{bmatrix} T_d \\ \dot{\delta}_w \\ \hline \delta_w \end{bmatrix} + \begin{bmatrix} 0 \\ \frac{kN}{J_{eq}} \\ \hline 0 \end{bmatrix} T_m \\ y = \left[\begin{array}{cc|c} 0 & 0 & 1 \end{array}\right] \begin{bmatrix} T_d \\ \dot{\delta}_w \\ \hline \delta_w \end{bmatrix} \end{cases} \quad (8)$$

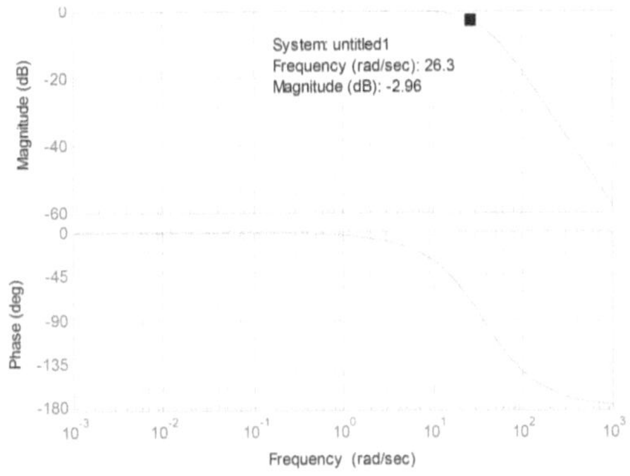

Figure 2. Frequency characteristic of observer.

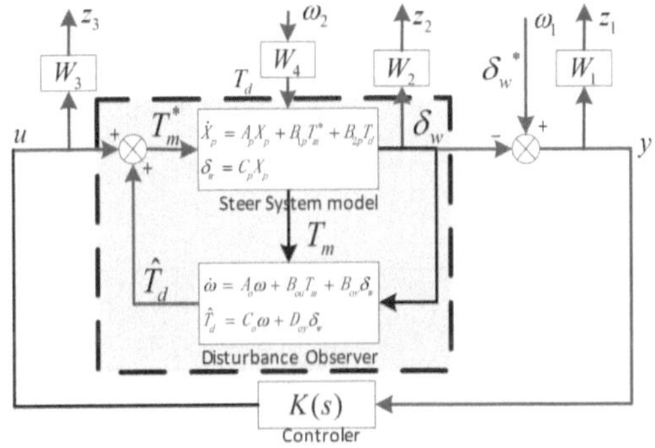

Figure 3. Control system architecture.

where T_m can be calculated directly from the motor current detection value, and δ_w is measurable. Observation model (8) is observable completely. Using pole placement method to design reduced-order observer, the observer expression is as follows:

$$\begin{cases} \dot{\omega} = \boldsymbol{A}_o \omega + \boldsymbol{B}_{ou} T_m + \boldsymbol{B}_{oy} \delta_w \\ \hat{T}_d = \boldsymbol{C}_o \omega + D_{oy} \delta_w \end{cases} \tag{9}$$

where

$$\boldsymbol{A}_o = \boldsymbol{A}_{11} - \boldsymbol{L}\boldsymbol{A}_{12} \qquad \boldsymbol{B}_{ou} = \boldsymbol{B}_1 - \boldsymbol{L}\boldsymbol{B}_2$$

$$\boldsymbol{B}_{oy} = \boldsymbol{A}_{12} - \boldsymbol{L}\boldsymbol{A}_{22} + (\boldsymbol{A}_{11} - \boldsymbol{L}\boldsymbol{A}_{21})\boldsymbol{L}$$

$$\boldsymbol{C}_o = \begin{bmatrix} 1 & 0 \end{bmatrix} \qquad D_{oy} = L_1$$

$\boldsymbol{A}_{11}, \boldsymbol{A}_{12}, \boldsymbol{A}_{21}, \boldsymbol{A}_{22}, \boldsymbol{B}_1, \boldsymbol{B}_2$ is the partitioned matrix in (9), and $\hat{T}_d$ is the Observed value of T_d. Controlled object parameters are shown in Table 1. Configure observer poles as $[-32-16i \quad -32+16i]$, then $\boldsymbol{L} = [-3.274 \times 10^4 \quad 57.40]^T$. The frequency characteristic from T_d to $\hat{T}_d$ is shown in Figure 2. The Observer Bandwidth is 26 rad/s. This bandwidth not only could observe T_d efficiently, but also has well noise suppression.

3.2 H_∞ *controller design*

Add $\hat{T}_d$ into T_m^* to compensate the road reaction torque:

$$T_m^* = u + \hat{T}_d \tag{10}$$

u is the new control input. Control system architecture after adding compensation is shown in Figure 3. New Controlled object is dashed circle section in Figure 3. Combined (3) and (9), its new state space description is as follows:

$$\begin{cases} \begin{bmatrix} \dot{\boldsymbol{X}}_p \\ \dot{\omega} \end{bmatrix} = \begin{bmatrix} \boldsymbol{A}_p + \boldsymbol{B}_{1p}\boldsymbol{D}_{oy}\boldsymbol{C}_p & \boldsymbol{B}_{1p}\boldsymbol{C}_o \\ \boldsymbol{B}_{ou}\boldsymbol{I}_o + \boldsymbol{B}_{oy}\boldsymbol{C}_p & \boldsymbol{A}_o \end{bmatrix} \begin{bmatrix} \boldsymbol{X}_p \\ \omega \end{bmatrix} \\ \qquad + \begin{bmatrix} \boldsymbol{B}_{1p} & \boldsymbol{B}_{2p} \\ \boldsymbol{O}_{2\times1} & \boldsymbol{O}_{2\times1} \end{bmatrix} \begin{bmatrix} u \\ T_d \end{bmatrix} \\ \delta_w = \begin{bmatrix} \boldsymbol{C}_p & \boldsymbol{O}_{1\times2} \end{bmatrix} \begin{bmatrix} \boldsymbol{X}_p \\ \omega \end{bmatrix} \\ \boldsymbol{I}_o = \begin{bmatrix} 1 & 0 & 0 \end{bmatrix} \end{cases} \tag{11}$$

Firstly, determine the uncertainty weighting function W_2 (high-pass). Taking into account the resonance characteristics and the inaccuracy of parameters, we get the multiplicative uncertainty description of the steering system:

$$L(s) = \frac{|\Delta|_{\max} (\tau_{21}{}^2 s^2 + 3\tau_{21} s + 1)^2}{(J_{eq} / B_{eq} s + 1)(\tau_{22}{}^2 s^2 + 0.2\tau_{22} s + 1)} \tag{12}$$

where $|\Delta|_{\max}$ is the maximum gain relative deviation caused by the inaccuracy of parameters, choose $|\Delta|_{max} = 1$. Thus the maximum gain is double of the standard value. Choose $\tau_{21} = \tau_m$. τ_{22} choose the time constant corresponding to the resonance frequency of the steer system, then $\tau_{22} = 0.006$ [8].

To ensure the amplitude-frequency curve of W_2 above that of multiplicative uncertainties on, select W_2 as follows:

$$W_2(s) = \frac{(0.02^2 s^2 + 0.06s + 1)(0.02s + 1)}{(0.1515s + 1)(0.006^2 s^2 + 0.0012s + 1)} \tag{13}$$

W_1 and W_4 affect tracking error and suppression of T_d. The frequency band of effective steering angle input δ_w^* is less then 1.5 Hz, T_d and δ_w have the same periodicity. Thus we choose 10 rad/s as a break angular frequency of W_1. As the disturbance caused by weight unevenness of the wheel is mainly on frequency band of 20 Hz–30 Hz [8], so choose 100 rad/s as another break angular frequency. Preliminarily determine

$$W_1(s) = \frac{k_1(0.003s + 1)^2}{(0.1s + 1)(0.01s + 1)} \tag{14}$$

The main purpose of W_4 is to adjust the system sensitivity to T_d. Select

$$W_4(s) = k_4 \tag{15}$$

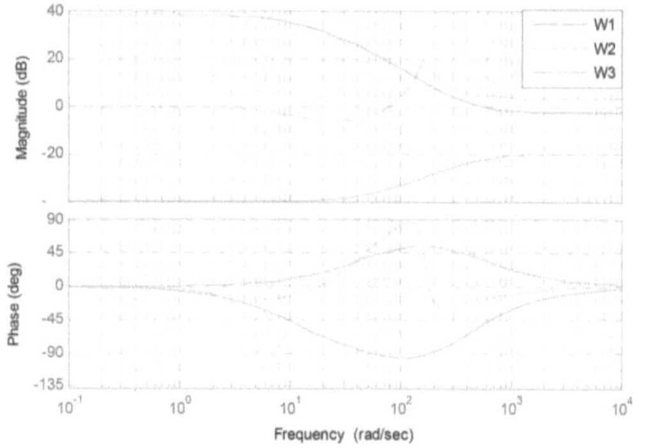

Figure 4. Bode diagram of weighting function.

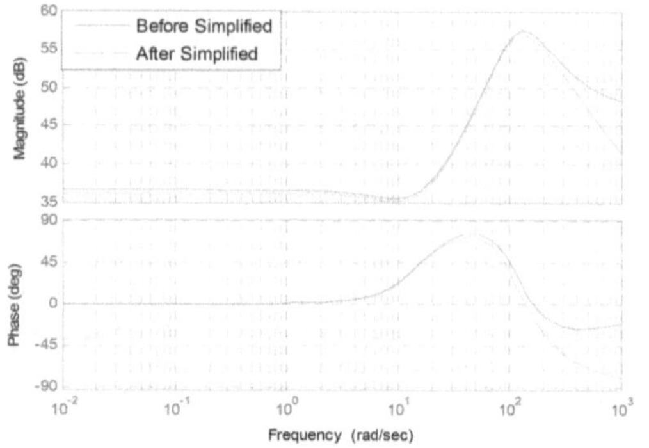

Figure 5. Bode diagram of controller.

In order to limit the bandwidth of the control signal within the motor torque control system, choose τ_m as a break angular frequency of W_3. To ensure W_3 regular transfer function, select

$$W_3(s) = \frac{k_3(0.02s + 1)}{(0.002s + 1)} \tag{16}$$

Parameter k_1 could be added to decrease the system sensitivity to T_d and δ_w^* and to make system dynamic response faster. Parameter k_4 could be added to further decrease the system sensitivity to T_d. Parameter k_3 could be added to decrease the amplitude of T_m^*.

MATLAB toolbox is used for solving this H_∞ problem. The solution method is based on Riccati equation, from [9] and [10]. Select $k_1 = 80$, $k_3 = 0.01$, $k_4 = 20$. Simplify the higher-order controller calculated by MATLAB as follows:

(1) Remove the zeros and poles above 200 rad/s.
(2) Eliminate similar zeros and poles.

The simplified controller is as follows:

$$K(s) = \frac{117453(s + 50)(s^2 + 21.73s + 198.5)}{(s + 10)(s + 136.2)(s^2 + 117.5s + 13180)} \tag{17}$$

The Bode diagram of W_1, W_2 and W_3 is shown in Figure 4. The Bode diagram of controller before and after simplified is shown in Figure 5. The controller has little change in the effective bandwidth after simplified in Figure 5. Thus, the simplified controller can replace the higher-order controller effectively.

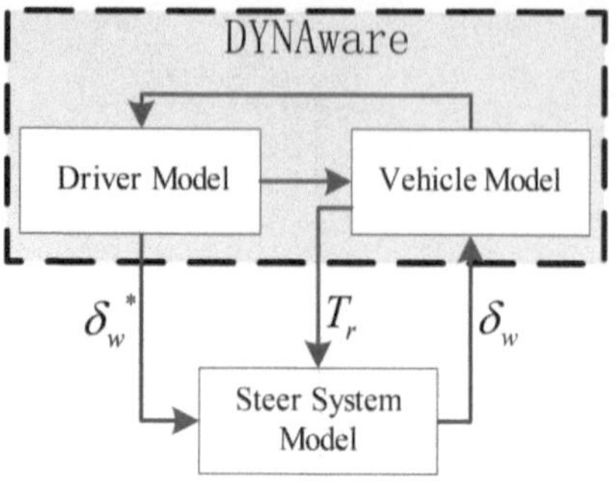

Figure 6. Simulation system architecture.

Table 1. Steering mechanism parameters.

Parameters	Value
τ_m	0.02 s
J_m	2.47×10^{-4} kg·m^2
J_w	3.8 kg·m^2
B_m	1.8×10^{-3} Nms/rad
B_w	10 Nms/rad
k	16.5
N	18

4 SIMULATION AND ANALYSIS

4.1 *Simulation environment*

The co-simulation platform is constructed by high-precision vehicle dynamics simulation software veDYNA and MATLAB Simulink. The architecture of co-simulation system is shown in Figure 6. Vehicle model is BMW_325i_88 which is built-in veDYNA. The steering mechanism parameters are shown in Table 1.

The vehicle model is given an input 0.4 Hz sine steering wheel angle of 90° in different vehicle speed. The relationship between T_r and δ_w is shown in Figure 7–8.

T_r extracted from veDYNA has the same periodicity with δ_w, and its basic tendency is consistent with (5), (6) & (7). However (7) is the description on steady state, and can't show the hysteresis characteristics of T_r. The veDYNA software can reflect the impact of vehicle load transfer, tire load changes, tire longitudinal force and other factors to T_r. T_r extracted from veDYNA is closer to the real vehicle, and can be used on control test.

4.2 *Simulation experiment*

Test Condition 1: vehicle speed 65 km/h, 0 degree road roughness, slalom test, distance between cones 30 m.

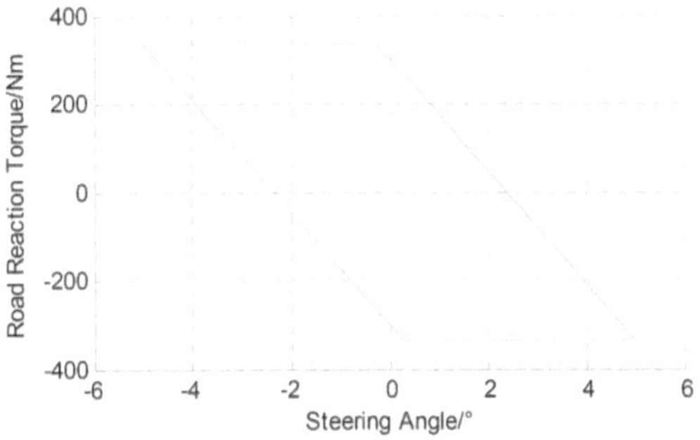

Figure 7. T_r–δ_w diagram during spot turn.

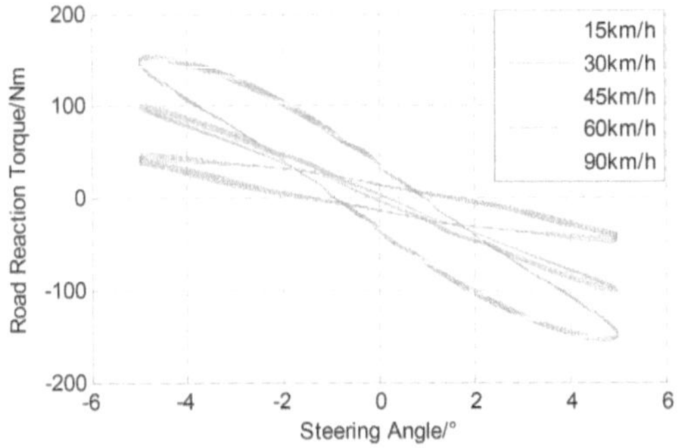

Figure 8. T_r–δ_w diagram in different vehicle speed.

Test Condition 2: vehicle speed 65 km/h, road roughness changes form 0 degree to 4 degree at 200 m, slalom test, distance between cones 30 m.

Test Condition 3: vehicle speed 80 km/h, 0 degree road roughness, double lane-change test.

Set the parameters of controlled object model is accurate and the above three test conditions are simulated. The observed result of T_d, steering angle tracking effect, tracking error and vehicle trajectory under each condition are shown in Figure 9–11. The observed result of T_d is comparable accuracy, and is not sensitive to high frequency vibrations caused by road surface roughness. The Tracking error is less than 0.2° in these conditions. At the same time, Vehicles can ensure good steering performance.

When the parameter B_{eq} is set double of regular value and zero separately, test condition 1 is simulated. The observed result of T_d, steering angle tracking effect, tracking error and vehicle trajectory under each condition are shown in Figure 12–13. Figure 13(a) shows that the observer's adaptability to parameter changing is not strong. However, Figure 13(b),(c) show that the tracking system is still able to guarantee well tracking performance. The reason is that the changes of model parameters were not considered when the observer was designed. But the controlled object used to design H_∞ controller is the system after compensated. So the affect to tracking error by the observation error caused by the changes of model parameters is well solved.

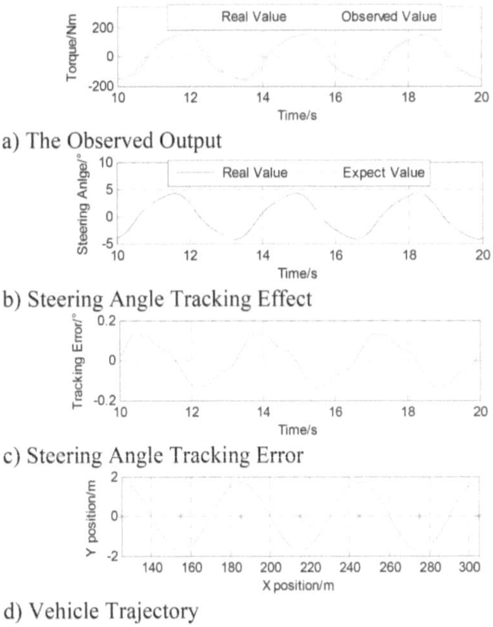

Figure 9. Simulation result on test condition 1.

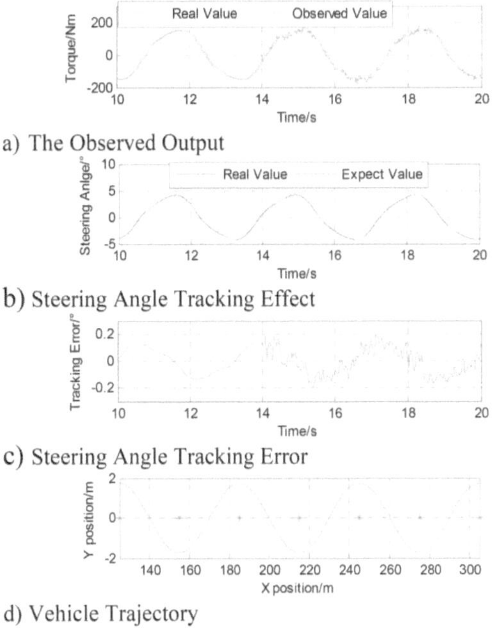

Figure 10. Simulation result on test condition 2.

5 CONCLUSION

In this paper, the steering angle tracking control approach using a linear observer and a H_∞ controller are presented for SBW system. In the approach, the observer is used to compensate the road reaction torque which is difficult to calculate directly, and the H_∞ controller ensures the robustness to the system parameters.

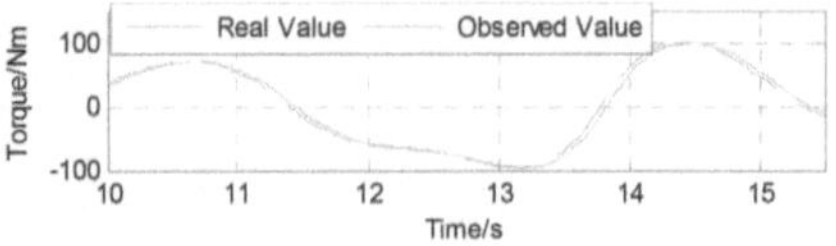

a) The Observed Output

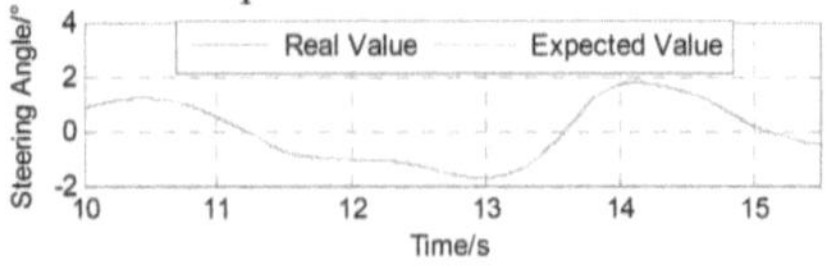

b) Steering Angle Tracking Effect

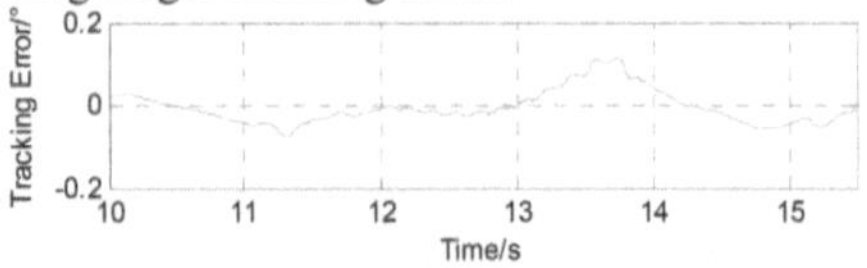

c) Steering Angle Tracking Error

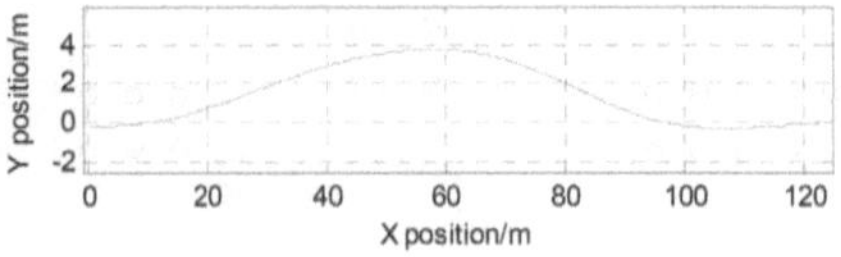

d) Vehicle Trajectory

Figure 11. Simulation result on test condition 3.

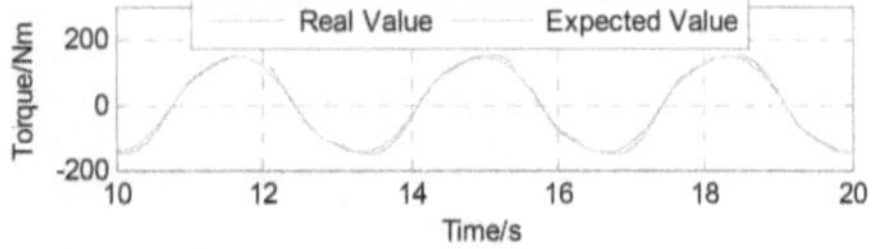

a) The Observer Output

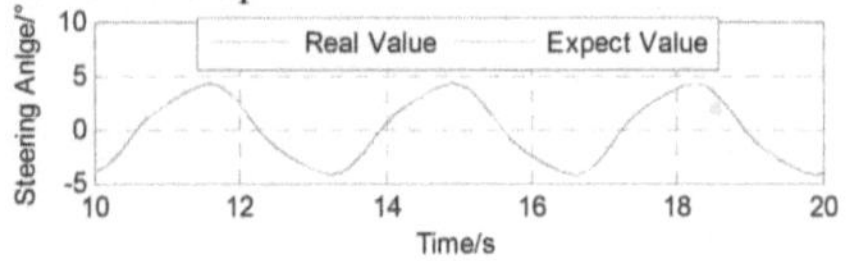

b) Steering Angle Tracking Effect

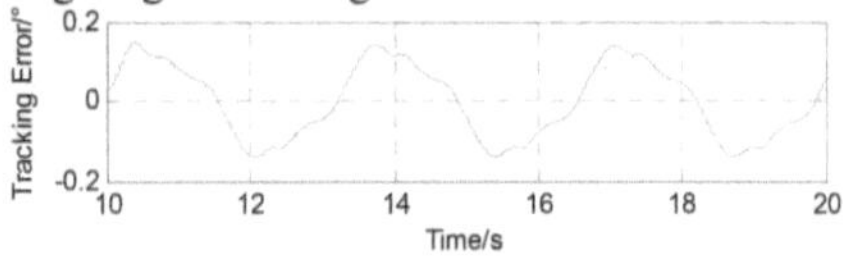

c) Steering Angle Tracking Error

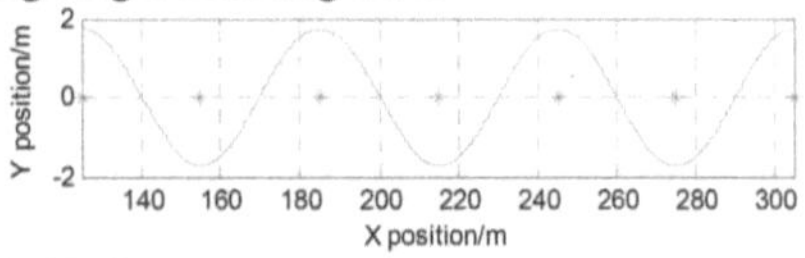

d) Vehicle Trajectory

Figure 12. Simulation result on Test Condition 1 (Double B_{eq}).

By targeted design observer bandwidth and weighting function, we get a second-order observer and a fourth-order controller. The simulation result shows that it can ensure well tracking performance and is not sensitive to system parameters in typical steering conditions. Moreover, the observer and controller is easy to implement on automobile electronic control platform.

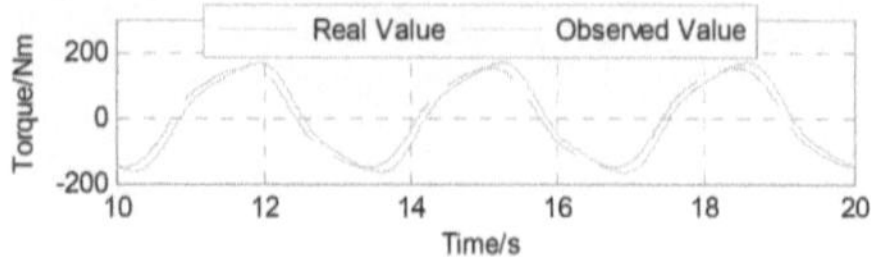

a) The Observer Output

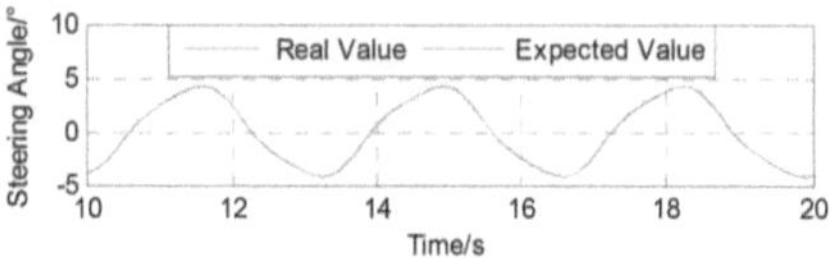

b) Steering Angle Tracking Effect

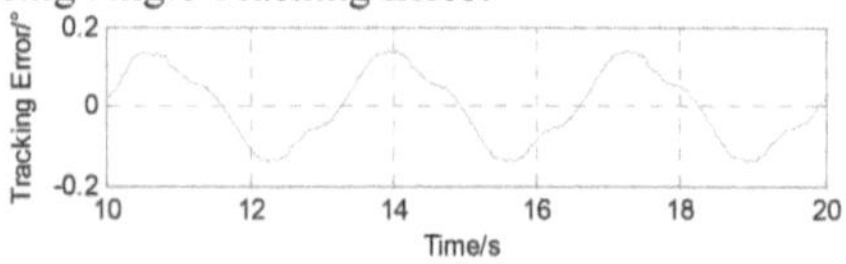

c) Steering Angle Tracking Error

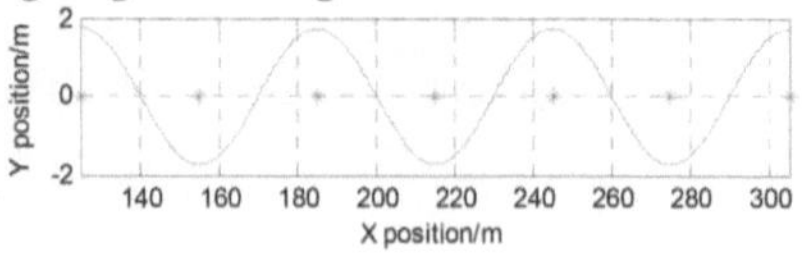

Figure 13. Simulation result on Test Condition 1 ($B_{eq} = 0$).

ACKNOWLEDGEMENT

Supported by National High Technology Research and Development Program of China (863 Program) (2012AA110701).

REFERENCES

[1] Bertoluzzo, M., Buja, G., & Menis, R. (2007). Control schemes for steer-by-wire systems. *Industrial Electronics Magazine, IEEE*, 1(1), 20–27.

[2] Im, J. S., Ozaki, F., Matsunaga, N., & Kawaji, S. (2007, October). Control of steering-by-wire system using bilateral control scheme with passivity approach. *Control, Automation and Systems, 2007. ICCAS'07. International Conference on* (pp. 1488–1493). IEEE.

[3] Zhai, P., Du, H., & Zheng, L. (2013). Bilateral Control of Vehicle Steer-by-Wire System with Variable Gear-Ratio. *Industrial Electronics and Applications, 2013 8th IEEE Conference on* (pp. 811–815). IEEE.

[4] Wang, X., Zong, C., Xing, H., & Hu, R. (2012). Bilateral Control Method of Torque Drive/Angle Feedback Used for Steer-by-Wire System. *SAE International Journal of Passenger Cars-Electronic and Electrical Systems*, 5(2), 479–485.

[5] Wang, H., Man, Z., Shen, W., & Zheng, J. (2013, June). Robust sliding mode control for Steer-by-Wire systems with AC motors in road vehicles. *Industrial Electronics and Applications (ICIEA), 2013 8th IEEE Conference on* (pp. 674–679). IEEE.

[6] Do, M., Man, Z., Zhang, C., Wang, H., & Tay, F. (2014). Robust Sliding Mode-Based Learning Control for Steer-by-Wire Systems in Modern Vehicles. *IEEE TRANSACTIONS ON VEHICULAR TECHNOLOGY*, 63(2), 580–590.

[7] Qiu, G., Luo, X., Wang, P., Wu, D., & Yang, Z. (2008). Design and Simulation of PID Controller Based on PMSM Servo System. *Journal of Chongqing University*, 3, 005.

[8] Mitschke, M., & Wallentowitz, H. (1972). *Dynamik der kraftfahrzeuge* (Vol. 4). Berlin: Springer.

[9] Doyle, J. C., Glover, K., Khargonekar, P. P., & Francis, B. A. (1989). State-space solutions to standard H_2 and H_∞ control problems. *Automatic Control, IEEE Transactions on*, 34(8), 831–847.

[10] Gahinet, P. M., Nemirovskii, A., Laub, A. J., & Chilali, M. (1994, December). The LMI control toolbox. In *IEEE Conference on Decision and Control* (Vol. 2, pp. 2038–2038). INSTITUTE OF ELECTRICAL ENGINEERS INC (IEE).

Future Mechatronics and Automation – Yang (Ed.)
© 2015 Taylor & Francis Group, London, ISBN 978-1-138-02648-3

Design and rendering of the 3D Lotus Pool by Moonlight

Yu-Xian Hui & Wei-Guang Liu
Zhong Yuan Institute of Technology, Zhengzhou, Henan, China

ABSTRACT: The Lotus Pool by Moonlight plays an important role in common natural beautiful scenery. In order to simulate this natural scene fast and realistically, new model and method are presented by this paper. First of all, the efficiency and vividness of rendering virtual scene can be improved by accelerating the optimization technology; then the lotus flowers and fish model can be reconstructed by reading the 3DS model of 3DSMAX, shinning stars can be rendered by using particle algorithm, and the sound of insect and cicadas are added to the virtual scene; finally, the interactive roaming can be realized by capturing the information of keyboard and mouse. Water spray splatters can be simulated when a collision occurs between fish and pond. The experiment shows that the roaming system given in this paper has the characteristic of timeliness and vividness.

1 INTRODUCTION

Virtual scene roaming technology is the integration of high-tech computer technology, including computer graphics, multimedia, and artificial intelligence and so on. It provides people with the support of experience in the virtual world, and is the combination of the virtual scene building technology and the virtual roaming technology. This paper will be based on the graphics; the relevant models are established by using 3DSMax, and rendered by using OpenGL, which finally realizes the lotus pond moonlight visual roaming on the platform of VS2010.

2 THE ARCHITECTURE DIAGRAM OF 3D MOONLIGHT

The architecture diagram of 3D moonlight is shown in Figure 1. It consists of optimized roaming technology module, 3D scenery construction module, and interactive roaming scenario module.

The scene can be browsed more smoothly and realistic because of using optimized roaming technology; the lotus flowers and fish model can be reconstructed by reading the 3DS model of 3DSMAX; shinning stars can be rendered by using particle algorithm, and the sound of insect and cicadas are added to the virtual scene by calling the function PlaySound(); the interactive roaming can be realized by capturing the information of keyboard and mouse, and by the rules of ray tracking, Water spray splatters can be simulated when a collision occurs between fish and pond.

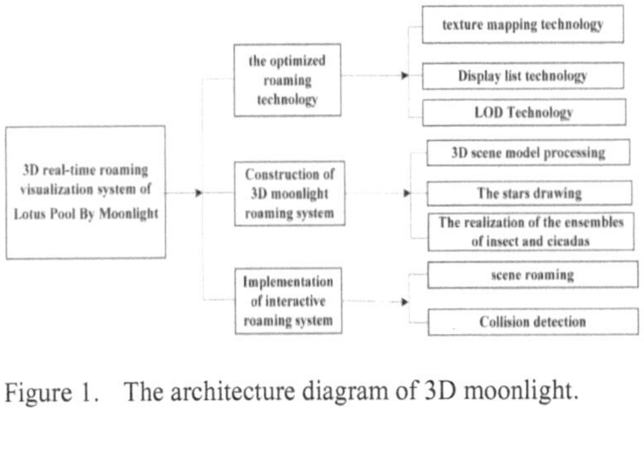

Figure 1. The architecture diagram of 3D moonlight.

3 THE OPTIMIZED ROAMING TECHNOLOGY

3.1 *Texture mapping technology*

The virtual scene models can be added texture such as photos as real appearance, without increasing the system memory and the quantity of polygon rendering. The pixels and geometric objects are combined by the texture mapping technology to create a vivid visual effect. Generally speaking, compared to the details of polygon features, texture features' occupy much less system resources, which can improve the real-time virtual scene roaming effectively.

Assuming that there is a pixel array $N \times M$, which is viewed as a continuous array instead of discrete elements. Any point in the array can be defined by the variables s and t. For each coordinate value, there is a pixel with the corresponding value. The continuous array is looked as a 2D texture mapping of which the key is how the 2D texture space will be mapped to actual scene. Because of different mapping definition function F, there are different effects of maps. So the mapping function F will directly affects the authenticity of the final virtual scene in the texture mapping algorithm.

Now considering a geometric object in 3D space, each point of its surface corresponds to the coordinate (x, y, z) of 3D world coordinates. From the

Figure 2. The pond lake.

mathematical point of view, mapping function F can be represented by equation (1)

$$(s,t)=F(x,y,z) \quad (1)$$

There are two steps of texture mapping technology as follows. 1) To set the texture attributes, 2) Mapping between texture and model space, model space and the screen space.

3.1.1 *Transparent texture technique*

Transparent texture technique is implemented by the texture technology and fusion technology, which can improve the scenery vividness. Fusion technology is fusion function combination of the source and destination color value, so that parts of the scene are displayed as transparent, usually using the color mode in A of RGBA, or Alpha to achieve.

3.1.2 *TSFS technology*

Although the texture mapping can greatly simplify the design of virtual scene, it occupies a certain system space. It will bring system heavy burden if a large number of higher resolution textures are used. While TSFS (Texture Synthesis from Samples) plays an important role in simulation of large-scale monotonous scenes, such as simulated lake, grass, clouds or flame. It can obtain good simulation results with the appropriate image processing. The pond lake of this procedure is as shown in Figure 2.

3.2 *Display list technology*

Display list technology is one of program optimization tool of OpenGL. It draws faster than instantaneous speed.

Once the display list is processed into a suitable for graphics hardware scheme, the degree of optimization is different because of different command of OpenGL. For example, the rotation matrix function glRotate* (). If it is put in the display list, which can greatly improve the performance of system. Because the calculation of rotation matrix is very complex, including a complex operation such as square, triangle functions. While it is stored as a final rotation matrix in the display list, so the execution speed can be as fast as the hardware implementation of glMultMatrix() function. In general, display list can matrix transform many adjacent combined into a single matrix multiplication, thus speeding up the speed.

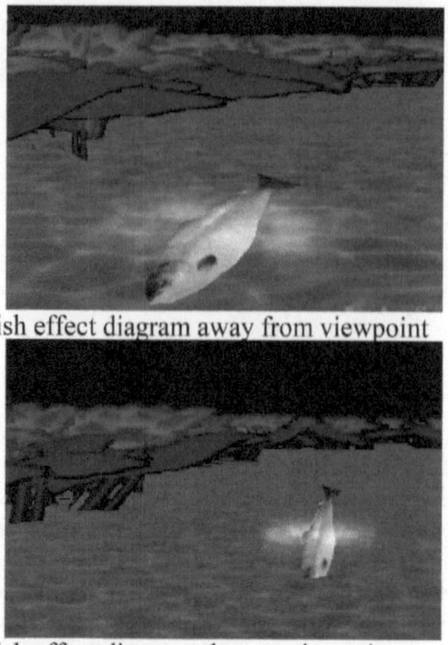

(a) The fish effect diagram away from viewpoint

(b) The fish effect diagram close to viewpoint

Figure 3. (a), (b) The fish effect diagram.

3.3 *LOD technology*

LOD (Level of Detail) technology can describe solid level of detail model by using a group of varying complexity. According to some objective criteria in the process of simulation, between these LOD models can be switched, which can change the complexity of scene in real time.

According to the viewpoint change, there are different levels of detail. As shown in Figure 3(a), (b), the closer the viewpoint, the bigger the fish, and fish details are more clearly; the farther the distance, the smaller the volume of fish, and the detail is fuzzier.

There are generally 3 kinds of standards to control LOD model switching: 1) distance standard, 2) eccentricity standard, 3) motion standard.

4 CONSTRUCTION OF 3D MOONLIGHT ROAMING SYSTEM

The development of moonlight scene roaming system is under the Window operating system, using VS2010 as the development platform, using 3DSMAX to construct the model, using PhotoShop to deal with texture mapping, and OpenGL is viewed as application program interface to construct 3D scene.

4.1 *3D scene model processing*

4.1.1 *PhotoShop texture processing*

Adobe PhotoShop referred to as "PS", it has a lot of functions, involved in various aspects of graphics, image, and text, video, publishing and other related. To add the Alpha channel using PS texture mapping

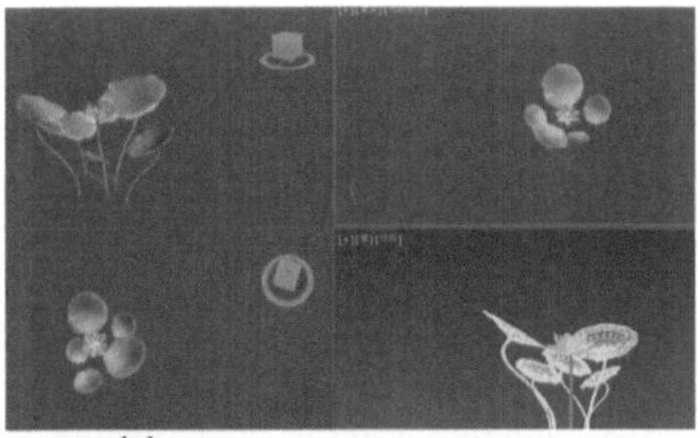

(a)3D lotus model.

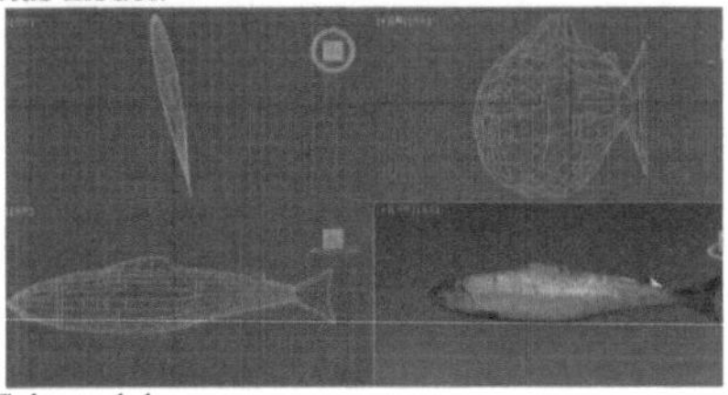

(b) 3D fish model.

Figure 4. 3D models diagram.

on the model, so that the channel selection is the black background. The black part will be transparent form when the screen displays during calling OpenGL function to render.

4.1.2 *3DSMAX modeling*

As shown in Figure 4, Lotus and fish models of the moonlight are constructed by using 3DSMAX software. 3DSMAX was launched by Autodesk, a set of 3D design software. The software has abundant functions of modeling. After the modeling is completed, it will be exported as 3DS file, then OpenGL reads these files and draws the corresponding model.

4.1.3 *OpenGL rendering*

In recent years, OpenGL is developed a high performance 3D graphics standard. It is launched by SGI Company, gradually mature, and a high performance software development of virtual reality, interactive scene simulation package. It has the advantages of high reliability, scalability, flexibility characteristics and has been widely applied in the military, architecture, products etc.

As shown in Figure 5, standard functions are provided by OpenGL, such as glTranslatef() and glRotatef(), to set virtual coordinate position in the screen and to render the 3DS model.

4.2 *The stars drawing*

Particle system can simulate some special fuzzy phenomenon in the 3D computer graphics while it is very hard to realize the realistic scene by other traditional technology.

The basic principle of particle algorithm is to simulate the fluid mechanics of objects using a large number of particles.

This paper uses particle algorithm to achieve the star twinkling drawing, as shown in Figure 6. There are thousands of irregular, random distributions of small stars in the sky, and each star has a certain life cycle,

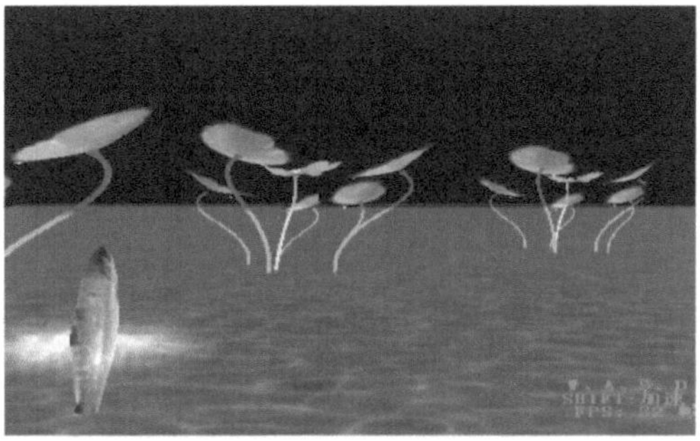

Figure 5. Using OpenGL to render 3DS model.

Figure 6. The stars drawing.

to experience the "birth", "movement and growth" and "death" three stages, reflecting the dynamic and stochastic.

4.3 *The realization of the ensembles of insect and cicadas*

In addition to the visual requirements, to create atmosphere, the sound is more important in virtual reality. In order to show the design ideas more fully and experience the beauty of the Lotus Pool by Moonlight, the insect sound and cicadas' ensembles are joined.

The design steps are as follows:

1) adding following codes to "stdAfx.h".

#include <mmsystem.h>

#pragma comment(lib, "winmm.lib")

2) adding function PlaySound() to "opengl.cpp" which can realize the scene rendering. The codes are as follows.

PlaySound("texture/concert.wav", NULL, SND_FILENAME);

5 IMPLEMENTATION OF INTERACTIVE ROAMING SYSTEM

5.1 *Scene roaming*

As for the virtual scene, if the viewers can only automatic roaming among them, but not interact with the surrounding environment, then it is difficult for users

Figure 7. The operation of scene roaming.

Figure 8a. The initial view scene.

Figure 8b. The view scene after switching of view.

to have the feeling of be personally on the scene. In order to make users control the scene and enhance the reality of scene animation, user actions need to be captured. So some input devices are depended on to achieve human-computer interaction, such as keyboard, mouse.

3D required positioning information function can be achieved in the scene by using the mouse. After Keyboard keystrokes are captured by application, they are converted into the corresponding control command, then the control transform are realized on the scene. In the actual operation, sometimes two or combinations of equipment are needed to complete scene roaming.

This paper uses GLUT utility library to carry out the corresponding message capture. Use of "W, A, S, D, shift, Ctrl" to control the forward, backward, to turn left, to turn right and the acceleration and deceleration function. In the fact, users' interaction with the scene displays the landscape screen redraw, as shown in Figure 7, Figure 8a,b.

Figure 8a,b the view scene the part screenshot of the Lotus Pool By Moonlight.

5.2 *Collision detection*

How to judge timely and accurately whether the collision occurs between objects is inevitable in the study of roaming system.

In the moonlight roaming system, in order to realize the fish fluttering water dynamic picture, this paper adopts a set of certain rules in the ray tracing algorithm.

Under normal circumstances, the light from a starting point is along the specified direction vector to proceed, the light can be expressed as:

$$\overrightarrow{Raystart \quad PointOnRay}$$

$$PointOnRay = Raystart + t * Raydirection \tag{2}$$

$t \in [0, \infty)$, *PointOnRay, Raydirection* are 3D vectors. The variable t represents distance between the light and the detection plane.

The detection plane can be represent using equation (3):

$$X_n \cdot X = d \tag{3}$$

X_n: The normal vector of the detection plane.
X: 3D coordinates of point inside the plane.
d: The dot product of X_n and X.

Then ray and plane equation are as follows:

$$\begin{cases} PointOnRay = Raystart + t * Raydirection \\ X_n \cdot X = d \end{cases} \tag{4}$$

If a light line and a plane intersect, there must be some points on the line which satisfy the formula (5):

$$X_n \cdot PointOnRay = d \tag{5}$$

Equation (2) is added to the above equation, the result is as follows:

$$X_n \cdot Raystart + t * (X_n \cdot Raydirection) = d \tag{6}$$

t is calculated using formula (7):

$$t = d - X_n \cdot Raystart / (X_n \cdot Raydirection) \tag{7}$$

After a serious of calculating, t can be expressed as formula (8):

$$t = (X_n \cdot (PointOnRay - Raystart)) / X_n \cdot Raydirection \tag{8}$$

Then the collision point can be calculated. There are two special situations that will don't bring collision:

$$\begin{cases} X_n \cdot Raydirection = 0; & a) \\ t < 0; & b) \end{cases}$$

Figure 9. The scene that fish thrashes water.

a) parallel between Light and plane
b) the collision point behind the starting point of light.

In addition to checking the collision point, whether the collision occurred in the current time slot should also be judged. Time slice (T) is unit step that the object from the current position along the direction of velocity displacement when the time is at a certain point. If the collision occurs, the collision point and starting distance will be calculated, and then a collision time can be obtained. A unit step is Dst, distance between the collision point and starting point is Dsc, time slice is T, the collision time (Tc) is:

$$Tc = Dsc * T / Dst \tag{9}$$

Assume that $T = 1.0(s)$; *Velocity* is roaming velocity; t has been calculated; then collision time can be calculated. If collision occurs after 0.7 seconds, it means that from the beginning of the time to 0.7 seconds after that a collision will occur. The coordinate's formula of collision point is as follows:

$$Collision \quad po\text{int} = Start + Velocity * Tc \tag{10}$$

For fish in the pond

{

When fish jumps, the distance between fish and water are calculated.

If the collision hasn't occurred, save and replace the nearer collision point.

}

If collision occurs (fish falling down the pond)

{

There are water spray splatters splashing

}

Otherwise, fish are jumping. Repeat the above steps.

The scene that fish thrashes water is shown in (Fig. 9).

By the above technique, the built models are imported into OpenGL, given corresponding texture and light. Then LOD scene display and collision detection are carried out. At the same time, the viewpoint is as virtual roaming system path.

The roaming system given in this paper has the characteristic of large scale, timeliness and vividness so that participants are personally on the scene of pleasure.

6 CONCLUSIONS

The 3D real-time roaming visualization system of Lotus Pool by Moonlight is developed on the platform of VS2010 combined with OpenGL. During the roaming, users can experience visual and auditory enjoyment. From the simulation results, the roaming system given in this paper has the characteristic of timeliness and vividness. In addition, it also has the very good reference value and practical significance.

But this system has also shortcomings. If 3D realistic model is called largely and repeatedly, rendering time will be longer, which affects the real-time performance of the whole scene roaming. The direction of future research is how to improve and optimize the model, which can achieve the fidelity and real-time balance.

REFERENCES

Li Ding-feng. 2012. The General development platform about 3D simulation system based on OpenGL. Railway Investigation and Surveying.

Li Qing-ling & Li Shuan. 2011. 3D simulation of human motion based on OpengGL. Computer Simulation 28(4):273.

Reeves W.T. & Blau R. 1985. Approximate and probabilistic algorithms for shading and rending structured particle system. In: Proceedings SIGGRAPH. New York: ACM Press, 313–322.

Wang Yan-an & Zeng Jun-feng. 2011. The exploration of garden design by virtual reality. Journal of Heilongjiang Vocational Institute of Ecological Engineering 22(5):158.

Wang Ying-ying. 2012. The Design and implementation of virtual roaming system of pastoral based on OpengGL. Computer and Information Technology 20(5):30–32.

Xie Shi-fu & Ma Li-yuan. 2013. The research and Realization of moving cable collision detection algorithm in Virtual environment. Journal of System Simulation 28(8): 1865–869.

Zhang Ming. 2013. The Research and Implementation of 3D dynamic flower simulation. Journal of Anhui Agricultural Sciences 41(3):1351–1353.

Zhou Xiao-jun & Wang Guang-xia. 2013. The fast organization and expression of multiple electronic map data based on OpenGL. Geo-Information Science 15(4):492–496.

Zhang Yu-jun. 2012. The Application of texture in visual simulation system based on OpenGL. Railway Investigation and Surveying 41(7):62–64.

Future Mechatronics and Automation – Yang (Ed.)
© 2015 Taylor & Francis Group, London, ISBN 978-1-138-02648-3

Finite element analysis and optimization of an economical welding robot

Shuwan Cui
Masters in Reading, Guangxi University of Science and Technology, College of Mechanical Engineering, Liuzhou, Guangxi, China

Jianjun Wei
Guangxi University of Science and Technology is Currently Dean of the Faculty of Mechanical Engineering, Liuzhou, Guangxi, China

ABSTRACT: The finite element analysis with ANSYS Workbench of the statics properties and whole structure of a welding robot indicates that the upper and lower arms bear the maximum stress and strain. Topological optimization of the upper and lower arm structure are carried out, and an optimized model is created and re-analyzed with ANSYS Workbench. The analysis results show that the weight of the optimized welding robot is reduced while the strength is enhanced, which means a better cost performance.

Keywords: Economical welding robot; Finite element analysis; Topological optimization

1 INTRODUCTION

Welding automation is an inevitable trend of advanced manufacturing technology development. It is also an effective way to improve welding productivity, ensure product's quality and reduce labor intensity[1]. As a family member of industrial robots, welding robots is one of the automated welding systems, which are flexible and intelligent, going beyond the traditional welding automation[2].

The applications of domestic welding robots are restricted in some cases because of their quality problems. The restrictions of their weight also limit their applications in a number of occasions. Meanwhile, they are also inhibited in SMEs because of expensive prices. So on the premise of ensuring its strength, lightening welding robot's weight can not only expand the scope of applying welding robots, but also reduce its costs. This paper mainly focuses on the welding robot in Figure 1, using its static analysis to find out a relatively intense and weak position, and then utilizing the topological optimization method to identify the surplus position of the robot's weight. Finally design the welding robots by the quality transferred optimization way[3]. The specific analysis process is shown in Figure 2.

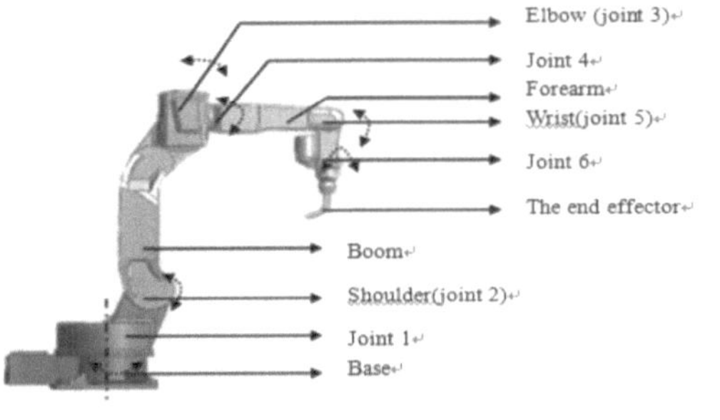

Figure 1. Schematic diagram of the overall economy robot welder.

2 ESTABLISHMENT OF ECONOMICAL FINITE ELEMENT MODEL OF THE WELDING ROBOT

Use three-dimensional software—Soild Works to establish welding robot solid model which is shown in Figure 3. The base of welding robot is fixed on the ground. When the welding robot's arm stretch to the maximum, that is, when the upper and lower arms in horizontal line, welding robot's weight is under the effect of gravity, load and other facts, and the displacement of the end effector is in maximum deformation.

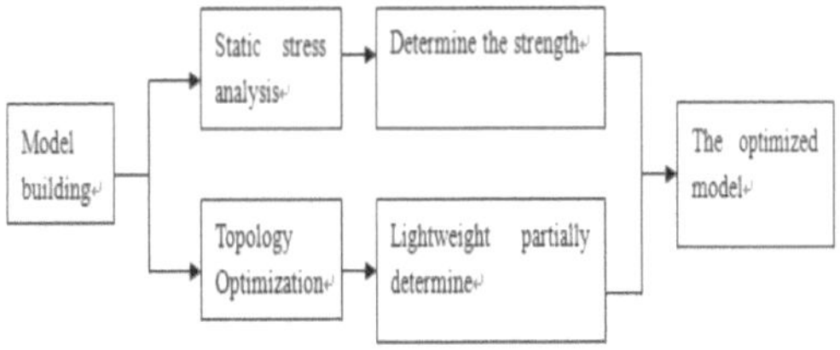

Figure 2. The overall flow chart.

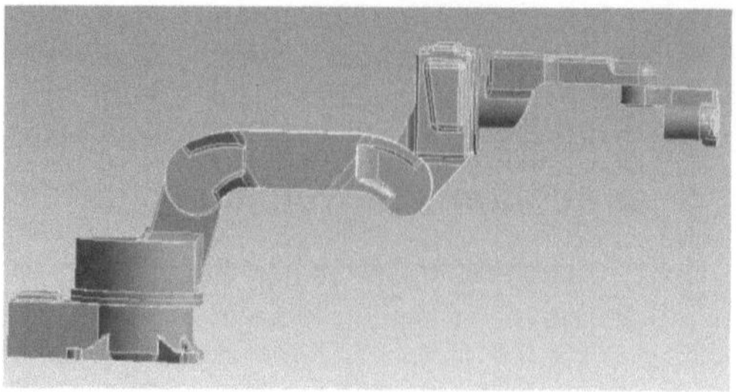

Figure 3. Welding robots in the most dangerous position and orientation.

Table 1. Welding robots Specifications.

Tectonic		Vertical multi-joint
Number of axes		6-axis
Maximum load capacity		4 kg
The body weight		170 kg
Range of motion	1-axis	±170° (±50°)
	2-axis	−155°∼90°
	3-axis	−170°∼180°
	4-axis	−155°∼155°
	5-axis	−45°∼255°
	6-axis	−205°∼205°

Table 2. Materials and attribute table.

	Value	
Material	Upper arm	Lower arm
Property	Malleable cast iron	Hard aluminum
Density (kg/cm^3)	7.3	2.7
Modulus of elasticity (GPa)	190	71
Poisson's ratio (μ)	0.3	0.3

In this case, the welding robot is in the most dangerous working conditions, so this paper focuses on the study of welding robot[4]. Specification parameter of welding robot are shown in Table 1.

3 FINITE ELEMENT ANALYSIS OF WELDING ROBOT

The CAD model is imported to ANSYS boom Workbench interface By Soildworks for analysis directly, it is good for data integrity[5–8].

3.1 *Definition of material*

The specific properties of the material of the Welding robot as shown in Table 2.

3.2 *Meshing*

After the material parameters are set, the overall cell meshing of welding robot is made by Soild 45, of the formation are 143,287 elements, 80,814 nodes after division. The meshing fig is shown in Figure 4.

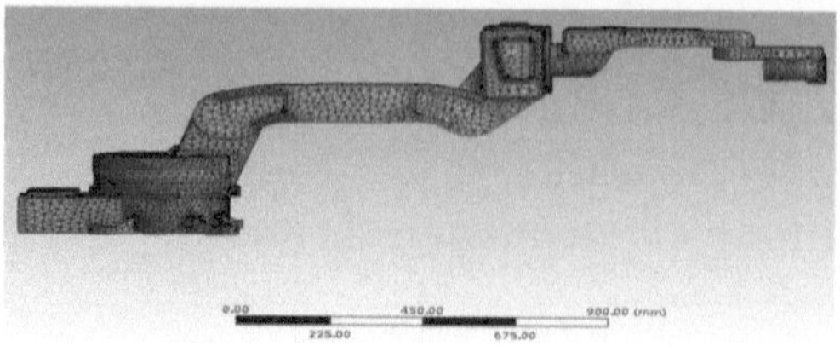

Figure 4. Overall mesh welding robot.

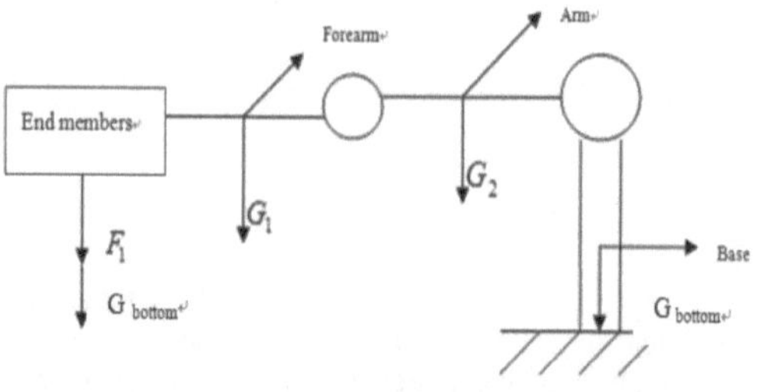

Figure 5. Force diagram welding robot.

3.3 *Impose constraints and load*

Since the physical model of the robot are connected with a rigid connection ministries, as the gravitational acceleration is applied, the Workbench software will automatically assembly and add gravity to the robotic of the various parts of the body[9].

The force of the Welding robot diagram as shown in Figure 5, first define four hole base as a fixed constraint, since the welding robot can carry a maximum of 4 kg parameters, so 4 kg of force is load to the welding robot at the end.

3.4 *Stress analysis of the structure*

Completion of the welding robot boundary conditions, the static analysis is applied to the welding robot overall showing the strain and stress cloud welding robot after processing, shown in Figure 6.

The analysis results can be seen in the robot arm the wrist stretch is the longest part of the load 4 kg of force. According to Figure 6, the large part of the upper and lower arms strain most likely to occur and should respectively analyzed on the upper and lower arms.

4 WELDING ROBOT ARM, THE ARM OF STRUCTURAL ANALYSIS AND OPTIMIZATION

4.1 *Structural analysis of the upper and lower arms*

The stress analysis of the welding robot are give the corresponding upper and lower arms According to the

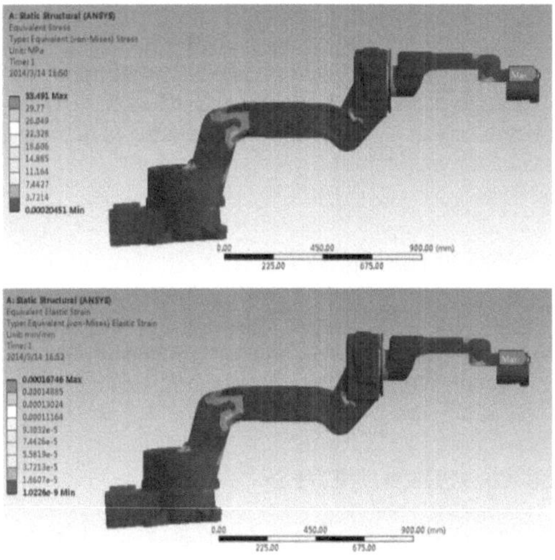

Figure 6. Welding robots overall stress and strain diagram.

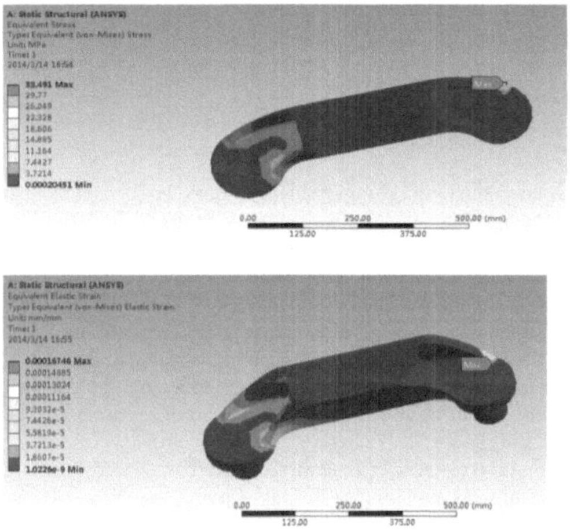

Figure 7. Welding robot arm stress, strain diagram.

overall structure of the strain contours in Figure 7, Figure 8.

The strain deformation can be seen through the upper and lower arms cloud of stress, it is mainly in the arm joints, we can design them.

4.2 *Upper and lower arms of topology optimization*

Topology optimization is also known as shape optimization can shape optimization, the goal of topology optimization is to find a single load or load objects bear the best material distribution program. This approach reflected in topology optimization as "the greatest stiffness" design[10–11], the difference of the traditional design and the topology optimization is the latter does not need optimized topology optimization and optimization of parameters defined. Upper and lower arms topology optimization mathematical model is used:

Solving: $x = \{x_1, x_2, \ldots\ldots x_n\}^T$

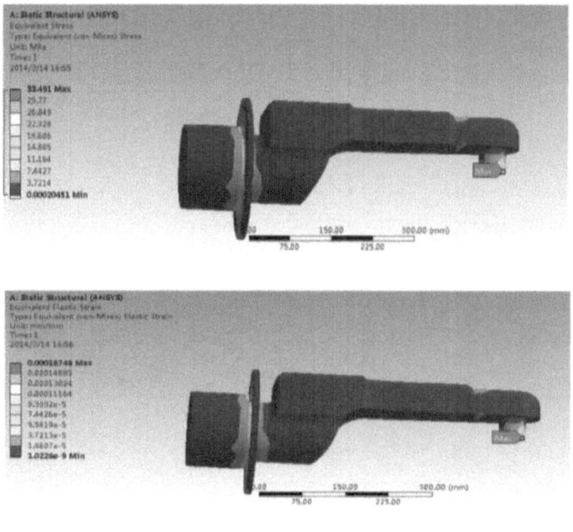

Figure 8. Welding robot forearm stress, strain diagram.

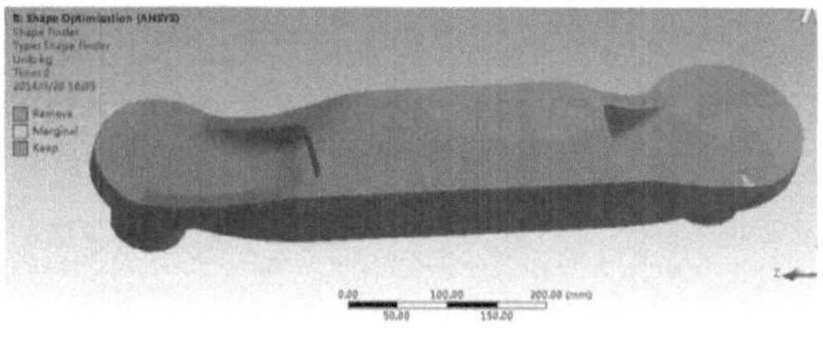

Figure 9(a). Mass reduction of 10%.

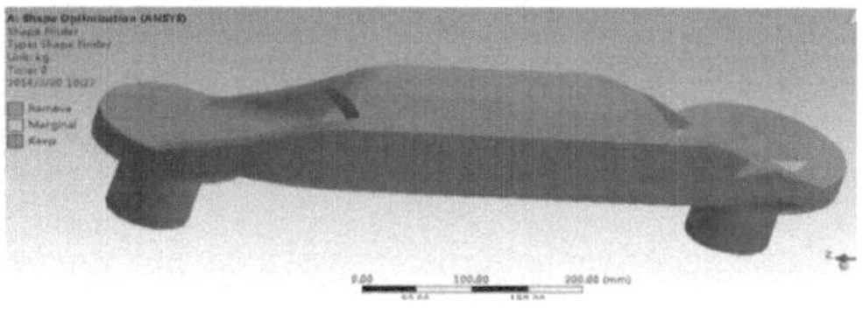

Figure 9(b). Mass reduction of 20%.

Where the objective function is: $C = F^T U$
Constraint:

$$\begin{cases} \dfrac{V_1}{V_0} \le e \\ 0 < x_{\min} < x_a < x_{\max} \\ F = kU \end{cases}$$

where C-type structure in flexibility, F-force vector, U-displacement vector, K-element stiffness matrix, e-optimized volume, xi-the effective density of unit, V1-the optimized volume remaining materials, V_0-the design area Volume, xmax, xmin-effective density unit off-line and on-line.

It is that upper arm is relatively dangerous than the lower arm's position According to the practice and analysis, so here we only make topology optimization for upper arm. the topology optimization reduction target is made of 10%, 20% to find the location and quality of surplus trend setting quality during, topology optimization results shown in Figure 9.

Figure 10(a). Improved front structure.

Figure 10(b). The improved structure.

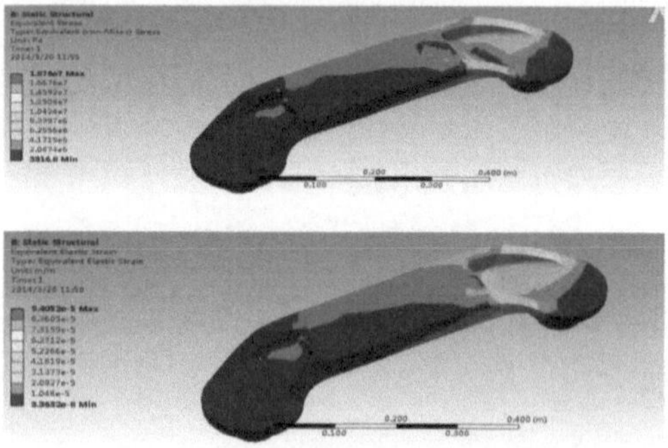

Figure 11. After optimization of the arm stress, strain cloud.

4.3 *Optimal design of arm*

It shows plate arm holes before assembly line can be appropriately removed to improve the program. By statics analysis and arm topology optimization thick cloud, the program and improve contrast as shown in Figure 10 with the actual conditions.

5 OPTIMIZATION RESULTS VERIFIED

Structure will be optimized again import FEM, stress analysis stress and strain contours shown in Figure 11. You can find structural stress is significantly cut, down 28.59%. At the same time, reducing the quality of 7.27%.

6 CONCLUSION

We found stress-strain large part based on the static analysis of the welding robot upper and lower arms, the proposed corresponding boom structure optimization design method is suggested on the basis of arm static analysis and topology optimization, By optimizing the stress makes welding robot arm slashed 28.59%, Quality reduced by 7.27%. The program targeted therefore has better applicability.

ACKNOWLEDGEMENT

Scientific and technological projects in Guangxi (Item Number: 12118007-11), Guangxi Graduate Education Innovation Program (Item Number: YCSZ2014198).

BIOGRAPHICAL NOTES

Cui Shuwan, born in 1989, Masters in Reading, Engaged in mechanical and electrical integration of research.

Wei Jianjun, born in 1964, Professor, Guangxi University of Science and Technology is currently Dean of the Faculty of Mechanical Engineering.

REFERENCES

Duan Jianzhong, Wang Xiaoyang. Soildworks mechanical design tutor-finite element static analysis [M]. Beijing: Electronic Industry Press, 2009.

Gong Shuguang editor, Xie Guilan, eds. ANSYS engineering application parsing [M]. Machinery Industry Press, 2003.

Hu Shengsun, welding automation technology and its application [M] Beijing: Mechanical Industry Press, 2007:2–4.

Lin Shangwu, Chen Shanben, Li Chengtong. Welding robot and its application [M]. Beijing Publishing House, 2000:2–4.

Pu Guangyi. ANSYS Workbench12 based tutorials and examples explain [M] Beijing: China Water Power Press, 2010:65–69.

Shi Guanglin, Zhu Lin, clutch cage positioning device design [J]. Mechanical design and manufacturing, 2012 (12): 137–139.

Shi Jiankui. Automobile link modal finite element analysis [J]. Equipment Manufacturing Technology. 2009(06).

Sigmund, O. A 99 line topology optimization code written in Matlb [J]. Educationalarticle, 2001(21): 120–127.

Xiao Zhiyong, Duan Jianzhong, Du Xinqiang, Li Jian. Motoman industrial robots static finite element analysis and design improvement [J]. Changsha University of Technology, 2011 (12):71–75.

Yang Guoliang, industrial robot dynamics simulation and finite element analysis [D], Wuhan: Huazhong University of Science and Technology, 2007:56–58.

Zhong Peisi, Zhao Dan, Xue Sunyan, Wei Qun analysis based on ANSYS modeling and automobile frame modal [J]. Mechanical design and manufacturing. 2008 (06).

Future Mechatronics and Automation – Yang (Ed.)
 ISBN 978-1-138-02648-3

Modal analysis on the instrument panel bracket of automotive

Shuwan Cui
Masters in Reading, Guangxi University of Science and Technology, College of Mechanical Engineering, Liuzhou, Guangxi, China

Jianjun Wei
Guangxi University of Science and Technology is Currently Dean of the Faculty of Mechanical Engineering, Liuzhou, Guangxi, China

ABSTRACT: In this paper, the author took the instrument panel bracket of light vehicle as the research object, focused on the vibration problem of the vehicle, adopted the ANSYS and Workbench software to do modal analysis on the instrument panel bracket for ensuring its low order vibration mode and frequency, and compared the analysis results with the permissible value, providing the theoretical basis and reference for improving and optimizing the structures followed.

Keywords: Dashboard bracket; Modal analysis; Topology optimization

1 INTRODUCTION

The instrument panel bracket is the key load-bearing part of the instrument panel assembly and attachments. Its general structure constitutes the frame beam from left to right and welding bracket for withstanding all kinds of electronics, air conditioning and steering module.

Modal analysis of structure not only reflects the rigidity of the whole vehicle body and the beam frame, but also the key index controlled the conventional vibration of vehicle[1]. The manufacturing process adopted for the instrument panel beam of automobile is welding. Its welding joints are quite more and the bearing performance is very crucial. Therefore, it is necessary to carry out the modal analysis which took the instrument panel bracket as research object.

In this paper, in order to ensure the accuracy of the analysis results, it's shown as the Fig. 1, the author took the light vehicle instrument as the research object, carried out modal analysis on the whole part by the FEA and compared the analysis results with the permissible value, providing the theoretical basis and reference for improving and optimizing the structures followed.

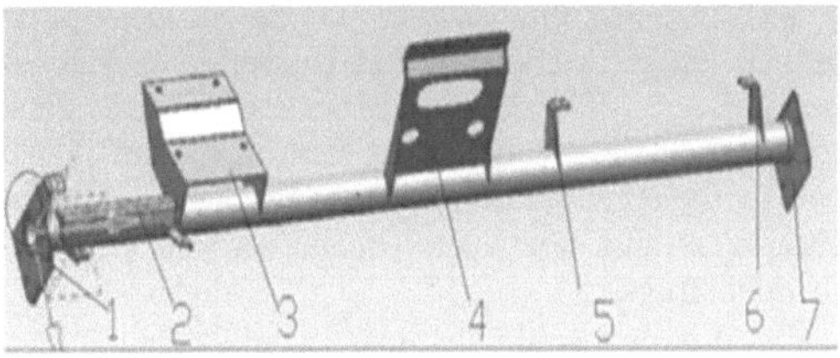

1-No. 1 end bracket
2-Big mounting bracket
3-Cassette player support
4-Steering column bracket
5-No.1 small mounting bracket
6-No.2 small mounting bracket
7-No.2 end bracket

Figure 1. The dashboard support overall structure diagram.

2 MODAL ANALYSIS

Modal analysis is an approximation method to study the characteristics of the structure. Its purpose is to describe the characteristics of the natural frequencies and mode of vibration[2]. The modal analysis can be achieved through the analytical analysis and experimental analysis these two methods. Analytical modal analysis, namely, according to the structural geometry, boundary condition and material properties, use the structure of the quality matrix, stiffness matrix and damping matrix to reflect the dynamic characteristics of quality, stiffness and damping distribution and so on. Through using CAE, establishing the finite element model of complex structures, and solving the structural dynamic parameters, it carried out the analysis of frequency response[3], and then it can be used to optimize and correction of the structure.

The structure of the instrument panel bracket model in this paper, its basic equation is as followed[4]:

$$[M]\{\ddot{X}\}+[K]\{X\}+[C]\{\dot{X}\}=\{F(t)\} \tag{1}$$

When the external force and the damping effect can be ignored, the mathematical model of the system can be used the following equation:

$$[M]\{\ddot{X}\}+[K]\{X\}=\{0\} \tag{2}$$

In the formula (2), [M] means the quality matrix, [K] is damping matrix, $\{\ddot{X}\}$, $\{\dot{X}\}$, $\{X\}$ are vector acceleration, velocity and displacement, while {F(t)} is the system suffered external load vector.

Supposed the answer to the above equation is:

$$\{X\}=\varphi \sin \omega t \tag{3}$$

In the formula (3), φ is the feature vector or model of vibration while ω is the frequency, apply the formula (3) into the formula (2) and make $\lambda=\omega^2$, and then it can get the following equation:

$$[K]\varphi-\lambda[M]\varphi=0 \tag{4}$$

In the formula (4), the condition of its answer is not zero is its matrix determinant is 0, that is:

$$\det([K]\varphi-\lambda[M])=\left|[K]-\lambda[M]\right|=0 \tag{5}$$

Expansion of the formula (5) can get an N polynomial equation about λ, the answer to this equation $\lambda_1, \lambda_2 \ldots, \lambda_n$, are the eigenvalues of the system. Applying $\lambda_1, \lambda_2 \ldots, \lambda_n$ into the formula (4), and then it can get the eigenvector φ_i corresponding to λ_i. Every eigenvalue and eigenvector decides a free vibration form of structure.

Because in real life, the frequency vibration higher than 10 is relatively rare, and the high frequency vibration influence on the dynamic characteristics of structure is quite small, in order to get the natural frequency in the first 7 step of the bracket, the number 7 is chose as the expansion module in the process of setting solution.

3 MODAL ANALYSIS ON THE INSTRUMENT PANEL BRACKET

Using Soildworks software to establish a model of the instrument panel bracket of light vehicle, in order to ensure the precision, the finite element model must be able to reflect the geometric and mechanical characteristics of the instrument panel assembly[5–6]. Through Soildworks and ANSYS seamless interface, it can directly apply the CAD model of the instrument panel bracket into ANSYS Workbench for analysis.

(1) Definition of materials: Beam frame and the instrument panel bracket are made from low carbon steel and low alloy steel mostly. In this paper, the instrument panel bracket material is CR340 and its components are connected by welding joints.

Table 1. The first seven order modal frequency.

Order number	Modal frequency/Hz
1	64.495
2	71.119
3	91.908
4	136.59
5	222.33
6	262.62
7	331.67

Figure 2(a). The dashboard stent first-order modal.

Figure 2(b). The dashboard stent second-order modal.

Figure 2(c). The dashboard stent third-order modal.

(2) Mesh: The overall structure of the instrument panel bracket mesh division, which consisted of 139965 nodes and 36523 units.

(3) Constraints and solution: Because there are multiple connections between connections the instrument panel bracket and body, testing the modal characteristics requires of constrained modal analysis. Constrained modal analysis is the modal analysis on the junction under the condition of setting boundary. Compared with the free modal analysis, it can truly reflect dynamic characteristics of the components. Therefore, the modal frequency value of the first 7 order before applying constraint at the connecting point showed in the Table 1, each mode is shown in Fig. 2.

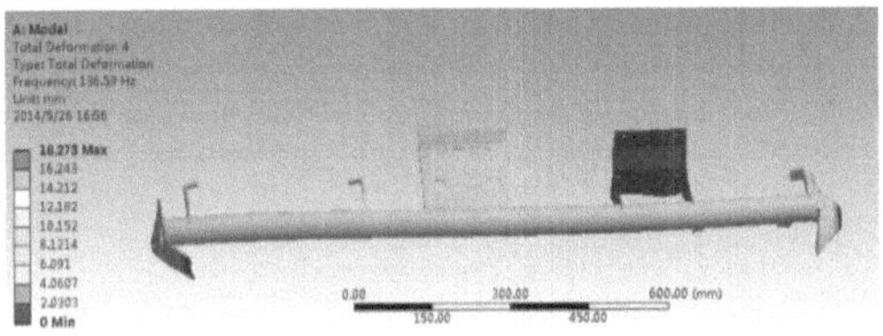

Figure 2(d). The dashboard stent fourth-order modal.

Figure 2(e). The dashboard stent fifth-order modal.

Figure 2(f). The dashboard stent sixth-order modal.

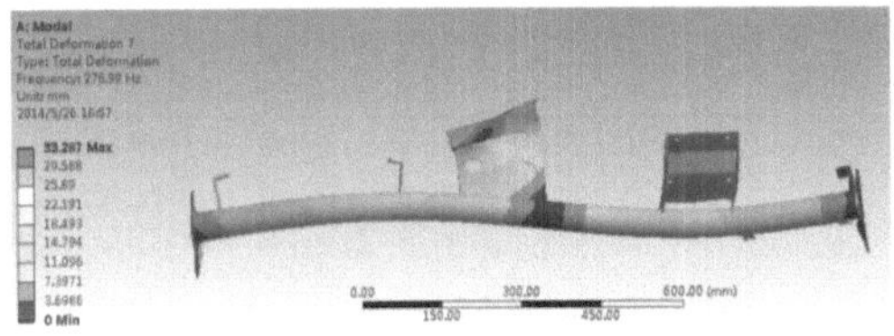

Figure 2(g). The dashboard stent seventh-order modal.

Modal analysis on the frame of automobile, the evaluation indexes are as followed[7–9]:

(1) The low frequency of the frame should be higher than the natural frequency of the structure under the suspension, but at the same time, it must avoid idling frequency of the engine for avoid the resonance phenomenon coming up;
(2) The frame elastic modal frequency should avoid the range of frequency during normal work of the engine as far as possible;
(3) The mode of vibration of the frame should be smooth, no mutation.

But in general, on the highway and better city road, the excitation frequency is of high 1–3 Hz, the vibration frequency caused by wheel imbalance is of low 11 Hz, the excitation frequency caused by engine idling is about 35 Hz, the excitation is quite larger. When the vehicle is in normal driving or driving at a high speed, the rotation rate of the engine is quite high, the excitation frequency is generally greater than 100 Hz. It can be seen from the table, modal frequency of beam bracket at the low order is within a proper range, which is far higher than the idling frequency of the engine, but also avoid the frequency range of the normal work of the engine.

Due to change frequency from the third step to fourth step, the fourth step and fifth step, the sixth step to the seventh step is big, therefore, during the permissible range of manufacturing processes, the appropriate adjustment on size, shape is allowed, which make the change of natural frequency of the frame be smooth and steady, no mutation.

4 CONCLUSION

Through the establishment of model and modal analysis on the instrument panel bracket, it can found that the low order modal frequency of the instrument panel bracket of automobile is 64.495 Hz, which is within the proper range and avoid the idling frequency of the engine. At the same time, because each modal frequency value is quite large, it is the local vibration and resonance will not occur, which provide the theoretical basis and reference for improvement and optimization of the structure followed.

ACKNOWLEDGEMENT

Scientific and technological projects in Guangxi (Item Number: 12118007-11), Guangxi Graduate Education Innovation Program (Item Number: YCSZ2014198).

BIOGRAPHICAL NOTES

Cui Shuwan, born in 1989, Masters in Reading, Engaged in mechanical and electrical integration of research.

Wei Jianjun, born in 1964, Professor, Guangxi University of Science and Technology is currently Dean of the Faculty of Mechanical Engineering.

REFERENCES

Chen Jiarui. Automobile structure [M]. Beijing: People's Traffic Press, 1993:247–285.

Gu Yan. Auto harshness (NVH) design CAE integrated solutions research [D]. Shanghai: Shanghai Jiaotong University, 2004:6–18.

Li Li. Van frame CAE engineering application research [D]. Wuhan: central China University of science and technology.

Li Yingping. Car body modal analysis case study [J]. Journal of automotive technology. 2007 (11).

Liu Weiping. Instrument panel performance analysis and optimization [D]. Wuhan: Wuhan University of technology, 2010-2-2.

Tang Yunjun, Liang Xuewen. Dashboard man-machine performance requirement and application [J]. Journal of business and the development of science and technology. 2009 (02).

Xiao Longxiang. Structure vibration modal analysis [M]. Tianjin: Tianjin Science Press, 1991:75–101.

Zhao Honghui. Car steering wheel vibration analysis and control [D]. Jilin: Jilin University, 2006-6-2.

Zhou Fangming, Yan Yi, Suchen, ChiJin Bo. Car dashboard beam welding based on ANSYS modal analysis. [J]. Journal of Wuhan university of science and technology, 2012 (6).

Future Mechatronics and Automation – Yang (Ed.)
 ISBN 978-1-138-02648-3

Preparing high aspect ratio sub-wavelength structures by X-ray lithography

Y.G. Li
School of Science, Shanghai Institute of Technology, Shanghai, China

S. Sugiyama
Department of Micro System, Ritsumeikan University, Shiga, Japan

ABSTRACT: A high aspect ratio sub-wavelength gratings is fabricated based on X-ray lithography. Polymethylmethacrylate (PMMA) was used as a positive resist in the X-ray lithography process. By using metal Ta as an absorber material, the X-ray lithography mask was fabricated by e-beam lithography and reactive ion etching process. As a result, different aspect ratio submicron structures can be achieved by changing the gap (or spacer) between the mask and the resist based on X-ray lithography. Measurements made by a scanning electric microscope indicate that the roughness of the gratings surface is about 20 nm; the PMMA gratings period is 300 nm and the aspect ratio is 3. The reasons why the fabricated grating shapes are obviously dependent on the gap between the mask and the substrate including the thickness of the resist layer is analyzed using X-ray diffraction.

1 INTRODUCTION

Three-dimensional nano structures with sloped sidewalls, nano gratings in curved surfaces or taped nano structures are key components in the optical system for use primarily in displays such as LCDs, energy fields such as solar cells. The semiconductor manufacturing process can meet nano fabrication demands, and the nano fabrication processs such as X-ray lithography process, electron beam lithography technique (Laddha 2001), and focus ion beam micromachining technology (Utke 2008) were developed. In 2007, high aspect ratio polymer structures fabrication technique is developed and high aspect ratio optical gratings have been created (Kato et al. 2007). The authors have developed high aspect ratio PMMA gratings with line as narrow as 150 nm using X-ray lithography. Based on the optical theoretical calculation, the sub-wavelength-structured (SWS) arrays has a low reflectivity over a wide spectral bandwidth. And sub-wavelength structured (SWS) surfaces can use as antireflection coating on in the solar cell devices (Zhao & Avrutsky 1999). Dr. Kanamori of Tohoku University has fabricated polymethyl methacrylate (PMMA) SWS grating by using a silicon mold technique (Kanamori et al. 2002). SWS surfaces typically behave as layers of single material with designable principal indices of refraction. It allows for the design of broadband wide field-of-view matching layers and artificial moth eyes for optical application. Ueno in Ritsumeikan University of Japan developed a high aspect ration X-ray lithography method called deep X-ray lithography for nano-structures fabrication. It is a form of non-contact lithography having the capability of fabricating high aspect ratio nano structures (Ueno 2001). In order to fabricate the anti-reflection film with moth eye structure, or use diffraction optical element (DOE) to adjust the wavelength of resonance reflection, the submicron structures are required. By using X-ray exposure, fabrication of the structure that its size is under less than the wavelength of the visible light is possible. A number of fabrication methods of diffractive optical elements have been reported so far, including electron beam lithography, FIB (focused ion beam) etching, and laser ablation. However, the resulting height of those structures is not sufficient. A fabrication technique for sub-wavelength two-dimensional arrays using X-ray lithography is introduced in the paper. The sub-wavelength gratings were fabricated by synchrotron radiation (SR) lithography by using SR source with the source wavelength range is from 0.1 nm to 1.0 nm, and metal Ta is used as X-ray absorber.

2 FABRICATION PROCESSES

2.1 *The X-ray mask*

A LIGA mask also called a X-ray lithography mask which included an metal absorber such as Au or Ta as a pattern layer, a membrane such as SiN or SiC and a substrate such as Al or Si. X-ray lithography uses X-rays to transfer a nano pattern from a mask to a light-sensitive positive PMMA resist which is coated on the silicon substrate. Our mask pattern for the submicron gratings fabrication is shown in Figure 1, the

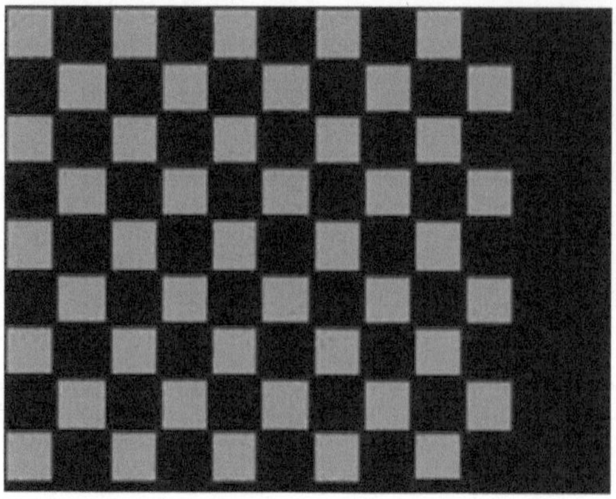

Figure 1. The mask pattern.

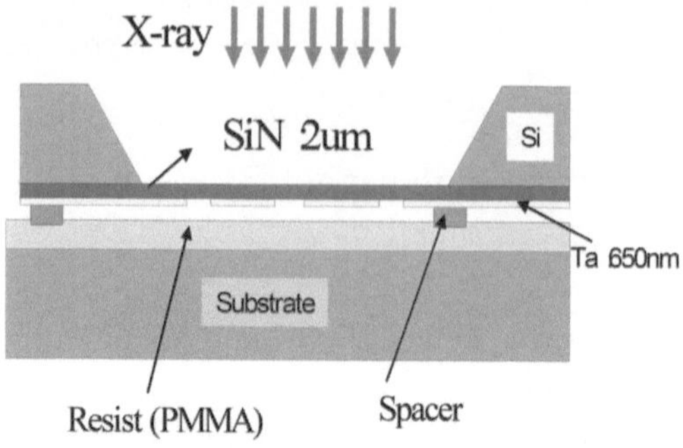

Figure 2. The mask configuration.

square size is 150 nm, and the pitch is 300 nm. And the mask configuration and PMMA resist (950PMMA-A11) for lithography is shown in Figure 2. A silicon nitride is suitable for a membrane of an X-ray mask due to its medium tensile stress and enough transparent and strength. In our work, a silicon nitride membrane is 2-μm in thickness and exhibits expected optical and mechanical properties. The silicon nitride film is produced on both 2-mm-thick silicon wafer surfaces by LPCVD. A central portion is RIE etched to create a silicon membrane by using a silicon nitride as etching mask on the backside. X-ray absorber is made of a heavy metal such as metal Au, W, and Ta. A 650 nm-thick Ta absorber was grown on the silicon nitride membrane by e-beam evaporation. The metal Ta is selected as the absorber because it has a high absorption capability for X-ray or a high etching accuracy and suitable for forming nano Ta pattern. The sample was then patterned by e-beam lithography to form the mask pattern. The final step in the fabrication of the LIGA mask is back etched the silicon by 30% KOH at 80°C. In order to protect the pattern on the other side, one side etching silicon tool is used in the KOH wet etching process. The metal Al frame is used because it has high mechanical strength and easy to fabricate. Fresnel diffraction is needed to consider duo to the X-ray mask pattern is nano scale in the near field of X-ray lithography. An X-ray radiation produces the resist and mask heat deformation and it also affects the X-ray lithography accuracy. Figure 3 shows the fabricated mask. The silicon wafer functions as a frame, the 650 nm in thickness metal Ta film acts as an absorber layer and 2 um in thickness SiN thin film layer as a membrane. The Ta absorbers on the mask are thick enough to keep enough contrast because the higher of the contrast of the X-ray, the thicker of the absorbers is needed. However, in order to save the noble metal and obtain the enough resolution, the thickness of Ta cannot be too thick because the higher pattern resolution is, the thinner of the absorbers is needed. In this experiment, the X-rays transparent contrast of the mask is 3 by considering to the required resolution.

Figure 3. The fabricated mask.

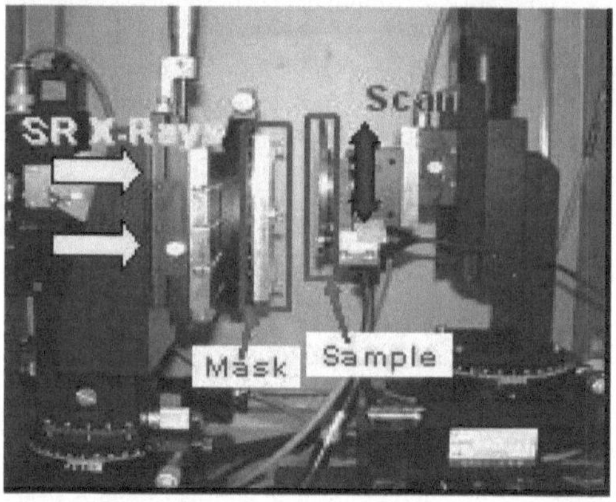

Figure 4. The exposure chamber.

2.2 *X-ray exposure and PMMA development*

The deep X-ray lithography has been becoming one of the most popular technology to fabricate sub-micron structures due to its ultra-short wavelength. To evaluate more accurately the relationship between the dose and the depth of the fabricated PMMA sample using the fabricated mask, test exposures were done. BL-6 was employed in the SR center of Ritsumeikan University in Japan. A theoretical and experimental data show that X-ray mask contrast is 30 when the PMMA

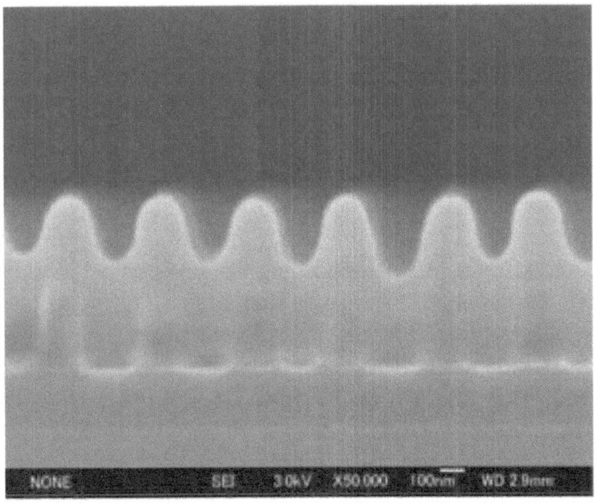

Figure 5. The fabricated gratings under the gap of 60 um.

resist is 1 um in thickness. It means that if the etching thickness under the Ta absorber is 0.1 μm, then etching depth under the not Ta area (X-ray exposed through this mask) is 3 μm thick PMMA. Figure4 shows the exposure chamber. The mask is set up in the mask holder and the sample is put in the sample holder. SR X-rays come from beamline and it is irradiated to the sample through the mask. The distribution of energy strength comes out from X rays that passed the mask further by the influence of the diffraction. AURORA is micro Synchrotron Radiation equipment and its operation electron energy is 575 MeV by 300 mA. The diameter of this compact SR equipment was 1 m. The optimized developer for X-ray lithography is a mixture of tetrahydro-1,4-oxazine (20%), 2-aminoethanol-1 (5%), 2-(2-butoxyethoxy)ethanol (60%), and water (15%). This developer can dissolute the exposed portions and remain unexposed portions of the resist. The improved PMMA developer can reduce crackings in the PMMA structures compared with conventional PMMA developers.

3 SEM ANALYSIS OF THE FABRICATED RESULTS

These nano structures were fabricated by a single mask process using LIGA. In fabrication of 1 um height PMMA structures, the X-ray exposure dose was about 0.05Ah. Developer fixing solution were used for rinsing or stabilizing process after development. The gap or space between X-ray mask and was the PMMA resist is adjustable, and two value of 60 um, 30 um is set in the experiment. The fabricated sub-micro gratings under the gap (or spacer) = 60 um and the dose is 0.0025Ah are shown in Figure 5. The 150 nm-width, 200 nm-height with 300 nm period structures has been successfully fabricated. The sub-micron gratings with the aspect ratio of 1.33 were achieved. Figure 6 shows the fabricated sub-micro gratings under the gap (or spacer) = 30 um and the dose is 0.0025Ah. The 150 nm-width, 450 nm-height

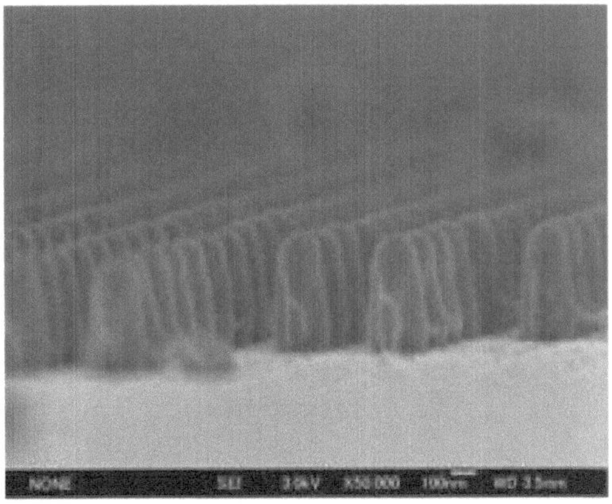

Figure 6. The fabricated gratings under the gap of 30 um.

with 300 nm period structures has been successfully fabricated. The sub-micron gratings with the aspect ratio of 3 were achieved. The experiment results show that the depth of the structures increases from 500 nm to 1000 nm when the dose is increased from 0.01Ah to 0.05Ah, in the same time, the gap of the gratings reduces from 150 nm to 120 nm. Therefore, the nearer gap yield better results than the far gap lithography. It is shown that the grating shapes become more complicated linked structures when the exposure dose decreases. To obtain the designed micro/nano structures, a enough exposure dose is needed. The gap between the mask and the PMMA is near enough and the theory limit value is 0. The gratings with 150 nm line width were fabricated when an exposure dose of 0.0025Ah is applied. When dose increases, the developed structures depth markedly increase. To obtain the desired structure, a compensation value must be considered. To avoid X-ray intensity variation to widen the line pattern, PMMA pattern SEM images is analyzed. A little undercut in the PMMA pattern can be observed. Measurements indicate that the roughness of the gratings surface is about 20 nm. The SEM images show good qualities and smooth sidewalls, the thickness of PMMA resist is up to 450 nm and the aspect ratio can reach 3.

Crack patterns on the PMMA surface is observed due to surface extension force after development. To avoid crack patterns, surface tension is need to reduce. A interlayer coupling agent such as silane is used, where PMMA structures remain after development.

The ultimate limit of resolution in Fresnel diffraction is defined by R_d,

$$R_d = 1.5(G^* \lambda)^{1/2}, \quad (1)$$

where λ is the wavelength of X-ray; G is the gap between the mask and the resist. The gap must be adjustable to near to zero for sub-micron grating fabrication based on the formula (1). Theoretical simulations based on the theory of the Fresnel diffraction show that the gap G have a heavy influence on the fabrication results. It is possible to fabricate the fine

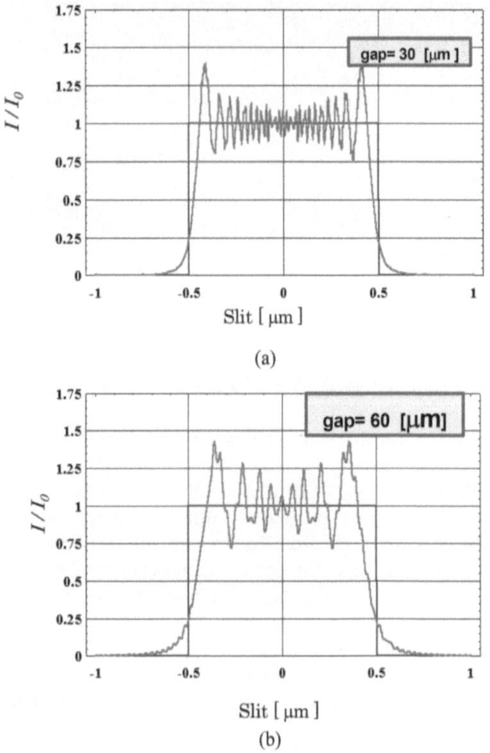

Figure 7. (a) Relation between X-ray intensity (a.u) and slit width when gap distance is 30 (μm); (b) Relation between X-ray intensity (a.u) and slit width when gap distance is 60 (μm).

nano gratings when the gap is less than 100 um. In this experiment, the gap is set in 30–60 um to assure the fine nano gratings. The effect of Fresnel diffraction on the shape of the grating is demonstrated by observing the top of the gratings [12]. The non-uniformity of the gap is produced by residual stress. To reduce the residual stress, a PMMA resist needs to be polymerized. The stresses also damage the bond strength between the PMMA and the wafer. In order to reduce areas of stress in the PMMA on the silicon wafer, a cooling system is also employed. Figure 7a shows the relation between X-ray intensity (a.u) and slit width when the proximity gap is assumed to be 30 μm, slit width is 1 μm. A blue line is the ideal intensity. The intensity distribution of light can be paraphrased as the resist shape because the shape of resist is decided by the amount of the irradiated rays. Figure 7b shows the relation between X-ray intensity (a.u) and slit width when the proximity gap is assumed to be 60 μm, slit width is 1 μm. It is shown that the intensity distribution vibrates drastically as the value of gap distance increase, and shape changes for the worse.

4 CONCLUSIONS

Fabrication technology for nano scale fine structures is developed. X-ray lithography using synchrotron radiation exposure has been employed to fabricate the PMMA nano gratings with a change in gap between the X-ray mask and the PMMA resist. The X-ray mask suspended on a silicon nitride membrane was fabricated utilizing an e-beam lithography system and a RIE process. The absorber is Ta metal film and the thickness was about 650 nm. Measurements made with a SEM indicate that the roughness of the gratings surface is about 20 nm; the PMMA gratings period is 300 nm and the aspect ratio is 3. SEM analysis indicates that the aspect ratio increases as the gap decreases under the same dosage condition. X-ray direct exposure of 1000 nm thick PMMA resist yielded submicron structures for anti-reflection application. Although the X-ray mask pattern is expensive due to complicated MEMS technology, this method seems to be useful for the fabrication nano-scale structures. By changing the lithography gap, the shape and the aspect ratio of the gratings can be changed using the same mask pattern. A challenging problem is how to keep fine submicron pattern. Silanes, as coupling agents, might be a solution.

ACKNOWLEDGEMENTS

This work was supported by "Professor Foundation" of Shanghai Institute of Technology (No. 10210Q140005).

REFERENCES

Laddha, S.L. 2001. Automatic determination of spatial dose distribution for improved accuracy in e-beam proximity effect correction. *Microelectronic Engineering* 57(8): 303–309.

Kanamori Y. & Hane K. 2002. Broadband antireflection subwavelength gratings for polymethyl methacrylate fabricated with molding technique, *Optical Review*, 9(5): 183–185.

Kato F. et al. 2007. Fabrication of high aspect ratio nano gratings using SR lithography, *Microsystem Technologies*, 13(3–4):221–225.

Ueno H. 2001. Ph.D Thesis, Ritsumeikan University, Japan.

Utke I. et al. 2008. Gas-assisted focused electron beam and ion beam processing and fabrication, *Journal of Vacuum Science & Technology B*, 26:1197–1201.

Zhao Y. & Avrutsky I. 1999. Two-dimensional colloidal crystal corrugated waveguides, *Optics Letters*, 24:817–819.

Future Mechatronics and Automation – Yang (Ed.)
© 2015 Taylor & Francis Group, London, ISBN 978-1-138-02648-3

Optimization and combination of machinery units for processing fish balls

J.M. Liu, G.R. Sun & F.G. Du
Forestry College of Beihua University, Jilin, China

X.R. Kong & K.J. Liu
Jilin WodaFood Co. Ltd., Jilin, China

X.S. Liu
Refrigeration Food Co. Ltd. of Jilin City, Jilin, China

ABSTRACT: To optimize and combine the machinery units for processing fish balls, the system of production line for processing fish balls, the equipment units composed in the processing of fish balls were all optimized by analysing the equipment and technologies. The equipment parameters are important in the production line of fish balls and these parameters were considered with a whole production line to achieve the effective convergence in different equipment units. Experimental analysis of fish processing equipment is available to determine reasonable device optimization parameters. The main processing methods are rinsing, grinding, forming, and cooking. The main production and processing equipment in the production line are meat grinders, beating machines, fish ball moulding machines, and digital boiling slots.

1 INTRODUCTION

With the improvement of human living standards, people prefer to eat some new kind of meat product. Thus making high quality meat products is becoming a necessary trend for the future. Fish balls are just a new kind of meat product and the most popular fish ball is the cuttlefish ball [1–3]. However, there is no auricular on sale. The components in fish balls are poor and the nutrition quality is poorer. Fish balls are important for food products in terms of nutritional properties [4], and they are a kind of flexible health care food. Mixing cuttlefish and auricular together in a fish ball could offer nutritional and health benefits as well as being delicious and also could enrich the quality of fish balls [5–7]. This thesis, through the equipment parameters optimization of the technology of fish balls, will ensure that the auricular has much better flavour than other fish.

2 MATERIALS AND METHODS

2.1 *Materials*

Fresh fish and fat were obtained from the market and stored 2°C until use. Tapioca starch, corn starch, soy protein isolate, carrageenan, compound phosphate, pieces of ice, salt, ginger, garlic, white granulated sugar were obtained from markets.

2.2 *Methods*

The processing method of fish balls involves three key steps:

(1) Rinsing is an important step in producing surimi, the purpose being to remove water-soluble protein which prevent surimi forming gel and to remove other matters that might harm the quality of the fish ball and hence all the sensory indexes of fish ball are improved.
(2) Grinding is divided into three steps: the first step is air blending; the second step is grinding with salt, the quantity of salt being 5% of raw material; the third step is grinding with condiment, and the purpose is to destroy fibres in the muscle and to make a salt soluble protein forming into gel.
(3) Heating is divided into two steps: the temperature and time are 45°C/35 min; 85°C/30 min, heating quickly through the temperature range of 55°C to 70°C, in order to avoid cracking of the fish ball gel. Heating and cooking in fish balls production in the process of the main role is to denature the protein coagulation; it transforms collosol to gel, gives the produce special flexibility, and also cooks the fish and sterilizes it.

The processing methods of a fish ball include three key steps, rinsing parameters, grinding parameters, and heating parameters. Every step of its attributes are listed as follows: Aroma (A) 20%, Colour (B) 20%,

Table 1. Standards of sensory evaluation on fish balls.

Item	Marking standard	Score
Aroma (A)	Rich flavour	16–20
	Fish flavour	12–16
	Fish aroma of plain	8–12
	A little fishy smell	4–8
	Thick fishy smell	0–4
Colour (B)	White	16–20
	Slightly yellow	12–16
	More yellow	8–12
	Grayish yellow	4–8
	Grey	0–4
Structural state (C)	Section dense, no large stomata, good elasticity	32–40
	Section dense, a little and small stomata, good elasticity	24–32
	Basic density section, a little and small stomata, good elasticity	16–24
	Section of loose, a little uneven hole, bad elasticity	8–16
	Section a slurry, soft non dense sense, bad elasticity	0–8
Flavour (D)	With fish unique flavour, fragrance, delicious	16–20
	Fish flavour, delicious	12–16
	Light flavour, taste general	8–12
	Almost no fish flavour, fishy smell slightly	4–8
	Strong fishy smell	0–4

Structural state (C) 40%, and Flavour (D) 20%. An evaluation index was used in evaluating fish balls in different processing stages and the evaluation standards are listed in Table 1.

Through the analysis of the three key steps of process parameter scoring, through SPSS (17.0) software processing of the three key steps of the fish's scoring tests were calculated by the method of LSD (Least Significant Difference).

3 RESULTS AND DISCUSSIONS

The effects of the rinsing temperature on the qualities of the fish balls are listed in Table 2. By the LSD analysis, there were great significant effects between two 7–9°C and 6–8°C, significant effects between 4–6°C and 6–8°C, 10–12°C and 6–8°C, 13–15°C and 6–8°C, 7–9°C and 13–15°C. The rinsing temperature had significant effect on the qualities of fish balls and the best treatment is at 7–9°C, the optimized parameter is 8°C.

3.1 *The effects of rinsing time on qualities of fish balls*

The effects of rinsing times on the qualities of the fish balls are shown in Table 3. By the LSD analysis, there were significantly different effects between 4 min and 1 min, 5 min and 1 min, 4 min and 2 min. The rinsing time had significantly different effects on the qualities of the fish balls and the best selection is 4 min, the optimized parameter is 4 min.

3.2 *The effects of rinsing times on the qualities of fish balls*

The effects of rinsing times on the qualities of fish balls are shown in Table 4. By the LSD analysis, rinsing times from 3 times to 5 times is the most significant numerical changes, significant effects between 3 times to 5 times. The rinsing times had significantly different effects on the qualities of the fish balls and the best selection is 5 times, the optimized parameter is 5 times.

3.3 *The effects of rinsing on the qualities of fish balls*

The effects of rinsing on the qualities of fish balls is the most important procedure of fish ball production and the main purpose is to remove water-soluble protein, pigment, taste, and fish fat. Through the rinse temperature, rinse time and rinse times, three factors experiment, the optimal value of the technological parameter is reached: rinsing temperature is 8 ± 1°C, the rinse times is 5, and the rinse time is 4 min. Through a contrasting experiment, the use of the texture analyzer to test folding degrees, by obtaining an optimal rinsing processing the conclusion is that with a dilute alkaline solution rinsing at 1634 g/cm^2, the gel strength is best, the gel strength by water rinsing is 1355 g/cm^2, the folding endurance is good, without rinsing, the gel strength is 1079 g/cm^2 and the folding endurance is poor. By comparison, the rinsing by a dilute alkaline solution is the best for the auricular cuttlefish ball's folding endurance and gel strength.

3.4 *The effects of grinding temperature on qualities of fish balls*

The effects of grinding temperature on the qualities of fish balls are shown in Table 5. By the LSD analysis, there were significantly different effects between 6–8°C and 12–14°C. The grinding temperature had significant effects on the qualities of fish balls and the best temperature was 6–8°C, while the optimized parameter was 7°C.

Table 2. Effect of rinsing temperatures on the qualities of fish balls (mean ± MSD, n = 3).

Rinse Temperature (°C)	4–6	7–9	10–12	13–15	16–18
Scores	85.7 ± 1.5	94.6 ± 2.1	85.6 ± 1.3	74.4 ± 1.5	65.4 ± 1.2

Note: $LSD_{0.01}$ (10) = 28.7, $LSD_{0.05}$(10) = 18.8

Table 3. Effect of rinsing times on the qualities of fish balls (mean ± MSD, n = 3).

Rinse Time (min)	1	2	3	4	5
Scores	69.1 ± 1.2	75.5 ± 1.3	86.1 ± 1.5	95.1 ± 1.5	87.4 ± 1.8

Note: $LSD_{0.01}$(10) = 26.2, $LSD_{0.05}$(10) = 17.2

Table 4. Effect of rinsing times on the qualities of fish balls (mean ± MSD, n = 3).

Rinse Times	3	4	5	6	7
Score	72.3 ± 1.2	84.1 ± 1.2	95.9 ± 1.3	86.7 ± 0.9	77.3 ± 1.2

Note: $LSD_{0.01}$(10) = 22.7, $LSD_{0.05}$(10) = 14.9

Table 5. Effect of grinding temperature on the qualities of fish balls (mean ± MSD, n = 3).

Grinding Temperature (°C)	0–2	3–5	6–8	9–11	12–14
Scores	83.9 ± 1.6	82.8 ± 1.6	94.9 ± 1.9	83.1 ± 2.8	73.3 ± 2.0

Note: $LSD_{0.01}$(10) = 20.0, $LSD_{0.05}$(10) = 13.1

Table 6. Effect of grinding types on the qualities of fish balls (mean ± MSD, n = 3).

	Grinding Time (min)				
Grinding types	1–3	4–6	7–9	10–12	13–15
Empty Grinding[1)]	83.3 ± 1.2	95.4 ± 1.4	82.1 ± 1.3	76.3 ± 1.5	71.7 ± 1.9
Salt Grinding[2)]	75.7 ± 1.3	81.7 ± 1.5	86.3 ± 1.6	96.1 ± 1.6	88.2 ± 1.3
Seasoning Grinding[3)]	79.3 ± 1.2	86.5 ± 1.5	97.1 ± 1.8	85.7 ± 1.7	83.2 ± 1.4

Note: 1) $LSD_{0.01}$(10) = 18.9, $LSD_{0.05}$(10) = 12.4; 2) $LSD_{0.01}$(10) = 17.6, $LSD_{0.05}$(10) = 11.5; 3) $LSD_{0.01}$(10) = 28.7, $LSD_{0.05}$(10) = 28.7

3.5 *The effect of grinding types on qualities of fish balls*

The effect of grinding types on the qualities of fish balls are shown in Table 6. By the LSD analysis:

For empty grinding, there were significantly different effects between 4–6 min and 13–15 min, significant effects between 4–6 min and 10–12 min, and the optimized grinding time was 5 min.

For salt grinding, there were significantly different effects between 10–12 min and 1–3 min, significant effects between 10–12 min and 4–6 min, and the optimized grinding time was 11 min.

For seasoning grinding, there were significantly different effects between 7–9 min and 1–3 min, significant effects between 7–9 min and 13–15 min, and the optimized grinding time was 8 min.

3.6 *The effects of grinding on qualities of fish balls*

Through a lot of grinding temperature experiments on the fish balls, it was concluded that the best grinding temperature is 7 ± 1°C. At this temperature and using a grinding time of about 5 min 11 min and 8 min, the strength of the gel of the fish balls is 1648 g/cm^2 and the folding endurance is at its best. It can avoid the shortcomings of other control groups and makes enough grinding, surimi gel property, and has good elasticity after heating. Using empty grinding 8 min, salt grinding 14 min, and seasoning grinding 11 min were adopted as a blending time to make a fish ball, and its gel strength and folding endurance respectively at 1426 g/cm^2 and its grade is good. Using empty grinding 14 min, seasoning grinding 2 min and salt grinding 2 min to make fish balls, the gel strength is 858 g/cm^2 and the folding endurance is of a poor grade. So, the comparison shows: adopting empty grinding 4 min, salt grinding 11 min, and seasoning grinding 9 min as blending times to make fish balls, gives the highest gel strength and the best folding endurance.

3.7 *The effects of heating methods on the gel strength and folding*

Heating methods consist of one-stage heating and two-stage heating: the former is when the fish ball is heated directly at 85°C for 30 min, the gel strength is 1239 g/cm^2 and the folding endurance is of good grade; the latter is firstly preheated to at 65°C for 25 min, and then heated at 90°C for 30 min; the gel strength is 668 g/cm^2 and the folding endurance grade is poor.

Because of alkaline protease, particularly sensitive to temperature in water-soluble protein in the fish. The activity is highest at 65°C. It can destroy the structure composed of actomyosin molecules, expose a hydrophobic group, lead to water loss, and deteriorate the gel strength.

The gel strength of fish balls obtained from the two-stage method which is firstly preheat at 45°C for 35 min, and then heated at 85°C for 30 min, is 1657 g/cm^2, higher than the one-stage method, and the folding endurance is of the best grade, because the fish ball has been through a slowly gelation temperature zone at 45°C , which contributes to the foundation of the net structure, It then goes through a gelation temperature zone quickly at 85°C, which avoids structural deterioration, thus enhancing the gel strength of the fish ball.

4 CONCLUSIONS

Fish balls with the highest gel strength and folding ability was produced under the following conditions respectively: washed with dilute alkaline solution temperature of rinsing is 8 ± 1°C, rinse 4 times and each time needs 5 min. Using the chopping machine to Chopping 3 sections, ground for 5 ± 1 min, 11 ± 1 min and 8 ± 1 min in order, its empty rolling needs 5 ± 1 min, chopping machine speed 500 rpm, salt needs 11 ± 1 min, chopping machine speed 1800 rpm and flavour needs 8 ± 1 min, chopping machine speed 1000 rpm. After cooking for with two stage heating process, will put the moulding of fish balls in digital boiled slots at 45°C temperature for 35 min, and then heated to 85°C temperature for 30 min. In this group of process parameters under the data of fish balls, the gel properties are the best.

To optimize the technological equipment parameters for the production of auricular fish ball and, if high quality fish balls are to be produced, the fish must first be rinsed in a dilute alkaline solution. This is because with dilute alkali brine is fulling of fish processing, can better promote the water-soluble protein dissolution and remove, and can make fish and pH increase to 6.8, close to neutral, thus improve the gel strength and folding of fish balls. The fish balls must be produced under the conditions of strict grinding times and heating controls in order to have good elasticity, a bright colour, and a good flavour. A grinding effect makes the fiber in fish tissue damage, again under the action of salt, salt soluble protein dissolution in muscle; Through mixing effect, the dissolution of salt soluble protein of fibrous interaction to form the high viscosity of the gel, and then in the subsequent degeneration in the heating process of coagulation, good protein mesh structure. Of course, the production of a fish ball is a complex process and there are other influential factors, which need to be studied further.

ACKNOWLEDGMENT

This research was financially supported by a five-year Science and Technology Research Support Program (No. 164, 2013) of the Department of Education of Jilin Province, China.

REFERENCES

[1] P.N. Forrester, P.N.W. Prinyawiwatkul, J.S. Godber & C. Phlakl, Treatment of catfish fillets with citric acid causes reduction of 2-methylisoborneol, but not musty flavor, J. Food Sci. 67 (2002)2615–2618.

[2] G.G. Kmamth, T.C. Lanier, E.A. Foegeding & D.D. Hamann, Non-disulfide covalent cross-linking of myosin heavy chain in "setting" of Alaska pollock and Atlantic croaker surimi, J. Food Biochem. 16 (1992)151–172.

[3] YC, Chung, L. Richardson & MT. Morrissey Effects of pH and NaCl on gel strength of Pacific whiting surimi, J. Aquatic Food Prod Technol. 2 (1993)52.

[4] Y. Kumazawa, T. Numazawa, K. Segurpand & M. Motoki: Suppression of surimi gel setting by transglutaminase inhibitors, J. Food Sci. 60 (1995) 715–717.

[5] C. C. Huang & C.F. Li: The effect of chitosan on thermal gelation of porcine salt proteins, Food Sci. 25 (1998) 150–162.

[6] Kocher, P.N. & E.A. Foegeding: Microcentrifuge-based method for measuring water-holding of protein gels, J. Food Sci. 58 (1993)1040–1046.

[7] S. Roller & N. Corvill, The antimicrobial properties of chitosan in mayonnaise and mayonnaise-based shrimp salads, J. Food, 63 (2000)202–209.

Section 2: Mechatronics

Future Mechatronics and Automation – Yang (Ed.)

Exploration on hospital strategy management based on niche theory

Chuan Zhu, Guo-wei Wang, Xiao-feng Xiong & Yuan Guo
Hospital of PLA, Chongqing, China

ABSTRACT: On basis of niche theory, the study put forward the concept of hospital niche, explained the formation mechanism of hospital niche from three aspects, analysed the connotation of hospital strategic management with synchronicity and diachronicity. The study realized that the essence of hospital strategic management was the process of constantly occupying advantage niche and dynamically maintaining their balance, which was based on correctly understanding and grasping hospitals' objective conditions and niche, with coordinating relationship between hospitals and environment through strategy implementation.

Keywords: Niche theory; Hospital niche; Hospital strategy management

1 INTRODUCTION

The concept of niche was first proposed by the ecologist Grinnell, whose originally meaning was "the final unit just occupied by one species of a subspecies." This definition emphasized the significance of the spatial distribution of species, highlighting the spatiality of niche (Luo et al. 2013). From the perspective of the community trophic relations between species, animal ecologist Charles Elton pointed out that niche was "the status and role of species in biomes", and this definition highlighted the functionality of niche. Hutchinson proposed that niche is "the sum of all relations of organisms and its biotic and abiotic environment", and he thought the actual niche of a species in the biomes will be smaller and smaller with the increase of competitive species (Li et al. 2010). To a certain extent, this definition explained why the living environment of species in community succession would become increasingly specialized (Zhao et al. 2010). Due to the different angles and starting points of ecologists, the definition of niche also had some differences. In this paper, it considers that niche refers to the sum of dependence on the resources and the adaptability of a biological unit (individual, population or species) in biomes, and also includes the functional relationships between its specific location and time and other relevant biological units. The niche of species not only depends on the region where it lives, but also is influenced and limited by its way of life and other biological units. As an objective existence in natural ecosystem, niche reflects that the subject and environment always exist information transformation conditions between matters, emerge and information, operating conditions of body living system, and mutual coordination operating conditions between the subject and environment (Li et al. 2012). We can say that, for all life phenomenons, niche phenomenon is a general principle with universality.

At present, due to the strong explanation and insight its basic principles, niche theory has been widely used in cultural, political, economic and other fields, as well as industries such as finance, education and urban planning, etc, forming the concepts of political niche, educational niche, business niche, and urban niche, which is becoming the strong theoretical analysis and practical guidance tools in the deep and comprehensive researches process of various fields and industries.

2 THE CONCEPT OF NICHE IN THE HOSPITAL

Currently, the studies of niche in the hospital are not many, and have not formed a unified or recognized concept of niche in the hospital. On the basis of the literature, this paper considers that niche in the hospital is that in the whole social environment and whole medical institutions, taking its own resources as basis and conditions, and in the process of medical activities and medical services, the hospital forms the survival and development space conductive to its own and the relative competitive advantages with the interactions between social environment and other medical institutional communities. Specially, if the hospital wants to remain its survival advantages and a good position in ecosystem, it must acquire the ecological resources needed as much as possible, and transform the input resources such as funds, talents, knowledge, information and technologies through the use of funds, consumption of goods and materials, transfer of knowledge, exchange of information, and updating of technologies, into resources for social needs, which transforms to be immediate interests and social benefits of the hospital, and in the process of transformation, the interactions in these resource elements constitute the broad living space of the hospital. While in a certain range of time and space to survive, the abilities of each hospital to obtain the input and output resources are relatively limited, which can only occupy a part of the broad living space, and this part of ecological space is the niche in the hospital. Niche in the hospital is the link between the survival development of a hospital itself and its survival environment, which reflects that in the living space, the roles played by the physical resources, funds, human resources, knowledge, information and technologies of a hospital in the process of flow, and manifests the competitiveness of a hospital (Geert et al. 2010). Niche in the hospital has three levels, first is the macro ecological environment such as politics, economy, culture and scientific technology; second is the intermediate ecological environment of a hospital and other medical institutional communities; and third is micro ecological environment within the hospital.

3 THE FORMATION MECHANISM OF NICHE IN THE HOSPITAL

First, the survival development of the hospital is inseparable from the space environment existed with a variety of resources. In the survival development of the hospital, the behaviors of the hospital affect the external environment, and at the same time, the external environment selectively affect the internal structure, operating mechanism, and specific behaviors of the hospital. Therefore, the survival development of the hospital is separable from the space environment existed with a variety of resources. Specially, the hospital absorbs various resources, which gradually forms its own core competitiveness and generates economic and social benefits, thus it forms an important driving factor of the development of the hospital and the changes of external environment, and gradually transforms into specific ecological characteristics to adapt to the environmental changes, then becomes a unique niche in the hospital. Moreover, if the hospital wants to avoid vicious competitions, it must maintain the collaborative symbiotic relationship with the outside world, which requires that the hospital must seek specific resources and clear space for development. When the hospital needs to re-select or screen various ecological factors from the outside world, or to change cooperative relationships with other medical institutions, it will lead to the adjustment and optimization of niche in the hospital (Gong et al. 2010).

Second, the sound balanced niche in the hospital in inseparable from the external environment and internal resources. The development of niche in the hospital not only depends on the adjustments of hospital's internal structure, operating mechanism, specific behaviors, and culture traditions, but also depends on the impact of sudden changes in the external environment. The most direct impact of sudden changes in the external environment is to break the original space environment existed with various resources, or emerges new ecological factors, which makes the hospital must adjust the internal structures, mechanism and traditions to absorb various new resources, so that to ensure the balance of niche. Therefore, if the niche in the hospital wants to establish a dynamic and balanced operating mechanism, it must be built on the basis of two ecological factors, which are internal resources and external environment. Otherwise, the hospital will face a huge ecological imbalance and the risk of ecological dislocation (Gong et al. 2010).

Third, the survival development of the hospital is inseparable from the symbiotic relationships between niches in the hospital. The complementarity and cooperation between niches in the hospital is beneficial to the normal and orderly exchange of matters, energies and information, and will weaken the functional hinders between each other. The symbiotic relationships between niches in the hospital have significant impacts on other niches in the hospital in the environment, and these important impacts will counteract on the niche in the hospital. The symbiotic relationships of niches in the hospital are the premise for the hospital to conduct the matter, energy and information exchanges, and are the intrinsic motivation to promote the sustainable development of the hospital, which is very important for the hospital to maintain core competitive advantages. Meanwhile, the symbiotic relationships are the foundation of the various resources shared by the hospital, and are the entry points for the hospital to express resource orientation and expand development space, which are helpful to stabilize the hospital population and ecological environment of the hospital.

4 THE ESSENCE OF THE HOSPITAL STRATEGY MANAGEMENT BASED ON NICHE THEORY

Hospital strategy refers to the overall planning of the hospital for the future development goals and the ways and measures to achieve the goals, in order to achieve its survival and sustainable development, and continuously acquire new competitive advantages on the basis of the hospital's external environment and the status of internal resources and abilities. While the hospital strategy management is the process for the internal resources and abilities of the hospital adapt to the demands of its external environment, and its purpose is to allow the management of the hospital to develop and implement the strategic plans of the hospital aiming at achieving its purposes and goals (Xue 2004).

Through the analysis of the formation mechanism of niche in the hospital, we can understand that the niche in the hospital is the results of self-initiatives and environmental selectivity. From the initiative of the hospital, it should be based on the actual situation to seek, occupy and pursue good niche, to actively change internal and external environment, and to pursue the sustainable development of the hospital; from the environmental selectivity, the ecological environment of the hospital has sudden changes, which is likely to lead that the original niche is not suitable for the need of new environment, hindering the survival and development of the hospital. This requires that the hospital must see the situation, timely select new development strategies, adjust the existing niche, and achieve the continuous balance of relationships between the hospital and the environment. Therefore, the hospital not only should live in harmony with the environment, but also achieves self-assertion by changing the environment, and the strategy management activities of the hospital have to deal with the dual problems of synchronicity and diachronism.

From the perspective of synchronicity, in a particular period or stage, the development of the hospital has relative stability. Hospital develops a reasonable strategic plan according to its own objective conditions and actual circumstances to achieve the adaption and development on the corresponding niche. Typically, the hospital strategy planning and management approach that the management of the hospital formulates strategies in accordance with certain procedures and methods, through the organizations within the hospital systematically to implement strategies changes to be the specific behaviors of the members within the hospital, and through strategic assessment to ensure the correct and reasonable strategy implementation, and at the same time to adjust the strategy planning based on the real situation. The survival activities and environment modes of the hospital will not change in a certain period, and the strategy of the hospital is relatively stable. However, if within this period, the survival environment of the hospital has significant changes or the value creation modes constituted by the hospital's internal structure, operating mechanism, specific behaviors and cultural traditions has changed, the strategy of the hospital should be changed. When the hospital facing with complex and changeable competitive environment, if it can adapt to the changes on the niche and create values, even is brave to use changes and create changes, it can achieve its own sustainable development. Therefore, during the process of conducting the strategy planning and implementation of the hospital, the hospital always adhere to the thinking mode of dynamic condition and contingency, and timely observe and grasp the change state of the internal and external environment. When the environment changes hinder the development of the hospital, the hospital should be good at making the best use of the circumstance to promote the sustainable development of the hospital through continuous learning and innovation; when the environment changes become the opportunity for the development of the hospital, the hospital will have the courage to compete and timely adjust the strategy of development to achieve long-term competitive advantages (Marjolein et al. 2008).

From the perspective of diachronism, each hospital has a certain life cycle and the hospital strategy management also will continue to evolve. In the early development of the hospital, it focus of strategy management is to design its own vision, actively cultivate the characteristics and culture of the hospital, fully grasp the various information in the external environment, and seek a breakthrough in the growth of the hospital. With the expansion of the hospital's scale and improvement of the organizational structure, the ability for the hospital to grasp the internal and external resources is enhanced, thus the hospital starts to initiatively adapt to the changes in the external environment and enters a rapid growth phase. During this period, most of the hospital's strategic choices are to compete and occupy more favorable ecological factors to achieve the development speed and efficiency, which reflect to be the fighting for the preferential policies, development of medical market and the competition between hospitals. At the same time, niches are quickly occupied, the capacities of favorable ecological factors tend to be saturated, and the development speed of the hospital slows down or even is suppressed, so the strategy of the hospital will gradually shift to be a compromise and breakthrough process towards the environment, such as seeking or initiatively creating new ecological factors. The symbiotic collaboration between hospitals enriches its ecological resources and construct ecosystem by strategic cooperation, thus improving survival and competitive environment, and occupying a good niche. For the complexity and uncertainty of the hospital, the response of the hospital towards the changes of environment will be slow, so at this time, the hospital strategy often can not support its continuous development, even is lag behind the development of the hospital. If the hospital can strengthen the strategy management and timely adjust strategy planning to maintain the advancement and guiding function, it enables the hospital to break out the limitations its life cycle, effectively prevents the hospital into a recession, and even usher into a new development peak (Marjolein et al. 2008).

CORRESPONDING AUTHOR:

Guo Yuan, 324 Hospital of PLA, Chongqing, China, 1121119659@qq.com

REFERENCES

Geert Verbong, Willem Christiaens, Rob Raven, Annelies Balkema. 2010. Strategic Niche Management in an unstable regime: Biomass gasification in India. *Environmental Science & Policy* 13(4): 272–281.

Gong Yi-zu, Xie Ling-ling. 2010. Niche Strategy: The new strategic development choice of undergraduate course colleges and universities. *Higher Education Exploration* 6: 10–15.

Li Hua-jun, Zhang Guang-yu, Liu Yi-xin. 2012. The Study on the Innovation System of Strategic and Emerging Industries Based on The SNM Theory. *Science & Technology Progress and Policy* 29(3): 61–64.

Li Ning, Lin Qing. 2010. Introduction to Niche Theory and Its Application. *The South of China Today* 7: 208–210.
Luo Jia-wen, Zhang Guang-yu, Tan Dan-dan. 2013. An Overview of the Research on the Strategic Niche Management. *Journal of Guangdong University of Technology (Social Sciences Edition)* 13(2): 84–89.
Marjolein C J Caniels, Henny A Romijn. 2008. Actor networks in Strategic Niche Management: Insights from social network theory. *Original Research Article Futures* 40: 613–629.
Xue Di. 2004. The Foundation of Chinese Hospital Strategy Management Practice. *Chinese Hospital Management* 24(7): 13–15.
Zhao Jia-jia. 2010. Analysis of Enterprise Niche Principle and Its Application in Enterprise Management. *Guide to Business* 9: 62–63.

Future Mechatronics and Automation – Yang (Ed.)
© 2015 Taylor & Francis Group, London, ISBN 978-1-138-02648-3

Noise adaptive UKF method used for boost trajectory tracking

Y. Wang, H. Chen, H. Zhao & W. Wu
School of Automation Science and Electrical Engineer of Beihang University, Beijing, China

ABSTRACT: To solve the problem of unknown system noise statistical characteristic in boost trajectory tracking, a noise adaptive UKF (AUKF) was proposed by introducing Sage-Husa noise estimator which is used for estimating the statistical characteristic of noise into UKF (Unscented Kalman Filter) to make a better trajectory tracking performance. This paper built a model for boost trajectory of missile and measurement at first. Then described the algorithm of AUKF and use it in trajectory tracking simulation. As for the problem of filtering divergence usually cause by noise adaptive UKF, an effective solution was presented to improve the numerical stability. Simulation results showed that AUKF has a better tracking performance than normal UKF.

1 INTRODUCTION

In missile's trajectory tracking, the SBIRS (Space-Based Infrared System) can gain azimuths and elevations by detecting infrared radiation of boost target using infrared sensor. Combine the measurements with filtering algorithm, the motion information of target can be estimated. Due to both the model for boost of motion and the model of angle-only measurement are nonlinear, nonlinear filtering algorithms are required in this situation.

Among nonlinear filtering algorithms, UKF (Unscented Kalman filter) uses unscented transformation to estimate the nonlinear probability density distribution instead of solving Jacobian matrix (Julier, S. J. & Uhlmann, J. K. 2004). It's widely used in trajectory tracking due to its advantage of high precision and well computational efficiency (Liang, X. et al. 2011).

In trajectory tracking, model error and inaccuracy statistical characteristic of noise will enlarge the estimated error, even result in filtering divergence (Zhou, H. et al. 1991). In general, the statistical characteristic of measurement noise can be known by the physical characteristics of sensors. While the statistical characteristic of system noise can't be estimated exactly because of much interference exists in maneuvering target.

To solve the problem, this paper introduces Sage-Husa noise estimator into UKF and proposes a solution to deal with the problem of filtering divergence usually cause by sub-optimal Sage AUKF (adaptive UKF) which improves the numerical stability of filter.

2 TRAJECTORY MODEL FOR BOOST OF MOTION

2.1 *Model for boost of motion*

The motion for missile's boost trajectory is affected by trust, gravity and atmospheric drag mainly. Consider the boost time is short, we take the flat earth model of gravity into account.

In ECI-CS (Earth-centered inertial coordinate systems), the model of boost trajectory can be described as follows (Sheng, W. 2011, Li, X. R. & Jilkov, V. P. 2001):

$$\begin{cases} \dfrac{dr}{dt} = \dot{r} \\ \dfrac{d\dot{r}}{dt} = a_c + a_a + a_g \end{cases} \tag{1}$$

where r, $\dot{r}$, a_c, a_a and a_g denote the position, velocity vector, the acceleration components included by thrust, aerodynamic drag and gravity respectively. And $a_g = -\mu_e r/\|r\|^3$, where $\mu_e = 3.98604418 \times 10^{14}\ \mathrm{m^3/s^2}$ denotes the constant of earth gravity.

Let $q = a_c/U$, where U denotes exhaust velocity and can be considered as a constant when the engine is stable. Through derivation, we acquire that

$$\frac{dq}{dt} = \frac{\dot{a}_c}{U} = q^2 \tag{2}$$

Meanwhile, according to dynamics principle, then

$$\frac{da_c}{dt} = \dot{a}_c e + \omega \times a_c \tag{3}$$

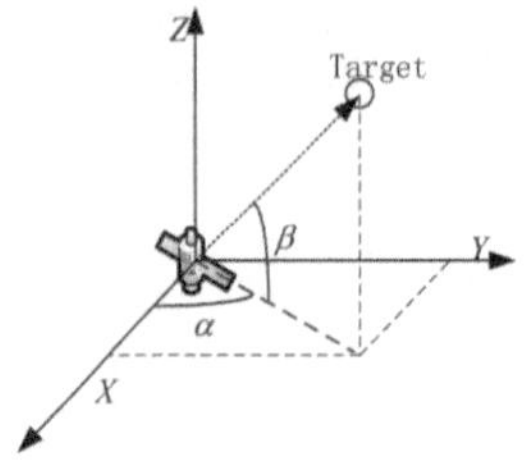

Figure 1. Diagrammatic map of measurement.

where e and ω denote unit vector of thrust acceleration and rotation angular velocity of e in ECI-CS. Combine formulas 1–3 with the unconsidered perturbation, the completed model of a boost target is given as follows:

$$\begin{cases} \dfrac{dr}{dt} = \dot{r} \\ \dfrac{d\dot{r}}{dt} = a_c - \dfrac{\mu_e}{\|r\|^3} r \\ \dfrac{da_c}{dt} = q a_c + \omega \times a_c \\ \dfrac{dq}{dt} = q^2 \end{cases} + W \tag{4}$$

where W denotes system noise with zero mean and covariance matrix Q. And

$$\omega \approx \frac{1}{\|u\|^2} u \times a_c - \frac{\mu_e}{\|u\|^2 \|r\|^3} u \times r \tag{5}$$

where u denote the velocity vector in ECF-CS (Earth-centered fixed coordinate systems).

2.2 *Measurement model*

The measurements of space based angle-only system contain azimuths and elevations as showed in Figure 1. The model can be described in ECI-CS as follows.

In the figure above, XYZ denote the axes of ECI-CS and α, β denote angle measurements of the system. The measurement model is given as follows (Sheng, W. 2011):

$$z_k = \begin{pmatrix} \alpha_k \\ \beta_k \end{pmatrix} = \begin{pmatrix} \tan^{-1} \dfrac{y_k - y_{s,k}}{x_k - x_{s,k}} \\ \sin^{-1} \dfrac{z_k - z_{s,k}}{\|r_k - r_{s,k}\|} \end{pmatrix} + v_k \tag{6}$$

where $r_k = (x_k, y_k, z_k)^T$, $r_{s,k} = (x_{s,k}, y_{s,k}, z_{s,k})^T$, $z_k = (\alpha_k, \beta_k)^T$ and v_k denote target position, satellite position, azimuths and elevations of measurement and measurement noise of sensors at the moment of k in ECI-CS. v_k is assumed to follow Gaussian distribution with zero mean and covariance matrix

$$R_k = \begin{pmatrix} \dfrac{\sigma_{LOS}^2}{\cos^2(\beta_k)} & 0 \\ 0 & \sigma_{LOS}^2 \end{pmatrix} \tag{7}$$

where σ_{LOS} denotes the error of Light of Sight (LOS).

3 UKF ALGORITHM

UKF (Unscented Kalman filter) (Unscented Kalman filter) uses unscented transformation to estimate the nonlinear probability density distribution instead of solving Jacobian matrix. It's effective in computation and has a high precision. The basic step of UKF is given as follows (Julier, S. J. & Uhlmann, J. K. 2004, Liang, X. et al. 2011):

Step 1 Give the filtering initial value of state-variable and covariance $\hat{X}_{0|0}$ and P_0.

Step 2 Compute the sigma points of state-variable.

$$\chi_{k-1} = \Big[\hat{X}_{k-1|k-1}, \hat{X}_{k-1|k-1} + \sqrt{(n+\lambda)P_{k-1|k-1}} \\ , \hat{X}_{k-1|k-1} - \sqrt{(n+\lambda)P_{k-1|k-1}} \Big] \tag{8}$$

Step 3 Predict the state-variable and measurement of next moment according to the nonlinear model based on the sigma points produced above.

$$\chi_{k|k-1} = f(\chi_{k-1}) \tag{9}$$

$$y_{k|k-1} = h(\chi_{k|k-1}) \tag{10}$$

$$\hat{X}_{k|k-1} = \sum_{i=0}^{2n} W_i^{(m)} \chi_{i,k|k-1} \tag{11}$$

$$\hat{Z}_{k|k-1} = \sum_{i=0}^{2n} W_i^{(m)} y_{i,k|k-1} \tag{12}$$

Step 4 Compute Kalman gain matrix, update state-variable $\hat{X}_{k|k}$ and covariance matrix $P_{k|k}$.

$$P_{k|k-1} = \sum_{i=0}^{2n} W_i^{(c)} \left(\chi_{i,k|k-1} - \hat{X}_{k|k-1}\right)\left(\chi_{i,k|k-1} - \hat{X}_{k|k-1}\right)^T + Q_{k-1} \tag{13}$$

$$P_{y_k y_k} = \sum_{i=0}^{2n} W_i^{(c)} \left(y_{i,k|k-1} - \hat{Z}_{k|k-1}\right)\left(y_{i,k|k-1} - \hat{Z}_{k|k-1}\right)^T + R_k \tag{14}$$

$$P_{x_k y_k} = \sum_{i=0}^{2n} W_i^{(c)} \left(\chi_{i,k|k-1} - \hat{X}_{k|k-1}\right)\left(y_{i,k|k-1} - \hat{Z}_{k|k-1}\right)^T \tag{15}$$

$$K_k = P_{x_k y_k} P_{y_k y_k}^{-1} \tag{16}$$

$$\hat{X}_{k|k} = \hat{X}_{k|k-1} + K_k \left(Z_k - \hat{Z}_{k|k-1}\right) \tag{17}$$

$$P_{k|k} = P_{k|k-1} - K_k P_{y_k y_k} K_k^T \tag{18}$$

Parameters used above are defined as follows:

$$\begin{cases} W_0^{(m)} = \lambda/(n+\lambda) \\ W_0^{(c)} = \lambda/(n+\lambda) + 1 - \alpha^2 + \beta \\ W_i^{(m)} = 1/\left[2(n+\lambda)\right], i = 1, \cdots, 2n \\ W_i^{(c)} = 1/\left[2(n+\lambda)\right], i = 1, \cdots, 2n \end{cases} \tag{19}$$

where n denotes the dimension of state. And $\lambda = \alpha^2(n+\kappa) - n$, where α is a scaling factor (which scales the spread of the distribution), usually set to a little value between $10^{-4} \sim 1$; β reflects quantity of priori information (under the assumption of no information in Gaussian distribution, $\beta = 2$); κ is a secondary scale parameter, usually set to 0.

4 NOISE ADAPTIVE UKF ALGORITHM (AUKF)

To solve the problem of model error or unknown system noise statistical characteristic in boost trajectory tracking, many adaptive filter algorithms were proposed. Among them, Sage-Husa noise estimator can estimate the noise statistical characteristic in real-time while filtering so that the model error can be corrected. By introducing it into UKF, the performance of estimated result can be improved.

4.1 *Sage-Husa noise estimator*

Sage-Husa sub-optimal maximum posteriori probability (MAP) noise estimator (Sage, A. P. & Husa, G. W. 1969) is used to solve the problem of unknown system noise statistical characteristic and estimate the noise statistical characteristic in real time while filtering. The main compute formulas for estimated system noise are given as follows (Shi, Y. & Han, C. 2011):

$$\hat{q}_k = (1-d_{k-1})\hat{q}_{k-1} + d_{k-1}\left(\hat{x}_{k|k} - \Phi_k \hat{x}_{k-1|k-1}\right) \quad (20)$$

$$\hat{Q}_k = (1-d_{k-1})\hat{Q}_{k-1} + d_{k-1}\left(K_k v_k v_k^T K_k^T + P_{k|k} - \Phi_k P_{k-1|k-1} \Phi_k^T\right) \quad (21)$$

where $\hat{q}_k$ and $\hat{Q}_k$ denote the mean and covariance matrix of system noise. And $d_{k-1} = (1-b)/(1-b^{k+1})$, $0 < b < 1$, where b is a forgetting factor which is used to control the memory length of filter, usually set to 0.95~0.99.

4.2 *Solution of filtering divergence*

Sub-optimal Sage-Husa filter usually causes filtering divergence due to $\hat{Q}_k$ becomes non-positive. One solution to deal with this problem is given as follows (Zhang, H. & Zhang, H. 1991):

By ignore $P_{k|k} - \Phi_k P_{k-1|k-1} \Phi_k^T$ item in formula (21), then

$$\hat{Q}_k = (1-d_{k-1})\hat{Q}_{k-1} + d_{k-1} K_k v_k v_k^T K_k^T \quad (22)$$

For finite time, $d_k < 1$, if $\hat{Q}_0$ is half positive, then $\hat{Q}_k$ is always half positive. So it can keep the filter stable.

However, the noise estimator given by Formula (22) is biased, it will enlarge the estimated error sometimes. To improve the precision, a better method is to make a judgment of $\hat{Q}_k$. If $\hat{Q}_k$ is half positive, use formula (21), otherwise use formula (22).

4.3 *Realization of AUKF*

The noise adaptive UKF is given by introducing the Sage-Husa noise estimator into UKF. Combine with formulas 20–22, some referenced changes are listed as follows.

Combine formula (20) with UKF, then

$$\hat{q}_k = (1-d_{k-1})\hat{q}_{k-1} + d_{k-1}\left(\hat{x}_{k|k} - \hat{X}_{k|k-1}\right) \quad (23)$$

Through formula (18) and the formula below

$$P_{k|k} - \Phi_k P_{k-1|k-1} \Phi_k^T = P_{k|k} - P_{k|k-1} \quad (24)$$

We see that

$$\hat{Q}_k = (1-d_{k-1})\hat{Q}_{k-1} + d_{k-1}\left(K_k v_k v_k^T K_k^T - K_k P_{y_k y_k} K_k^T\right) = \left(K_k\left(v_k v_k^T - P_{y_k y_k}\right)K_k^T\right) \quad (25)$$

And formula (9) should change to the following one due to the mean of system noise $\hat{q}_k$ is not zero when Sage-Husa noise estimator was introduced.

$$\chi_{k|k-1} = f(\chi_{k-1}) + \hat{q}_{k-1} \quad (26)$$

However, the convergence rate of AUKF is slower than normal UKF due to the influence of inaccurate initial value and covariance matrix. To overcome this problem, we can use UKF at first and then switch to AUKF while UKF converges.

5 SIMULATION

In order to verify the performance of noise adaptive UKF, we used normal UKF and AUKF to make a simulation under the condition of unknown system noise statistics respectively. As a contrast, a referenced UKF filter whose initial covariance matrix of system noise is exact was introduced.

5.1 *Simulation scene and referenced parameter configuration*

To insure observability, measurements were acquired from two GEO (Geostationary earth orbit) satellites which locate in 10°E (East) and 69°E. Sampling frequency was 10 Hz and measurement error was set to $\sigma_{LOS} = 10$ urad. Boost trajectory used in simulation came by integrating with formula (4) using fourth rank Runge-Kutta method and adding sys em noise with zero mean and covariance matrix $Q = diag\ (10^2, 10^2, 10^2, 10, 10, 10, 10^{-1}, 10^{-1}, 10^{-1}, 10^{-8})$ to insure tracking performance. The total time of boost trajectory lasted 67s in simulation. Initial position of missile located in 30°N(North), 50°E, altitude of 16 km, the azimuth and elevation is 20°NE and 45°NE (Northeast) respectively and initial velocity was 3 km/s in ECF-CS. To verify the efficiency of AUKF algorithm, two simulation scenes with different initial covariance matrix of system noise Q were designed as follows.

Table 1. Compare statistics of position and velocity estimated error of UKF and AUKF.

Algorithm	$E(\Delta r_x)$ m	$\sigma(\Delta r_x)$ m	$E(\Delta v_x)$ m/s	$\sigma(\Delta v_x)$ m/s
UKF	8.517	94.65	1.717	26.34
AUKF	9.710	82.65	−1.313	22.46

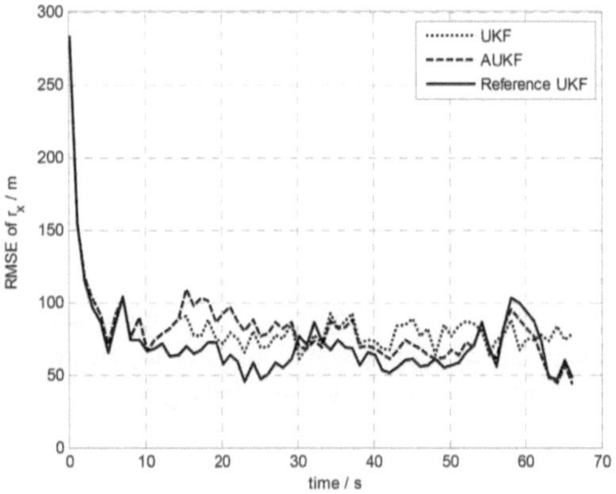

Figure 2. Position RMSE of x-axis.

5.2 *Simulation and results analysis*

Simulation results were got by making 50 times of Monte-Carlo simulation using UKF and AUKF respectively. Forgetting factor b was set to 0.98. And define RMSE (Root Mean Square Error) as follows:

$$E = \left(\frac{1}{N} \sum_{i=1}^{N} \left\| \left(\hat{x}_k^i - x_k^i \right) \right\|_2 \right)^{\frac{1}{2}} \tag{27}$$

where N is the simulation times and $\hat{x}_k^i - x_k^i$ denotes the estimated error of state variable.

The following results were obtained from two different simulation scenes with different initial covariance matrix of system noise Q.

5.2.1 *Initial covariance matrix of system noise is larger than the true one*

In this scene, initial Q was set to $Q_0 = diag(40^2, 40^2, 40^2, 30, 30, 30, 1, 1, 1, 10^{-6})$. By using UKF and AUKF respectively, we got the statistics of position and velocity estimated error of x-axis in one Monte-Carlo simulation as showed in Table 1. It showed that AUKF has a less standard deviation than UKF that indicate AUKF has better performance.

Figures 2–3 represent the RMSE of estimated error of r_x (position of x-axis) and v_x (velocity of x-axis) respectively. And Figure 4 represents the changing curve of the second norm of covariance matrix P. We can see that AUKF has a less RMSE than UKF from Figures 2–3. Also, AUKF has a less norm of covariance matrix P and its result is closer to reference UKF than normal UKF eventually. It showed that AUKF performs better than UKF.

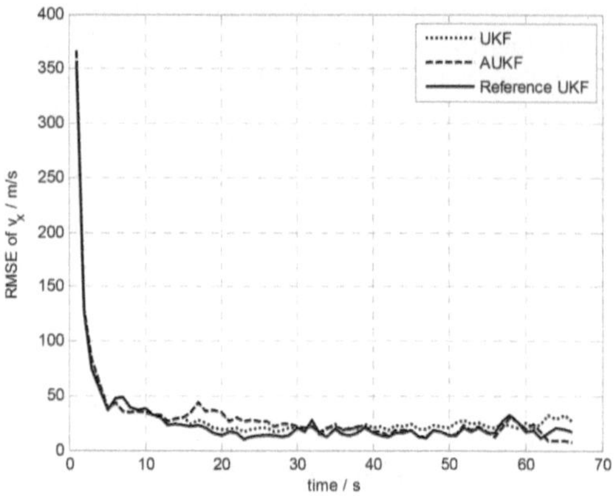

Figure 3. Velocity RMSE of x-axis.

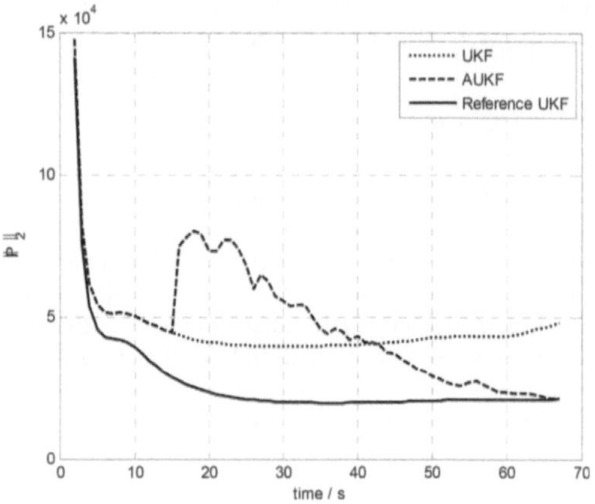

Figure 4. $\|P\|_2$.

Table 2. Compare statistics of position and velocity estimated error of UKF and AUKF.

Algorithm	$E(\Delta r_x)$ m	$\sigma(\Delta r_x)$ m	$E(\Delta v_x)$ m/s	$\sigma(\Delta v_x)$ m/s
UKF	−42.49	87.80	−10.33	25.12
AUKF	24.68	66.85	−6.609	17.71

5.2.2 *Initial covariance matrix of system noise is smaller than the true one*

In this scene, initial Q was set to $Q_0 = diag(1, 1, 1, 0.1, 0.1, 0.1, 10^{-2}, 10^{-2}, 10^{-2}, 10^{-10})$. Table 2 is the statistics of the estimated error. We can see from Table 2 that AUKF has a less estimated error than UKF.

Like Figure 2 and Figure 3, we can see that AUKF has a less RMSE than UKF from Figures 5–6. And Figure 7 indicated that UKF has a less norm of covariance matrix P than reference UKF due to the initial covariance matrix is set smaller than the true one. However, it's no meet the actual situation. While AUKF can update the covariance matrix P through a real-time

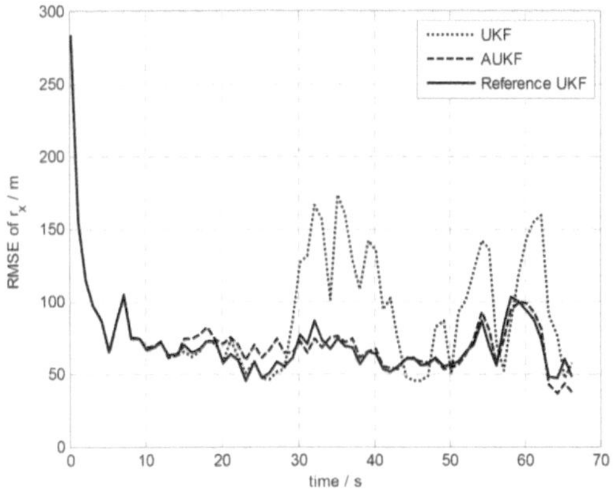

Figure 5. Velocity RMSE of x-axis.

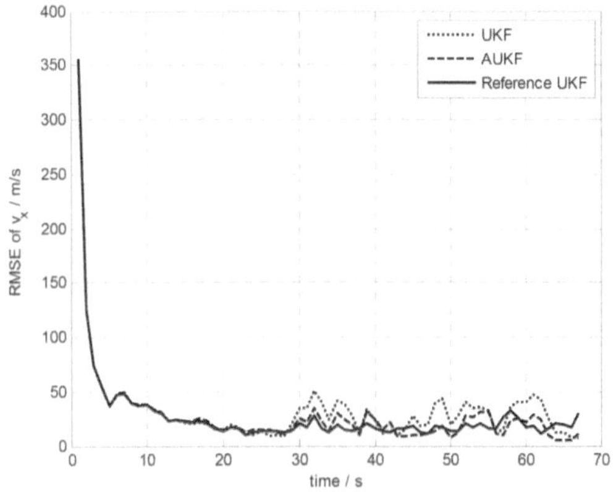

Figure 6. Position RMSE of x-axis.

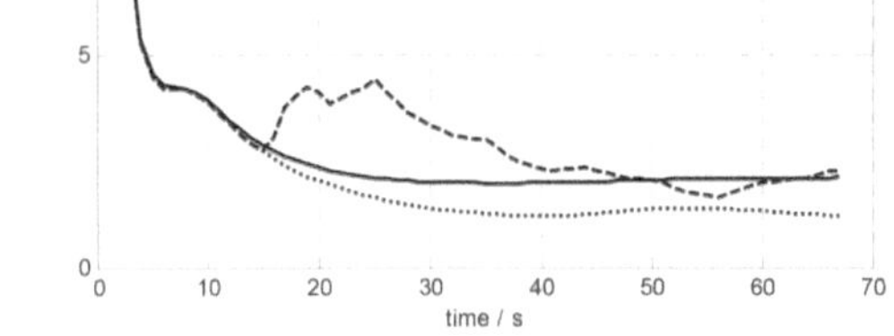

Figure 7. $\|P\|_2$.

correction of covariance matrix Q and make the filtering curve more close to the one of referenced UKF eventually. As a result, the estimated precision can be improved under the condition of unknown system noise.

6 CONCLUSION

Due to the exact statistics characteristic of boost trajectory is usually unknown, this paper proposed an AUKF algorithm that can estimate the statistic characteristic of the system noise in real-time by introducing Sage-Husa noise estimator into UKF to make a better tracking performance. Meanwhile, this paper gave a solution to overcome the situation of filtering divergence that insures the numerical stability of filter. At last, we make simulations under two different initial conditions and analyzed the results. Simulation results showed that AUKF performs better than normal UKF under the condition of unknown system noise and it can be used for trajectory tracking effectively.

REFERENCES

Julier, S. J. & Uhlmann, J. K. 2004. Unscented filtering and nonlinear estimation. In Proceedings of the IEEE 92(3): 401–422.

Li, X. R. & Jilkov, V. P. 2001. A survey of maneuvering target tracking—Part II: Ballistic target models. In Proc. 2001 SPIE Conf. on Signal and Data Processing of Small Targets 4473: 559–581.

Liang, X., Zhou, Z. & Qu, G. 2011. Comparison of nonlinear filters in angles-only boost trajectory estimation based on SBIRS. Spacecraft Engineering 20(3): 56–63.

Sage, A. P. & Husa, G. W. 1969. Adaptive filtering with unknown prior statistics. In Proceedings of joint automatic control conference 1969: 760–769, Boulder: CO.

Sheng, W. 2011. Research on target tracking technologies for space-based optical surveillance system, Changsha: National University of Defense Technology.

Shi, Y. & Han, C. 2011. Adaptive UKF method with Applications to Target tracking. Acta Automatica Sinica 6(37): 755–759.

Zhang, H. & Zhang, H. 1991. Method using for preventing divergence of adaptive Kalman filter. Control and Decision 6(1): 53–56.

Zhou, H., Jing, Z. & Wang, P. 1991. Tracking of maneuvering targets. Beijing: National Defence Industry Press.

Future Mechatronics and Automation – Yang (Ed.)
© 2015 Taylor & Francis Group, London, ISBN 978-1-138-02648-3

Geometric orbit determination of GEO satellites based on dynamics

Yandong Wang, Huan Zhao, Huiyong Chen & Wenyi Wu
School of Automation Science and Electrical Engineering, Beijing University of Aeronautics and Astronautics, Beijing, China

ABSTRACT: The geometric method of orbit determination based on dynamics has been widely applied in LEO satellites orbit determination. But, this method is seldom used in the GEO satellite's orbit determination. So, in this paper, we consider using a similar approach to achieve the GEO satellite's orbit. First, we use the result of the dynamic orbit determination as the priori orbit, secondly we use the geometric method to correct the position of the satellite, and then we can get the precise orbit of the satellite. The main factors such as the quality of the observation data, the length of the baseline and the number of the ground stations which affect the geometric orbit determination are analyzed. So we can provide some references for orbit determination of GEO satellites.

1 INTRODUCTION

With the rapid development of space technology, GEO orbit, because of its geostationary orbit and high altitude, has become an important earth satellite orbit in communications, meteorology, reconnaissance, navigation, precision timing, tracking and data relay and other important scientific research. So the timeliness and accuracy of its orbit have become more and more important[2]. Geometric method, dynamic method and Kalman filter method are commonly used in GEO satellite orbit determination. Geometric method is not affected by it's dynamic model, but it can't predict the orbit either. The precision of orbit determined by dynamic method is vulnerable to mechanical model. While the GEO satellites often adjust it's orbit, so it is difficult to get it's accurate mechanical model[4].

In this paper, we consider combining the dynamic method and geometric method effectively so that we can obtain the precise orbit of GEO satellites. Computing the integral of the orbit dynamics model equation to get the general information of orbit. After we get the general position of the satellite, using it as the priori orbit. Then using the geometric method to achieve the precise orbit. The integral for the orbit dynamic model equation is applied to get the general positions of satellites, so there is no need to establish the accurate dynamic model of GEO satellites. Therefore, the key is to improve the precision of the orbit determined by the geometric method. The main factors affecting the accuracy of orbit calculated by geometric method are the quality of observation data and the tracking geometry for GEO. For we can only use the several existing ground stations within our national border, so we can only improve the tracking geometry for GEO by changing the length of the baseline or the number of the ground stations. It is necessary to analyze how these factors affected the precision of the orbit determined by the geometric method. So we can provide some references for orbit determination of GEO satellites[1].

2 DYNAMIC MODEL

Taking account of the perturbation of the earth's shape, the gravitation of the sun and the moon and the radiation pressure from the sun light, an orbital dynamic model of GEO satellites is established in the inertial coordinate system:

$$\ddot{r} = -\frac{GM}{r^3}r + \ddot{R}(t, r, \dot{r}) \tag{1}$$

On the right side of the equation, the first part is the two-body motion which is the main part of the equation, and the second part is the sum of the perturbations. r, $\dot{r}$ and $\ddot{r}$ respectively donates the position, the velocity and the acceleration of the satellite at the moment of t in the inertial coordinate system. G is earth's gravitational constant, and M is the mass of the earth[5].

For the GEO satellite is far from the earth, the orbital radius of it is about 6.6 times of the Earth's equatorial radius. So, we just take second-order item of the perturbation of the earth's shape and the gravitation of the sun and the moon into account, ignoring other perturbations.

3 MEASUREMENT MODEL

At the moment of t, ground station i (i = 1, 2, 3...) observed the GEO satellite. Then we can get the

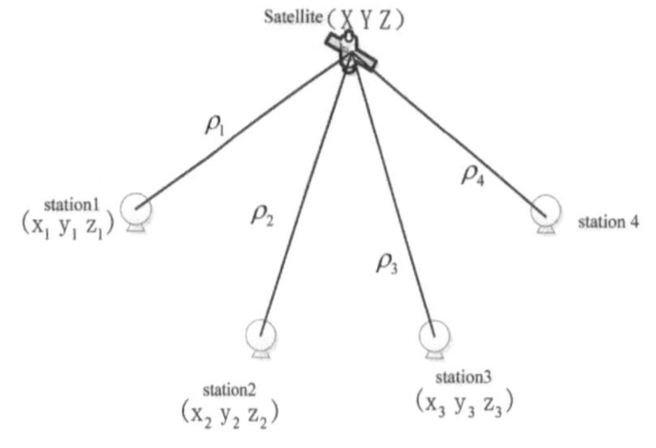

Figure 1. Measurement model.

pseudo-range between ground station i and the GEO satellite. Figure 1 shows the pseudo-range measurement. Then we can get the measurement model, the equation is as follows[3]:

$$\rho_i = \sqrt{(X-x_i)^2+(Y-y_i)^2+(Z-z_i)^2} + d_{trop} + d_{ion} + \varepsilon \tag{2}$$

$$\rho_i^{'} = \sqrt{(X-x_i)^2+(Y-y_i)^2+(Z-z_i)^2} + \delta \tag{3}$$

(X, Y, Z) and (x_i, y_i, z_i) respectively donates the position coordinates of the satellite and the ground station i at the moment of t in the inertial coordinate system, ρ_i donates the pseudo-range between ground station i and the satellite at the moment of t, ρ_i' donates the compensated pseudo-range at the moment of t, σ donates observation noise.

Expanding the equation (3) in a taylor series about $(\hat{X}, \hat{Y}, \hat{Z})$, the result is as follows[6]:

$$\frac{\hat{X}-x_i}{\hat{\rho}_i}\Delta X + \frac{\hat{Y}-y_i}{\hat{\rho}_i}\Delta Y + \frac{\hat{Z}-z_i}{\hat{\rho}_i}\Delta Z - \Delta\rho_i = \varepsilon_i \tag{4}$$

While,

$$\hat{\rho}_i = \sqrt{(\hat{X}-x_i)^2+(\hat{Y}-y_i)^2+(\hat{Z}-z_i)^2} \tag{5}$$

$$\Delta\rho_i = \rho_i^{'} - \hat{\rho}_i \tag{6}$$

$$a_{xi} = \frac{\hat{X}-x_i}{\hat{\rho}_i},\quad a_{yi} = \frac{\hat{Y}-y_i}{\hat{\rho}_i},\quad a_{zi} = \frac{\hat{Z}-z_i}{\hat{\rho}_i} \tag{7}$$

So,

$$\varepsilon_i = a_{xi}\Delta X + a_{yi}\Delta Y + a_{zi}\Delta Z - \Delta\rho_i \tag{8}$$

So, observation equation is as follows:

$$H = \begin{bmatrix} a_{x1} & a_{y1} & a_{z1} \\ a_{x2} & a_{y2} & a_{z2} \\ \cdots & \cdots & \cdots \\ a_{xk} & a_{yk} & a_{zk} \end{bmatrix} \tag{9}$$

Table 1. Results of the orbit determination with different observation noise.

error		noise (0,3.3)	(0,1)	(0,0.3)	(0,0.1)
X error	mean (m)	−0.07	−0.23	−0.01	0.01
	sigma (m)	40.38	13.26	3.83	1.27
Y error	mean (m)	0.02	0.22	−0.01	0.01
	sigma (m)	40.17	10.97	3.81	1.28
Z error	mean (m)	0.02	−0.14	−0.03	0.01
	sigma (m)	61.90	18.63	5.83	1.95
position error	mean (m)	74.21	23.49	7.01	2.34
	sigma (m)	39.57	12.50	3.73	1.25

where k is the number of the ground stations. Then we can obtain the increment of the satellite's position by adopting least square method. The results are as follows:

$$\Delta x = (H^T H)^{-1} H^T \Delta\rho \tag{10}$$

The position of the satellite consists of the approximate part and increment part:

$$X = \hat{X} + \Delta X,\quad Y = \hat{Y} + \Delta Y,\quad Z = \hat{Z} + \Delta Z \tag{11}$$

Iterate until the results meet the orbit accuracy requirement.

4 SIMULATION ANALYSIS

In the simulation, the eccentricity of the GEO satellite we used in this simulation is 0.024, the orbit inclination angle is 2.4°, the right ascension of ascending node is 117.25°E. STK simulation is used to get the normal orbit of the GEO satellite which is used to evaluate the orbit calculated by the geometric method based on dynamics.

Then we use the geometric method based on dynamics to achieve the orbit of the GEO satellite. In the simulation we analyze how the factors such as the quality of the observation data, the length of the baseline and the number of the ground stations affect the precision of the orbit determination. The result is as follows.

4.1 *The influence of observation data*

The ground stations which we used in orbit determinations are Beijing, Harbin, Sanya and Urumqi. During the simulation, we add different observation noise in different cases. The mean of the observation noise is 0 m, while the standard deviations are 3.3 m, 1 m, 0.3 m, 0.1 m respectively. Then we can get the error of the satellite's position. The mean and standard deviation of the error of the satellite's position under the four different cases are shown in table 1 (simulation time 86400 s).

Table 2. Results of the orbit determination with different base length.

error		baseline length (km) 2000	3000
X error	mean (m)	−0.07	−0.01
	sigma (m)	6.97	3.83
Y error	mean (m)	0.02	−0.01
	sigma (m)	7.00	3.81
Z error	mean (m)	−0.05	−0.03
	sigma (m)	11.52	5.83
position error	mean (m)	13.27	7.01
	sigma (m)	7.34	3.73

Table 3. Results of the orbit determination with different number of stations.

noise		number 3	4	5	6
X error	mean (m)	−0.01	−0.01	0.01	−0.01
	sigma (m)	3.96	3.83	3.14	2.83
Y error	mean (m)	−0.02	−0.01	0.01	0.02
	sigma (m)	3.95	3.81	3.13	2.82
Z error	mean (m)	−0.01	−0.03	0.03	−0.01
	sigma (m)	5.95	5.82	5.83	5.34
position error	mean (m)	7.22	7.01	6.41	5.87
	sigma (m)	3.82	3.74	3.53	3.17

From the simulation results we can get that when the standard deviation of observation noise changed from 3.3 m to 0.1 m, the precision of orbit determination is significantly improved. So, it's important to improve the quality of the observation data and reduce the observation error.

4.2 *The influence of baseline length*

We selected two groups of ground stations with different baseline length to achieve the orbit. The two groups of ground stations include Beijing-Harbin-Sanya-Urumqi, which has the base line length of about 3000 km, and Hohhot-Kunming-Fuzhou-Wuhan, which has the base line length of 2000 km. Respectively using these two pairs of ground stations to achieve the orbit of the satellite. The mean of the observation noise is 0 m, and the standard deviation is 0.3 m. The mean and standard deviation of the error of the satellite's position under the two different cases are shown in table 2 (simulation time 86400 s).

As can be seen from the simulation results, when the baseline length between ground stations is 3000 km the accuracy of the orbit determination is better orbit than it is 2000 km.

4.3 *The influence of the number of the ground stations*

Respectively using different numbers (three, four, five, six) of ground stations to obtain the orbit of the satellite. The mean of the observation noise is 0 m, and the standard deviation is 0.3 m. The mean and standard deviation of the error of the satellite's position under the two different cases are shown in table 3 (simulation time 86400 s).

We can see from the simulation data that with the increase of number of ground station the orbit determination precision is increased.

5 CONCLUSION

From simulation results, we conclude that the quality of the observation data has great influence on the precision of orbit determination. So, in order to improve the precision of the GEO satellite orbit we should try to improve the quality of the observation data and reduce the observation error. As for the ground stations, they should be distributed as evenly as possible in the territory, and the baseline length between the ground stations should be as long as possible. With the number of ground stations increased, the accuracy of orbit can be improved, but maintenance costs of the equipment required is correspondingly increased. Considering all the factors, we think that we should use four ground stations to achieve the orbit.

REFERENCES

Baomin Han & Yuanxi Yang. 2007. Kinematic orbit determination of low-earth orbiters based on GPS precise point positioning technique [J]. Journal of southwest jiaotong university.

Lan Du. 2006. A study of the precise orbit determination of geostationary satellites [D]. The PLA information engineering university.

Suiqiang. 2012. Zhi. Research on Real-Time Orbit Determination for Navigation Satellite [D]. The PLA information engineering university

Tianhe Xu & Kaifei He. 2009. Geometry orbit determination of GEO satellite attending to systematic errors [J]. Journal of geodesy and geodynamics.

Xiaolin Jia, WenHai Jiao, Gang Wang & Xianbing Wu. 2004. Research on dynamic orbit determination foe geostationary satellite using two-way pseudorange. Chinese journal of space science.

Xinhai Chen. 1987. Optimal estimate theory [J]. Beijing: Beijing Aviation Institute.

Future Mechatronics and Automation – Yang (Ed.)
© 2015 Taylor & Francis Group, London, ISBN 978-1-138-02648-3

The design of an intelligent hydropower station operation simulation model

Tie Chen & XiChun Wu
Electrical Engineering & Renewable Energy School of CTGU, Yichang, Hubei Province, China

ABSTRACT: By aiming at how to reflect dynamic process under full behaviour and full process, this paper proposes new model designs to resolve this problem. First, it reveals the hierarchical characteristics of hydropower plant production behaviour in its structure. Then, it proposes an operational behaviour conceptual model of a hydropower plant for the first time. Key elements such as entities, roles, and activities are abstracted from production details and defined to achieve a conceptual model. Combined with the production structure, these key elements are described in the perspective of entities based on agent-oriented concepts. General intelligent model templates are designed and established to realize simulation design to achieve function emergence from microscopic details to macroscopic overall. This model design has been proved in operation simulation in some hydropower plants. It can lay the foundation for intelligent hydropower plant operation simulation.

Keywords: hydropower plant; operation simulation; conceptual model; agent-oriented

1 INTRODUCTION

Hydropower plant production system is a complex system with a large number of hydraulic, mechanical, and electrical physical devices. Interaction among these devices cause joint dynamics to achieve production goals through all kinds of production activities. Interaction includes a continuous physical dynamic system, a discrete event dynamic system, and a process dynamic system between continuous and discrete. Interaction is so tight that its process to its goals shows functions that gradually emerge from micro to macro systems of activity. In newly built hydropower plants, the wide use of automation devices with distributive intelligence can cause unexpected actions to destructive results. It is necessary to research and simulate production behaviours. There is very little research in this area at present.

Figure 1 proposes object-oriented models to simulate the typical operation process of a hydropower plant. Models are programmed by the method of "top to bottom" to achieve typical functions and fails to accomplish subsequent dynamic process caused by unexpected actions. Intelligent models based on multi-agents are proposed to simulate all production behaviour and its dynamic process.

Agent-oriented models are valuable research tools for the analysis of dynamic and emergent phenomena in large-scale complex systems. They have been widely applied in ecological systems, in social systems, in economic systems, in human organizations, the military, and in other fields (Figs. 2). Figure 3 proposes the new concept of a "relay agent." Figure 4 develops a decision support system for the modern power system operating with agent technology.

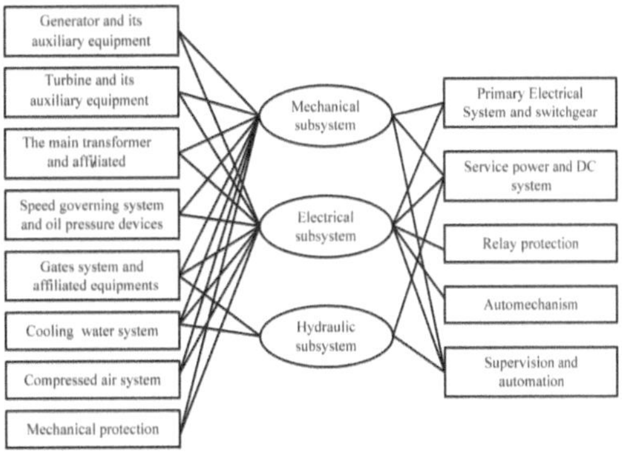

Figure 1. Production entity and its function.

The process of agent-oriented modelling contains the following five main stages: observing the practical real system, building the conceptual model, establishing the simulation model, simulating the operation, and analysing the results (Figs. 5–7). To describe the practical system by the conception model is not only the foundation of modelling, but also the key to estimate the models.

2 ANALYSIS OF PRODUCTION ACTION

A hydropower plant converts the energy in flowing water into electricity. It happens under the coupling effect among the different sub-systems of hydraulic, mechanical and electrical. The effect is reflected in three levels.

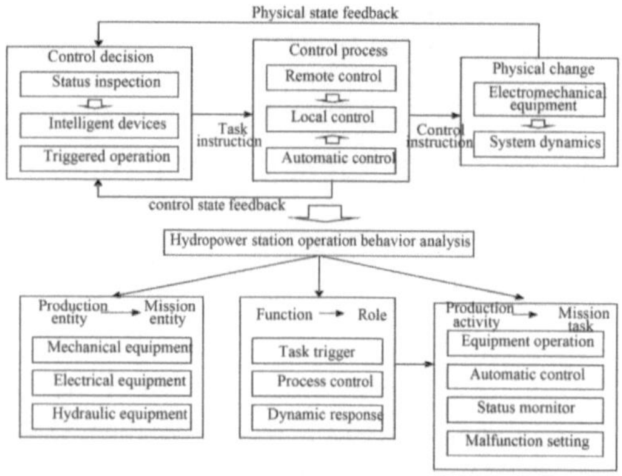

Figure 2. Analysis of production behaviour.

The first level is the core of energy conversion and its distribution in the different main physical components. A typical hydroelectric power plant consists of a reservoir and a diversion system, a hydraulic turbine, a generator, a main transformer, an electrical network, and plant power. Hydropower can be obtained in a reservoir and be sent to a hydraulic turbine by a diversion system where it is converted to mechanical energy in a hydraulic turbine by a speed governor. It is then converted into electricity in a generator. Most of this is transferred to an electrical network, but small amounts are used in the plant power for power supply during energy conversion. It includes hydro-mechanical-electric dynamics with a closed-loop, which is always operating unless the generator stops completely.

The second level is the controlling, monitoring, measuring, regulating, and protecting of the auxiliary systems of the state and function of the main physical equipment.

The third level is the interaction inside the components of the equipment. Their states are decided by a control process which determines their functions. Action methods decide how to call for corresponding processes, which may be scheduled in production behaviour or accidental action.

The devices' state changing causes a dynamic process. The control process determines the devices' state. The production behaviour in a hydropower plant operation show obviously hierarchical relationships in its working structure.

With a complex system perspective, we can separate production behaviour into different process stages. It follows as "decision," "process," and "physics" in logical and functional relationships. "Decision" represents a behavioural pattern of action modes, "process" represents a behavioural process of action results with their process in different modes, and "physics" represents a behaviour response of dynamic changes caused by the operation. Each stage is constructed by different production entities to meet corresponding functions as Figure 2 shows.

High-stage entities may be triggered by existing production tasks to form task instruction, which can influence the process states of entities in the middle-stage. Then the control instruction must be formed to change the physics entities' state in low-stage. The entities' state changes in a low-stage may trigger a new task instruction in a high-stage by an intelligent device, and then have a further effect on the changing process of entities in a low-stage. It constitutes a close-loop feedback mechanism. In this feedback mechanism, behaviour details are reflected as a functional emergence of production entity under some task instruction, which should come from scheduled production behaviour or accidental actions.

3 OPERATION CONCEPTUAL MODEL

3.1 *Basic concept*

Within the conceptual models' perspective, an operation conceptual model is established by abstracting the production process for the first time. It provides a standardized description in detailed model framework with its composition, entity, activity, function, and interaction. It can be described as establishing a simulation context based on authoritative information, the decomposition of mission space, the abstraction of mission entities, tasks, activities, environments, and algorithms, and describing the relationships between these elements (Figs. 8).

As Figure 2 shows by decomposing the production process in its structure, key elements of production entities, functions, production activities as mission entities, roles, and mission tasks are abstracted, and determine their relationship in agent concepts to get an operation conceptual model.

3.2 *Establishing key elements*

In this paper, hydropower station production entities are divided into three categories: the trigger entity ("trigger"), the process entities ("process"), and ("physics") the physical entities. They belong to relevant levels in the hierarchical structure to realize the functions of each stage in the production process.

Trigger entities are production entities achieving different human-computer interactive operation modes and form task instruction. They contain total existing triggering modes of production tasks. Process entities are production entities achieving different operation modes, process, logics, operation results, and they form the control instructions. They contain the total control process of the production tasks. Physics entities are production entities achieving behavioural dynamic responses in hydraulic, mechanical, and electrical sub-systems. They contain the total behavioural results of the production tasks.

A certain mission entity must correspond to a certain production entity in every production process, and vice versa. A mission entity come from a production entity, and mission entities are also divided into three categories. They can be divided into lower groups for demands of simulation particle size.

3.2.1 *Trigger entities*

We define trigger entities as four classes: operation, monitoring, automation, and setting. They are divided downwards to the third points and include twenty-three groups.

Operation classes include 6 general groups of switches, buttons, linking pieces, control widgets, fuses and valves. It can be triggered by human-computer interactive operation modes which include remote, field, auto, and hand modes.

The monitor class includes 2 general groups of signal lamp and bulletins. It can monitor total entities' states.

The automation class includes 11 general groups of the PLC for auxiliary devices, protections, emergency power supply auto restart, AGC, AVR. It can be triggered by physics entities.

The setting class includes 2 general groups of electrical fault setting, mechanical fault setting. They are set as the specialized demands for fault training. It can be triggered by human-computer interactive operation.

3.2.2 *Process entities*

We define process entities as three classes: system, procedure, and components. They are divided downwards to the third points and include 13 groups.

The procedure class includes "turn-on/off", breaker operation, auxiliary equipment and gate operation; it can be triggered to inflect operation procedure by task instruction from automatic (remote/local) operation.

The component class includes 5 general groups of switching devices and operating mechanism. It can change its switching states in automatic (remote/local) mode by instruction from procedure and hand (local) mode by task instruction from field operation.

The system class includes 4 general groups of secondary electrical systems. It reflects the states of monitoring systems, synchronization devices, excitation, and governing systems as constraint conditions for an operation and procedure entity. Components entity triggered this class.

3.2.3 *Physics entities*

We define physics entities into two classes: a complex class and a logic class. They are divided downwards to the third points and include 19 groups.

The logic class includes the network topology of electrical and mechanical sub-systems, and seven general groups. It can be triggered by process entity for dynamic calculation. This type of entity is a topology structure for dynamic computation.

The complex class includes the main equipment in electrical, mechanical, and hydraulic sub-systems, and twelve general groups. It reflects the system's dynamic states.

According to the classification above, the mission entity dominates the production activity by its role, which can be described as a role that highlights and functions anything emerging by triggered entities. An activity is driven by a triggered entity, and is controlled by a process entity, changes the state of a physics entity, and is detected by a triggered entity. Considering subsequent activity may be caused by a certain activity, the numbers of production activity are as many as the numbers of triggered entities.

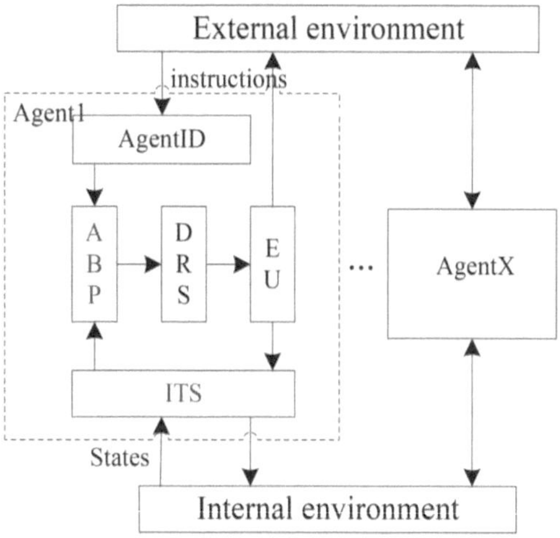

Figure 3. Agent-model structure.

A mission task comes from an activity, and a certain mission task must correspond to a certain activity in every production process, and vice versa. Therefore, we define a mission task as one of four kinds: operation, automatic control, monitoring, and fault setting. With definition triggering orders, numbers, interaction, and corresponding algorithm in mission tasks, we can form the operational concept model of a hydropower plant.

4 AGENT-ORIENTED MODEL

4.1 *General agent model templates*

An agent class is devised according to an entity class, and each agent class consists of several sub-agents with similar attributes but with different technical details which correspond to entity groups.

In order to realize the interactions and repeated use of agents, we define the general agent model template as follows:

$$Agent = < Time, AgentID, ITS, ABP, KRS, EU > \quad (1)$$

Agent ID is the agent logo. It can perceive the environment, and forwardly receive behaviour instructions of triggering conditions, and transform these messages into recognizable variables which represent behaviour modes. Because of internal variables being defined in agent models, an agent ID structure can also be designed as a unity which includes information interface between agent and environment, instruction rules, and agent identification. Agent ID can be classified by its types of groups.

ITS: Information Transfer System. ITS can forwardly achieve internal information interaction among agents. Because interaction cannot occur directly among agents, the ITS structure can also be designed and unified which includes the information interface, rules, and the interaction object. Input information dates from internal environments can be transformed into recognizable variables by independent input interfaces which may be used to form constraint conditions in the ABP. Constraint conditions include behaviour constrains and states' constraints. The former is decided by the executing priority of agents; the latter is decided by the executing condition of the agent. For example, we can operate a breaker successfully only under certain conditions of the correct operational modes, the current operates status and the power supply.

Output information, which represents agent states from the EU, can be transformed into information data by the independent output interface to the internal environment. There is only input interface in monitor agents because of their roles which are the ultimate link in production activity.

ABP: Agent Behaviour Property. ABP is a status library of agents. It can be defined as multi-attribute variables in behaviour modes. Behaviour constraints and states constraints have action results in different behaviour modes.

DRS: Decision Reasoning System. It can be described as logic functions:

$$\psi : B \times A \times C \times S \rightarrow S \tag{2}$$

where B is behaviour modes library; A is action function of behaviour modes; C is constrains library; S is status library.

EU: Execution Unit. It can transform action results into interaction information and send them to external environment.

4.2 *Establishing multi-agent models*

We construct four kinds of agent-class model templates at the decision level, four kinds at the process level, and two kinds at the physics level. Models are constructed in single step execution.

4.2.1 *Define environment*

Decide the agent link reaction, and establish the input and output information library. Link the input information and output information and the instruction process is formed as Figure 4 shows.

Behaviour constraints come from status interaction between similar agents. State constraints come from status in different levels as Figure 5 shows.

Establish internal environment and external environment. Send instruction library to external environment. Send status library to internal environment.

Each agent can be triggered by the unique instruction from an external environment and sends new instructions back after the agent has executed it. Internal environment is a vector of agent states. Because some agent status information may be also used as a production instruction, which causes information to be repeated, a real-time database can be devised to solve these two problems.

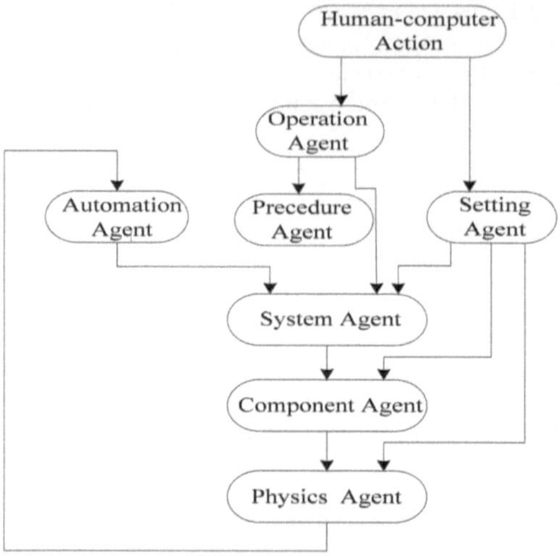

Figure 4. Instruction process.

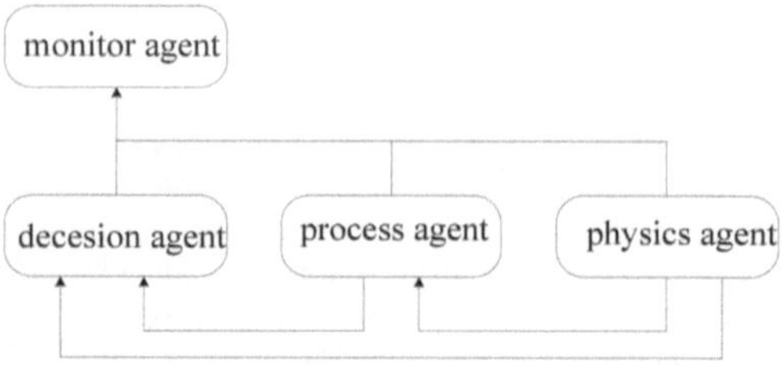

Figure 5. Status process.

4.2.2 *Agent models*

Define structure variables for agent execution. Establish an interface between structure variables and the real-time database.

Establish agent model by structure variables. Complex-class agents include dynamic functions and hydraulic, mechanical, and electrical systems. Subsystem agent models for system decoupling need to be established as Figure 6 shows.

4.2.3 *Multi-agent models*

Multi-agent models are constructed as Figure 7 shows. Simulation commences and the simulation step can be set to adapt dynamic electrical models. When simulation starts, the models can be triggered by their own orders.

5 CONCLUSIONS

This article proposed an operation of a conceptual model of a hydropower station for the first time based

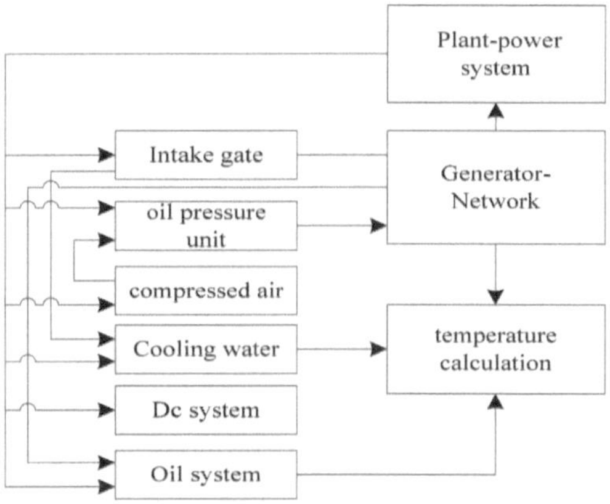

Figure 6. System decoupling model.

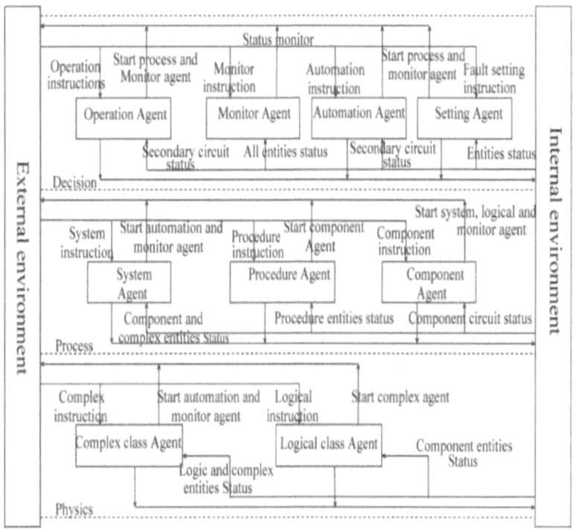

Figure 7. System decoupling model.

on agents to achieve the functions emerging in operation simulation. It illustrates the design and establishes a standardized agent-oriented model structure. It provides a technical foundation to achieve full range and full scope models in a hydropower station operation simulation system.

There are some key problems to be solved in modelling. It mainly focuses on how to determine model algorithms to suit the demands of real-time and accuracy. We have established multi-agent models for some large hydropower plants. The models have been used in this hydropower plant. Simulation objects contain mainly hydraulic, mechanical, and electrical equipment and their auxiliary control, monitoring, and protection devices. The system has been used for operating personnel training.

REFERENCES

Fang Hong-qing, Chen Long, Shen Zu-yi, Wu Kai. Hydropower station digital simulation based on objet-oriented technology. Journal of Yangzhou University (Natural Science Edition), vol. 6 No. 4, June, 2003, pp. 64–67.

[Full Scope Real-Time Simulation of Hydropower Plant for a Training and Research Simulator. Xianshan Li, Chengming Wu, and Xiangyong Hu, 2005 IEEE/PES Transmission and Distribution Conference & Exhibition]

Gilbert N, Troitzsch K G. Simulation for the Social Scientist 2nd Edition [M]. USA: Open University Press, 2005.

Law A M, Kelton W D. Simulation Modeling and Analysis Third Edition [M]. USA: The McGraw-Hill Companies, Inc., 2000. Fishwick P. Simulation Model Design and Execution [M].

Liu Xiao-ping, Tang Yi-ming, Zheng Li-ping. Survey of Complex System and Complex System Simulation. Journal of System Simulation. Vol. 20 No. 23, Dec., 2008, pp. 6303–6315.

Wang Zi-cai, Wang Yong. Research Evolvement and Direction of Complex System Simulation Conceptual Model, Journal of Astronautics, Vol. 28, No. 4, July 2007, pp. 779–785.

Yang Xusheng, Y. Sheng Wanxing, Y. Wang Sun'an. Study On multi-agent architecture based decision support system for Power System, Automation of Electric Power System, Vol. 26, No. 18, Sept 25, 2002, pp. 45–49.

Yasushi Tomita, Chihiro Fukui, Hiroyuki Kudo. Acooprative Protection System with an Agent Model. IEEE Transactions on Power Delivery, Vol. 13, No. 4, October 1998, pp. 1060–1066.

Zhang Fa, Xuan Hui-yu, Zhao Qiao-xia. Methodology of Multi-Agent Based Simulation for Complex Systems, Journal of System Simulation, Vol. 21, No. 8, Apr., 2009, pp. 2386–2390.

Future Mechatronics and Automation – Yang (Ed.)
© 2015 Taylor & Francis Group, London, ISBN 978-1-138-02648-3

The development of an intelligent portable fumigation treatment bed

Dong-Liang Zhao & Yan-Xing Guo
Orthopaedic Institute of Henan Province, Luoyang Orthopaedic Hospital of Henan Province, Luoyang, Henan, China

ABSTRACT: The objective is to develop a kind of portable fumigation treatment bed, which has a folding body and a treatment device for easy disassembly. It is convenient for carrying and transportation. It can treat the shoulders, waist, knees and other parts of the body by means of fumigation. Methods: The bed frame is made of 201 stainless steel pieces. The treatment device adopts stainless steel whose grade is 304. Most electrical components adopt ABB, Schneider, and other famous brands. The treatment temperature is controlled automatically through microcomputer technology and a control algorithm of Proportion Integration Differentiation. Results: The developed prototype has the advantages of having a compact structure, is lightweight, and is easy to carry and transport. It is very safe and highly reliable. The temperature control precision is ±1°C.

1 INTRODUCTION

Fumigation treatment is an important part of traditional medicine in China's core theory. The dialectical therapy includes selecting a herbal formula according to the illness and fumigating the skin in order to enable the drug ion to penetrate through the skin. It has a good effect in the treatment of various diseases such as all types of neck, shoulder, waist, and leg pain disorders, fracture postoperative rehabilitation, treatment in the area of the anus and intestinal surgery, and for gynaecological diseases etc. In recent years, some domestic manufacturers have improved the traditional fumigation equipment and produced many fumigation beds. However, the fumigation beds on the market are very large, have complicated structures, are heavy, and inconvenient to transport etc. In order to overcome the shortcomings of the existing fumigation beds, We invented a type of portable fumigation therapeutic bed, which has a folding bed body, and a treatment device for easy disassembly. It is convenient for carrying and transportation. It can give heated fumigation treatment to the shoulders, waist, knees, and other parts of the body.

2 DESIGN OF THE MECHANICAL STRUCTURE

The instrument is mainly composed of a bed surface, a treatment device, and the bedstead. The bed surface is composed of a fixed bed plate and a movable bed plate. The bed has a moisture-proof board. Each plank is covered with sponge and leather. The bed planks above the treatment tank are movable. They can be easily removed when the neck, waist, knees, and other parts are treated. The heating box and the control box are connected into a whole through the bolts to form a treatment device. The treatment device is hung on a folding orbit. The control box is closed, which can prevent the steam getting into it. When carrying out treatment, the bed is opened and the treatment device is moved above the treatment site by pulleys. The movable bed plate is removed. The bed head is provided with a massage hole to facilitate the treatment of patient's knee joints while they are in a prone position. The fumigation bed will become a massage bed when all the movable bed plates are put into place. The bed body has three sections to its folding frame structure. When transporting or storing, first open the medicine-discharging valve and drain the liquid medicine in the heating box. Then remove the whole treatment device from the folding orbit and take the movable bed planks away. The fixed bed legs in the middle of the bed are not moved; both ends of the foldable bed legs are folded inwards into the middle. The folding orbit and the bedstead are folded up together. The advantages of a folding bed are that it is a small, compact structure, and is lightweight. The bed structure is folded, as shown in Figure 2. The folded bed can be directly placed in a car trunk and is easy to carry and transport. The bed frame is made of 201 stainless steel pieces. It is lightweight and can be moisture-proof. The treatment device consists of 304 stainless steel pieces in order to prevent the traditional Chinese medicine from corroding it.

3 HARDWARE DESIGN OF THE CONTROL SYSTEM

The block diagram of the control system is shown in Figure 4. The running process of the system is as follows: first, the Chinese liquid medicine is added to

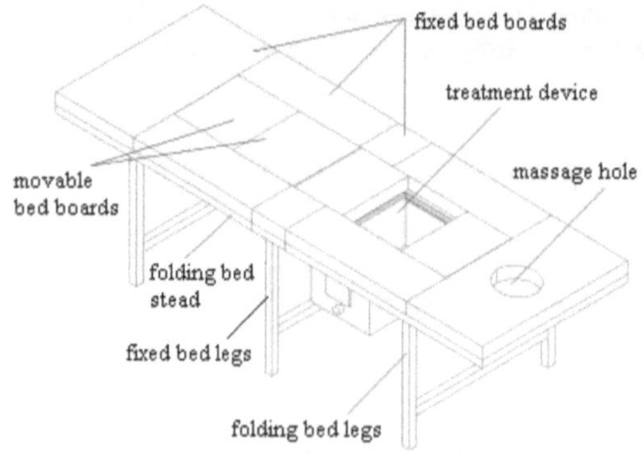

Figure 1. The schematic diagram of the fumigation bed machine.

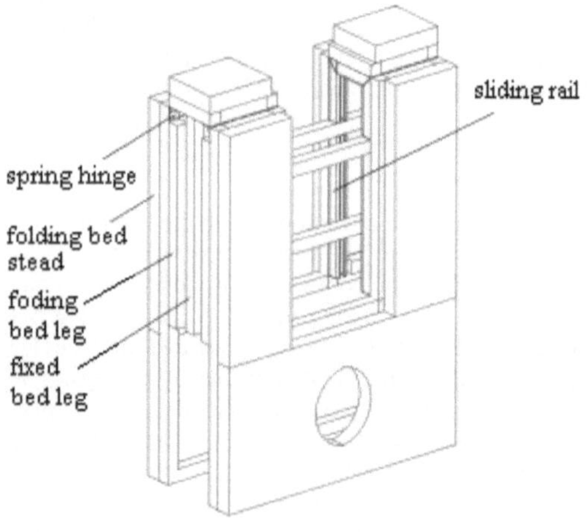

Figure 2. The schematic diagram of the folded fumigation bedstead.

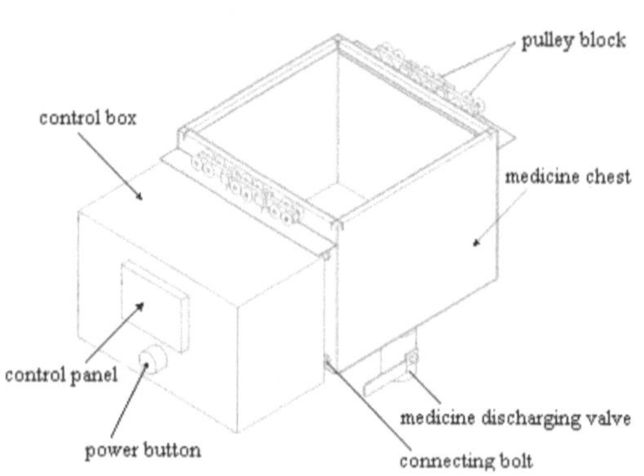

Figure 3. The schematic diagram of the treatment device.

the medicine chest. When the liquid medicine in the medicine chest reaches a certain amount, the liquid level switch closes. Then the system can work normally. This function prevents the heater from being damaged. The input of the treatment temperature and the time setting value is through the metal dome

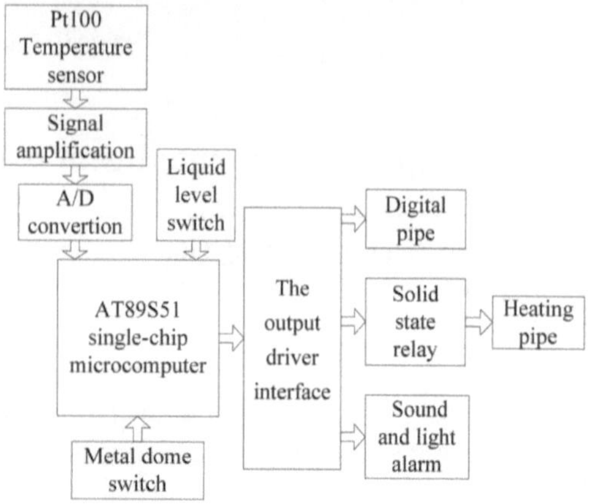

Figure 4. Block diagram of the control system.

switch. The equipment goes into an automatic operation state when the start button is pressed. The temperature of the liquid medicine is conversed into mv level signals by the Pt100 temperature sensor. It is converted into 0–5V voltage signal through the amplification circuit. Then it is converted into digital quantity through the A/D converter and is sent to an MCU. At first, the single chip microcomputer carries on the scale transformation and registers the marked temperature value and operates the digital pipe to display it. Then, according to the given process parameters and the feedback value of temperature, the device adjusts the operation in accordance with the reservation control algorithm. It determines the output control quantity to drive the solid state relay and controls the input energy of the heater. It makes the temperature of the heater change according to the prescriptive process curve to achieve a constant temperature control. When the treatment time is over, the sound and light alarm will sound automatically to remind the user. The leakage protection air switch is installed in the main loop. As long as the leakage current is detected, the power supply is automatically cut off to prevent the user from getting an electric shock.

4 DESIGN OF A DIGITAL TEMPERATURE CONTROLLER

a) The mathematical model of the heater

The theoretical analysis and experimental results illustrate that the heater can be used in a first order inertial link plus an approximate delay link. The heater in this system is a sealed heating source and there are no disturbance factors. The energy of the power supply is solely used for raising the temperature of the liquid medicine. Therefore, its mathematical model is:

$$\frac{\theta(s)}{U(s)} = \frac{Ke^{-\tau s}}{T_p S + 1} \tag{1}$$

Table 1. Data of Heater Model Detected by Flying curve.

Time (min)	0.5	1.0	1.5	2.0	2.5	3.0	3.5	4.0	4.5	5.0
TEMP (°C)	23	28	34	41	46	51	57	62	66	69

where $\theta(s)$ = temperature of medicine liquid; $U(s)$ = input voltage; K = the static gain of this target; T_p = the time constant of this target; and τ = the pure delay time of this target. The current methods commonly used in engineering are applying the step input signal to the process object and measuring the step response of the process object. Then the step response curve is used to determine the approximate transfer function of the process. Specifically, the approximate transfer function is determined by the Cohen-Coon formula. The given input step signal is 70°C, and a thermometer was used to measure the temperature of the traditional Chinese liquid medicine in the medicine tank. We measured once every half minute.

The experimental data is as shown in Table 1.

The Cohen-Coon formula is shown below:

$$\begin{cases} K = \dfrac{\Delta C}{\Delta M} \\ T_p = 1.5(t_{0.632} - t_{0.28}) \\ \tau = 1.5(t_{0.28} - \frac{1}{3} t_{0.632}) \end{cases} \quad (2)$$

where ΔM = step input of the system; ΔC = output response of the system; $t_{0.28}$ = the time (min) when flying curve = $0.28\Delta C$; and $t_{0.632}$ = the time (min) when flying curve = $0.632\Delta C$. Thus $K = 0.92$; $T_p = 144$ s; $\tau = 30$ s. Therefore, the model heater is:

$$\frac{\theta(s)}{U(s)} = \frac{0.92e^{-30s}}{144s+1} \quad (3)$$

b) Determining the temperature control algorithm

In order to reduce the overshoot and eliminate oscillation, the system begins to control the temperature consistently when the temperature freely rises to a given value of 85%. Temperature control adopts the integral separation PID adjustment. The position control equation of PID regulation is:

$$u_{ct}(t) = K_p[e(t) + \frac{1}{T_i}\int e(t)dt + T_D\frac{de(t)}{dt}] \quad (4)$$

where $u_{ct}(t)$ = output of controller, and $e(t)$ = the difference of given constant temperature value (θ_0) on t time and the measured liquid temperature (θ). It is discredited for:

$$u_{ct}(k) = u_{ct}(k-1) + K_p\{[x(k) - x(k-1)] + \frac{T}{T_i}X(k) + \frac{T_D}{T}[X(k) - 2x(k-1) + x(k-2)]\} \quad (5)$$

where T = sampling period; T_D = differentiating time constant; T_i = integration time constant; K_p = coefficients of proportionality; and $x(k)$ = the difference ($\Delta\theta$) of this sampling θ_0 and θ. If the control degree = 1.2, then $T = 0.16\tau = 4.8$ s; $K_p = T_p/\tau = 4.8$; $T_i = 1.9\tau = 57$ s; $T_D = 0.55\tau = 16.5$ s. Substitute for the above formulate to obtain:

$$u(k) = u(k-1) + 4.8[x(k) - x(k-1)] + 0.404X(k) + 16.5[X(k) - 2x(k-1) + x(k-2)] \quad (6)$$

5 DESIGN OF THE CONTROL SYSTEM SOFTWARE

The software of the system adopts the modularized and structured concept of a program design, which is convenient for the maintenance and transplantation of the program. The main program includes system initialization, a constant temperature and a constant temperature time input processing, the display of temperature and time of constant temperature, judgements and treatment of constant temperature at the beginning, and a constant temperature at the completion of the treatment and judgements and treatment whether the sampling period reaches, and so on. The flow chart of the main program is as shown in Figure 5.

For "Input and treatment of the constant temperature and constant temperature time," complete the following functions: Receive the user key-press and key-press identification, convert the input number to a binary number, and send it to the corresponding given value unit after inputting the constant temperature and time.

"Sending the actual temperature and the time of constant temperature to 8279" completes the display of liquid temperature and constant temperature time. Because there is a display buffer in the 8279, it can automatically and dynamically send the code of display buffer to the digital tube and display. Therefore, it is sufficient to convert the display value to display code and then send it to 8279.

"End of constant temperature time" is comparing the calculation time of the constant temperature according to the timing program with the given constant time. If the time of calculation equals the given time, the time of constant temperature is up. This temperature control process is over. The system will stop detecting and controlling the temperature and only carry out displaying and alarming functions until the system restarts. The constant temperature-timing program is actually a clock interrupt service program. The time of the software timer one we select is 100 ms. When the time is up, the interruption times increase

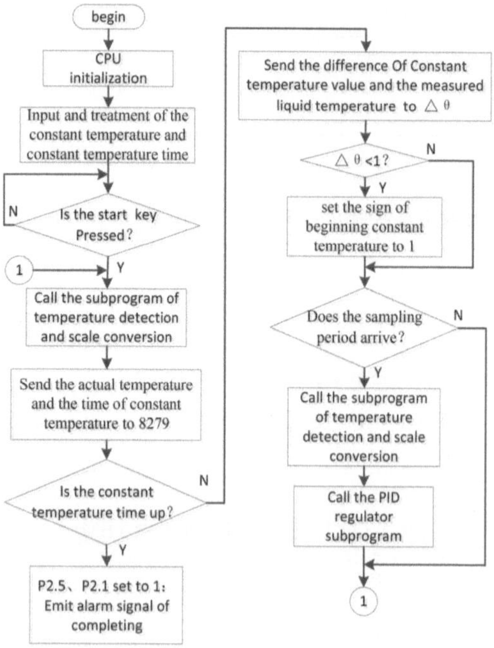

Figure 5. The flow chart of the main program.

by one and the time of the timer is resetted as 100 ms. The second unit increases by one every 10 times. When the values of the second unit and the unit of the given time of constant temperature are equal, it shows that the time of constant temperature is up.

"$|\Delta\theta| < 1°C$" is used to determine whether the liquid temperature reaches the constant temperature value. Because the constant temperature error given by the design requirements is $\pm 1°C$, so we think that the liquid temperature has reached the constant temperature value when $|\Delta\theta| < 1°C$. The MCU will set the sign of beginning a constant temperature to 1 and begin to reckon by keeping time of the constant temperature.

"The sampling period arriving" is to determine whether the sampling cycle timing counter is 0. The timing of the sampling period is realized by the power supply frequency. When a power cycle arrives, the counter will be -1. The initial value of the counter is the power supply frequency corresponding to a sampling period. When the counter reaches the value 0, it shows that a new sampling period is beginning.

6 CONCLUSIONS

After this intelligent portable fumigation treatment bed has been manufactured, by means of repeatedly testing and modifying the program, all the functions, such as constant temperature control and the automatic timing alarm, have achieved the expected design requirements. The precision of temperature control is $\pm 1°C$. All the safety measures of protecting the heater from electric leakage, dry heating, and current overload, etc. achieve the expected effect. This fumigation bed has the characteristics of an intelligent instrument for the safety and convenience of the user. The treatment device and the bed body can be assembled together conveniently. The treatment device can be smoothly moved to below those parts of the body needing treatment along the treatment rail. The bed can be folded conveniently and is light in weight. It can be lifted by one person after being folded, and can be put in the trunk of a car. It is easy to carry and transport.

REFERENCES

Chen Zhi-huang, He Dan-dan, Shen Yan. Research summaries about the traditional Chinese medicine fumigation treatment application in arthromyodynia [J]. Zhong Guo Zhong Yi Ji Zheng, 2011, Feb; 20(02):282–283.

Deng W, Zhao CJ, etc. Study on spectral detection of green plant target [J]. Guang Pu Xue Yu Guang Pu Fen Xi. 2010 Aug; 30(8):2179–83.

Lu H, Feng X, Chen Y, Chen Y, Ni Y. Sheng Wu Yi Xue Gong Cheng Xue Za Zhi [J]. An improved PID algorithm for temperature control used by tumor combined therapeutic instrument. 2003 Sep; 20(3):521–3.

McCracken M[1], Mayer M, Jourard I, Moon JT, Persic J. Symmetric miniaturized heating system for active microelectronic devices [J]. Rev Sci Instrum. 2010 Jul; 81(7):075112.

Tang Li-ming, Liu Tie-bing, Wu Min-qi, Shi Tao. Design and manufacture of folding and mobile hospital bed [J]. Beijing Biomedical Engineering. 2008, Jun; 27(3): 286–288.

Zhang J[1], Zhang F, Ren M, Hou G, Fang F. Cascade control of superheated steam temperature with neuro-PID controller [J]. ISA Trans. 2012 Nov; 51(6):778–85.

Future Mechatronics and Automation – Yang (Ed.)
© 2015 Taylor & Francis Group, London, ISBN 978-1-138-02648-3

The design of a controller with Smith predictor for networked control systems with long time delay

Yongguang Ma, Junru Jia & Junqing Bo
Control and computer engineering, North China Electric Power University, Baoding, China

ABSTRACT: In the networked control systems, the network random delay and the lag of nonlinear system itself have a serious impact on control performance of the system and reduce the real-time performance of the system. For the long time delay problem of the closed loop control system, the method that regards the network delay and the object as a time-varying generalized controlled object is proposed. At the same time,in order to estimate the parameters of the Smith predictive controller, the recursive least square method is used to identify the generalized controlled object online. Further, the fuzzy PID controller is introduced to make up for the negative effects that model mismatch causes and optimize the control performance. Simulation results demonstrate the effectiveness of the proposed method.

Keywords: networked control system; recursive least square method; Smith predictive controll; fuzzy PID

1 INTRODUCTION

With the development of computer, network, communication technology and control theory, networked control systems(NCSs) are being applied to industrial control field on a daily broadening scale, as in [1]. Networked control systems bring many advantages over the point-to-point wired conventional control systems, including stronger diagnostic capacity simpler installation, higher reliability and better interactivity, etc and can be applied to complex large system, as in [2]. It's easy to realize remote control and resource sharing as well. At the same time, due to the introduction of the network in the closed loop control system, many new problems have arisen in the networked control system, such as sampling period, clock synchronization mechanism, transmission mode, node-driven means, network delay, network scheduling, packet loss, packet sequential disorder and communication constraints, etc, as in [3]. The random network delay makes it harder to analyze or synthesize the system. Overlong time delay will even make the system lose stability. Therefore, for the analysis and design of the network control system, network time delay compensation is an important factor and receive wide concern from the experts and scholars.

Take the network random delay and the lag of nonlinear system itself in the networked control systems into account, in this paper, the introduction of Smith predictive control to realize the network time delay compensation is put forward. Since the design of Smith predictor is based on the accurate model of controlled object and network time delay, the method that regards the network delay and the controlled plant as a time-varying generalized controlled object is proposed. At the same time, the recursive least square method is adopted to identify the parameters of the Smith predictive controller online. For the better robustness of the closed loop control system, the design combines Fuzzy self-adaptive PID control with Smith predictive controll to achieve time delay compensation networked control systems with long time delay.

2 THE PROBLEM OF TIME DELAY IN NETWORKED CONTROL SYSTEMS

The networked control system is a fully distributed, networked real-time feedback control system. It refers to the set of field sensor, controller, actuator and communication network in one area and is available for data transmission between devices to make it possible for users located in different positions in this area to achieve resource sharing and coordinated operation, as in [4]. The typical structure of networked control systems is described as in figure 1.

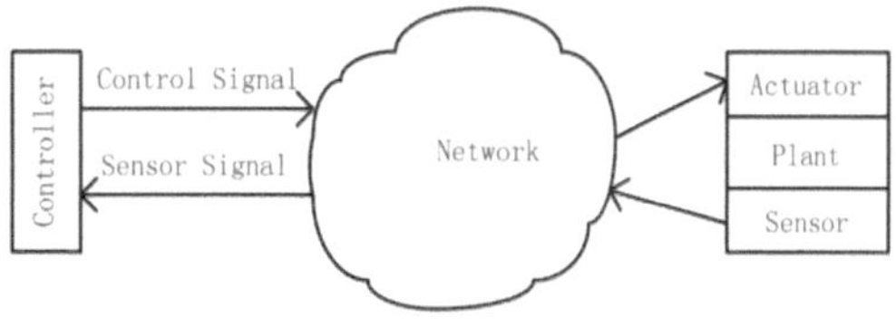

Figure 1. The typical structure of networked control systems.

As a general rule, network control systems consist of sensor, controller, controlled object and communication network. The output of the controlled object is taken sample by the sensor first, then being sent to the input end of the controller through the network. After finishing calculation and processing of the received messages, the controller will send the output to the input end of the controlled object through the communication network again, as in [5]. In view of the limited network bandwidth and irregular changes of network data flow, it is inevitable to cause data collision, multipath transmission, network interruption and network congestion, etc. Thus information exchange time delay will occur, which is also referred to as network time delay, as in [6]. The network time delay is characterized by random timely variation or uncertain timely variation.

3 RECURSIVEL EAST-SQUARES ESTIMATION

3.1 *Least squares*

The least square method has been widely used in parameter estimation, system identification and prediction, etc. Least squares estimation refers to calculating estimated values of parameters that minimize the square of the difference between actual value and discreet value.

In terms of SISO system, assume that the transfer function of the controlled plant identified is defined as follows:

$$\frac{Y(z)}{U(z)} = \frac{b_1 z^{-1} + b_2 z^{-2} + \cdots b_m z^{-m}}{1 + a_1 z^{-1} + a_2 z^{-2} + \cdots a_n z^{-n}} \tag{1}$$

Assume $m = n$ (In the following reasoning process, in case of $m \leq n$, set $b_{m+1} = b_{m+2} = \cdots b_n = 0$). Then the transfer function in (1) can be translated into

$$\begin{aligned} y(n+k) &= -a_1 y(n+k-1) - \cdots - a_n y(k) \\ &\quad + b_1 y(n+k-1) + \cdots + b_n y(k) \end{aligned} \tag{2}$$

Simplify (2) as

$$y(n+k) = -\sum_{i=1}^{n} a_i y(n+k-i) + \sum_{i=1}^{n} b_i u(n+k-i) \tag{3}$$

Assume

$$\Phi_N = \begin{bmatrix} -y(n) & -y(n-1) & \cdots & -y(1) & u(n) & u(n-1) & \cdots & u(1) \\ -y(n+1) & -y(n) & \cdots & -y(2) & u(n+1) & u(n) & \cdots & u(2) \\ \vdots & \vdots & \vdots & \vdots & \vdots & \vdots & \vdots & \vdots \\ -y(n+N-1) & -y(n+N-2) & \cdots & -y(N) & u(n+N-1) & u(n+N-2) & \cdots & u(N) \end{bmatrix}$$

$$\theta = \begin{bmatrix} a_1 & a_2 & \cdots & a_n & b_1 & b_2 & \cdots & b_n \end{bmatrix}^T$$

$$y_N = \begin{bmatrix} y_{n+1} & y_{n+2} & \cdots & y_{n+N} \end{bmatrix}^T$$

Where $N \geq 3n$.

In general, select

$J = (y_N - \Phi_N\theta)^T (y_N - \Phi_N\theta)$ as the object function. Demand first order partial derivative of J with respect to independent variable θ and assume $\frac{\partial J}{\partial \theta} = 0$. Then obtain

$$\hat{\theta}_N = (\Phi_N^T \Phi_N)\backslash \Phi_N^T y_N \tag{4}$$

At this point, J achieve the minimum. $\hat{\theta}_N$ is just the least square estimation value of θ.

3.2 *Recursive least square methodology*

Define estimated value of parameter based on sampling value at the $n+N+1$ time as $\hat{\theta}_{N+1}$. The following equality can be obtained from (4)

$$\hat{\theta}_{N+1} = (\Phi_{N+1}^T \Phi_{N+1})\backslash \Phi_{N+1}^T y_{N+1} \tag{5}$$

Where

$$\Phi_{N+1} = \begin{bmatrix} \Phi_N \\ \varphi_{N+1}^T \end{bmatrix}, \quad y_{N+1} = \begin{bmatrix} y_N \\ y(n+N+1) \end{bmatrix}$$

Assume

$$P_N = \left[\Phi_N^T \Phi_N\right]^{-1} \tag{6}$$

From (6), one can obtain

$$\begin{aligned} P_{N+1} &= [\Phi_{N+1}^T \Phi_{N+1}]^{-1} = [\Phi_N^T \Phi_N + \varphi_{N+1}\varphi_{N+1}^T]^{-1} \\ &= \left[P_N^{-1} + \varphi_{N+1}\varphi_{N+1}^T\right]^{-1} \end{aligned} \tag{7}$$

Thus, the estimated value of parameter based on sampling value at the $n+N+1$ time is just

$$\begin{aligned} \hat{\theta}_{N+1} &= P_{N+1}\Phi_{N+1}^T y_{N+1} \\ &= P_{N+1}[\Phi_N^T y_N + \varphi_{N+1} y(n+N+1)] \end{aligned} \tag{8}$$

Add (7) into (8), obtain

$$\hat{\theta}_{N+1} = \hat{\theta}_N + P_{N+1}\varphi_{N+1}[y(n+N+1) - \varphi_{N+1}^T \hat{\theta}_N] \tag{9}$$

Inversion formula of matrix is introduced to simplify reasoning process of recursion formula of inverse matrix. The formula is as follows:

$$(A+BCD)^{-1} = A^{-1} - A^{-1}B\,(C^{-1} + DA^{-1}B)^{-1}DA^{-1} \tag{10}$$

(7) can be translated into

$$P_{N+1} = P_N - P_N\varphi_{N+1}(\varphi_{N+1}^T P_N \varphi_{N+1} + 1)^{-1}\varphi_{N+1}^T P_N \tag{11}$$

or

$$P_{N+1} = P_N - \frac{P_N\varphi_{N+1}}{\varphi_{N+1}^T P_N \varphi_{N+1} + 1}[y(n+N+1) - \varphi_{N+1}^T \hat{\theta}_N] \tag{12}$$

Since $\varphi_{N+1}^T P_N \varphi_{N+1}$ is a scalar, one can obtain

$$\hat{\theta}_{N+1} = \hat{\theta}_N + \frac{P_N\varphi_{N+1}}{\varphi_{N+1}^T P_N \varphi_{N+1} + 1}[y(n+N+1) - \varphi_{N+1}^T \hat{\theta}_N] \tag{13}$$

4 THE DESIGN OF FUZZY PID CONTROLLER

The design of fuzzy controller is mainly composed of the following five parts,

- The determination of input variables and output variables.
- The design of control rules.
- The determination of the domain of fuzzy sets and parameters, such as quantification factors and scaling factors.
- The design of programs of fuzzy control algorithm.
- Reasonable selection of sampling time, as in [7].

4.1 *The design of fuzzy partition of input variables*

Define variation range of deviation e and deviation changes ec as the domain of fuzzy sets. Select e,ec = [−3,3] as master domain of accurate amount. The fuzzy reasoning is based on fuzzy data. Therefore, the input must be firstly fuzzed by the controller. The domain of fuzzy sets are:

E,EC = {−3, −2, −1,0,1,2,3}. Thus, quantification factors are as follows: Ke = 1, Kec = 1. Fuzzy subset of fuzzy controller are :

e,ec = {NB,NM,NS,ZO,PS,PM,PB}. Elements of the subset represent negative big, negative middle, negative small, zero, positive small, positive middle, positive big. The membership functions consist of triangle function and normal distribution function.

4.2 *The design of fuzzy partition of output variables*

Define ΔKp, ΔKi, ΔKd as the domain of fuzzy sets. Select ΔKp = [−3,3], $\Delta K_i = [-1.08, 1.08]$ $\Delta Kd = [-0.0003, 0.0003]$ as master domain of accurate amount. The domain of fuzzy sets are:

ΔKp,ΔKi,ΔKd = {−3,−2,−1,0,1,2,3}.

Scaling factors are as follows:

Kp0 = 1, Ki0 = 0.36, Kd0 = 0.0001 kd0 = 0.0001. Fuzzy subset of Fuzzy controller are:

ΔKp, ΔKi, ΔKd = {NB, NM, NS, ZO, PS, PM, PB}.

The membership functions consist of triangle function and normal distribution function.

4.3 *The design of control rules*

On the basis of technical expertise and practical operation experience, the design of fuzzy rule table is the core of fuzzy controller, as in [8]. The parameter adjustment rules Kp, Ki, Kd are summarized as follows:

4.4 *The selection of defuzzification method*

The centroid method is often applied to defuzzification of fuzzy output. It is also referred to as weighted average method, which regards the basic variable value corresponding to centre of area that encircled by membership function curve and basic variable axes as articulate value, as in [9].

Table 1. Fuzzy tuning rules of Kp.

ec Kp e	NB	NM	NS	ZO	PS	PM	PB
NB	NB	NB	NM	NM	NS	ZO	ZO
NM	NB	NB	NM	NS	NS	ZO	ZO
NS	NB	NM	NS	NS	ZO	PS	PS
ZO	NM	NM	NS	ZO	PS	PM	PM
PS	NM	NS	ZO	PS	PS	PM	PB
PM	ZO	ZO	PS	PS	PM	PB	PB
PB	ZO	ZO	PS	PM	PM	PB	PB

Table 2. Fuzzy tuning rules of Ki.

ec Kp e	NB	NM	NS	ZO	PS	PM	PB
NB	PB	PB	PM	PM	PS	ZO	ZO
NM	PB	PB	PM	PS	PS	ZO	NS
NS	PM	PM	PM	PS	ZO	NS	NS
ZO	PM	PM	PS	ZO	NS	NM	NM
PS	PS	PS	ZO	NS	NS	NM	NM
PM	PS	ZO	NS	NM	NM	NM	NB
PB	ZO	ZO	NM	NM	NM	NB	NB

Table 3. Fuzzy tuning rules of Kd.

ec Kp e	NB	NM	NS	ZO	PS	PM	PB
NB	PS	NS	NB	NB	NB	NM	PS
NM	PS	NS	NB	NM	NM	NS	ZO
NS	ZO	NS	NM	NM	NS	NS	ZO
ZO	ZO	NS	NS	NS	NS	NS	ZO
PS	ZO	ZO	ZO	ZO	ZO	ZO	ZO
PM	PB	NS	PS	PS	PS	PS	PB
PB	PB	PM	PM	PM	PS	PS	PB

5 FUZZY SMITH PREDICTIVE CONTROL BASED ON RECURSIVE LEAST SQUARE METHODOLOGY

In order to reduce the influence of network time delay on control system, the recursive least square method is adopted to identify the generalized controlled object online to adjust parameters of Smith predictive model constantly. Fuzzy adaptive PID control is mainly used to control the generalized controlled object of the system and is able to adjust control parameters by itself in line with random network delay or parameters changes of the controlled object, which can reduce the influence of model mismatch greatly, quicken the response and improve the control performance of the system.

6 SIMULATION EXPERIMENT

The transfer function of controlled object is given as

$$G(s) = \frac{1}{100s+1}e^{-2s}$$

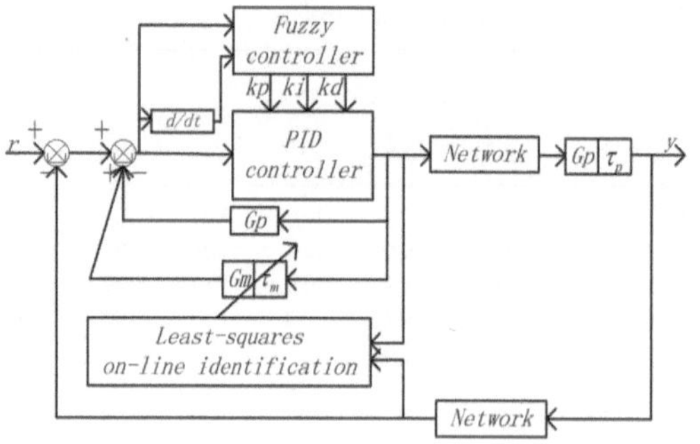

Figure 2. Fuzzy Smith predictive control structure based on recursive least square methodology.

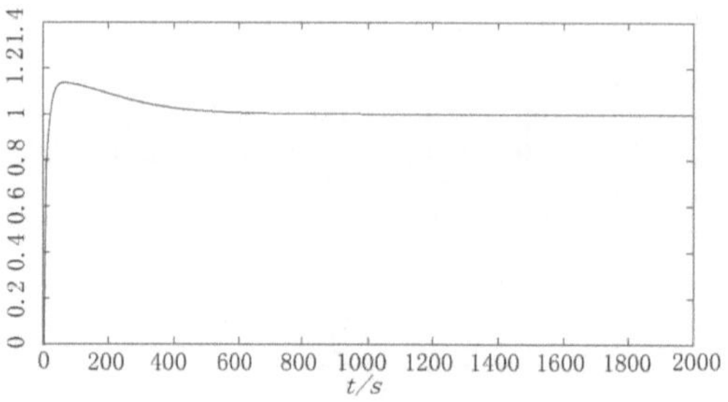

Figure 3. $\tau = 5$s Smith predictive control method based on recursive least square methodology on-line identification.

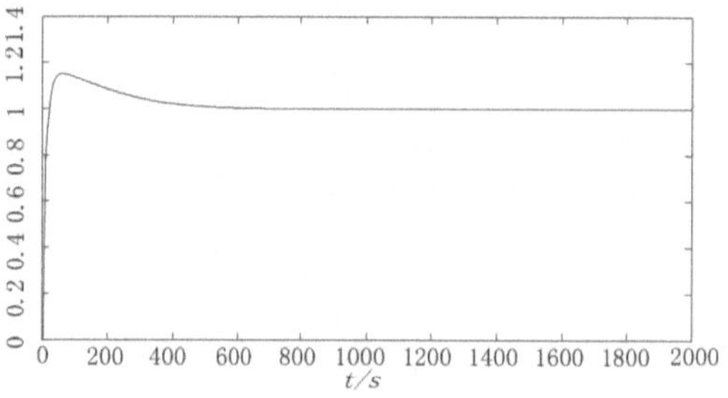

Figure 4. $\tau = 5$s Fuzzy Smith predictive control method based on recursive least square methodology on-line identification.

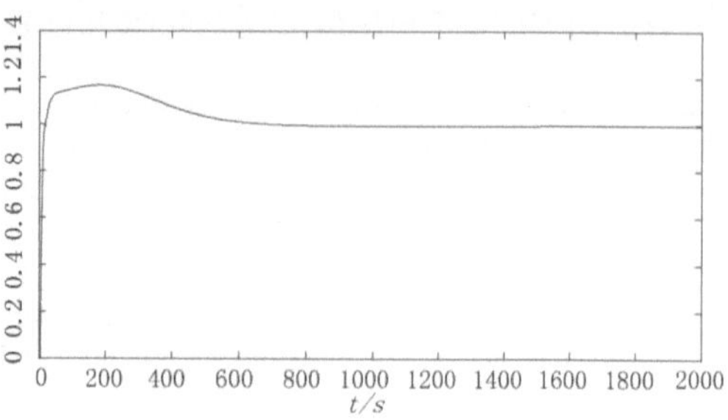

Figure 5. $\tau = 10$s Smith predictive control method based on recursive least square methodology on-line identification.

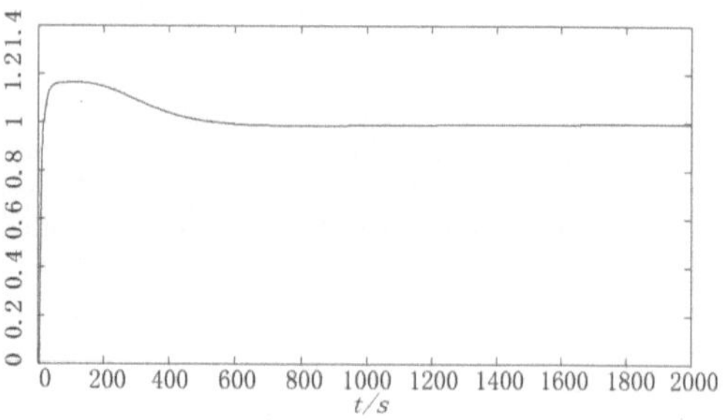

Figure 6. $\tau = 10$s Fuzzy Smith predictive control method based on recursive least square methodology on-line identification.

Define Ts = 1s as the sampling period and kp = 9.2, ki = 24, kd = 0.2 as initial values PID controller parameters. When network time delay changes, the step response corresponding to Smith predictive control method based on recursive least square methodology on-line identification and the step response on condition that the fuzzy controller is introduced are shown in figures 3–6.

It can be seen from step response curves of the system with long time delay that the system has longer adjustment time, when fuzzy Smith predictive control method based on recursive least square methodology on-line identification is adopted. The control performance is not so satisfactory. The introduction of fuzzy adaptive PID control can shorten the adjustment time definitely and its control performance is superior to the former.

7 CONCLUSIONS

The network random delay and the lag of nonlinear system itself exert serious negative effects. Given networked control systems with long time delay, the method that regards the communication network and the controlled object as a generalized controlled object and introduces Smith predictor whose parameters are obtained on the basis of recursive least square methodology is proposed. Fuzzy adaptive PID control is introduced simultaneously. By utilizing the proposed delay compensation strategy, the performance of systems is optimized.

REFERENCES

Du Feng & Qian Qingquan. Study of networked control systems based on modified Smith prediction compensation. Systems Engineering and Electronics 31(3): 661–665.

Duan Hongxia & Li Hongxin & Wang Longgui. Application of a new Smith fuzzy PID controller in networked control systems. Gansu Science and Technology 29(11): 12–15.

Feng Jianzhou. Research of networked control system based on stochastic delay process, Ph D Thesis. Qingdao: China University of Petroleum, 2008.

Li Zhen. Research on delay compensation of network control system. Luoyang: Henan University of Science and Technology, 2011.

Lu An. Simulation Analysis on the predictive control for networked control system with double-sided time-delay, Ph D Thesis. Chongqing: Chongqing University, 2012.
Wang Sujuan. Fuzzy immune PI control of networked control system based on prediction compensation, Ph D Thesis. Tianjin: Tianjin University, 2008.
Zhang Haitao & Li Zhen. Simulation of networked control system based on Smith prediction compensation. *Computer Engineering and Applications* 48(8): 243–245.
Wen Yangdong & Li Yu. Network control system of the time delay estimation and compensation. Microcomputer and Its Applications 30(13): 75–79.
Zhang Ya & Tian Yuping. Predictor-based compensator of networked control systems with random distributed delays. Journal of Southeast University(Natural Science Edition) 39(2): 309–314.

Future Mechatronics and Automation – Yang (Ed.)
© 2015 Taylor & Francis Group, London, ISBN 978-1-138-02648-3

The behavioural identified technology of drivers based on mechanical vision

Jia Lin Tang, Guang Li Zhuang, Bing Hua Su, Shu Fen Chen & Xi Ying Li
School of Information Technology, ZHBIT, Zhuhai, China
Sun Yat-sen University, SYSU, Guangzhou, China

ABSTRACT: By means of introducing a behavioural identified technology of drivers based on mechanical vision in depth, it is known that this technology can analyse the non-contact ground image of drivers. Therefore, it can extract the effective characteristic information and, as a result, it can distinguish whether the person who is driving is tired, is violating the traffic regulations, or has been drinking or taking medication. This technology sends out some relevant voice warnings. In this way, it could reduce the probability of traffic accidents. Even more, it would protect drivers and passengers' lives and property. The experimental result shows that this measure would achieve the identification of retrieving the target on a video sequence with real-time capabilities, as well as, achieving good recognition results.

Keywords: Machine vision, facial image, traffic accident, real-time.

1 INTRODUCTION

With the constant popularization of cars, traffic accidents have grown in great numbers. Every year there are about 20 million people who are injured or disabled in traffic accidents every year; over 1million people have died, which results in more than 500 billion dollars of financial loss every year all over the world. It poses a big threat to people's lives and property. Nowadays, traffic accidents have become the focus of the whole of society. In recent years, more than 100,000 people died because of traffic accidents each year in our country, ranking it the highest in the world, and the financial loss has reached tens of billions of dollars.

One of the most important reasons for serious traffic accidents is that drivers are driving while lacking sleep, have been drinking, are tired, are taking medication, or driving badly. Previously, there have been many experts who have taken part in the research of monitoring driving. However, our market has no such a successful and reliable product, Therefore, producing a system based on mechanical vision to monitor drivers' performance which gives an alarm if a problem is detected, has become an urgent need for the population. Overall, exploiting a system to identify the vehicular drivers' behaviour and then give an alarm has a very intensive social and economic significance.

2 APPROACH

2.1 *Face recognition technology*

A face recognition system process is shown in Figure 1-1, and generally involves three steps: head detection, face detection, facial feature extraction, face recognition, and verification. On a specific implementation process, it is necessary to first capture the image set, and then to use the face detection module for face detection. If a facial image is detected, then a feature point is located, generally based between the two eyes, according to the distance of the two eyes; face images are normalized in processing. Normalization processing includes image preprocessing, image scaling, and effective facial region selection and other operations. Finally, a normalized facial image is extracted, and then sent to the classifier so the recognition result can be obtained.

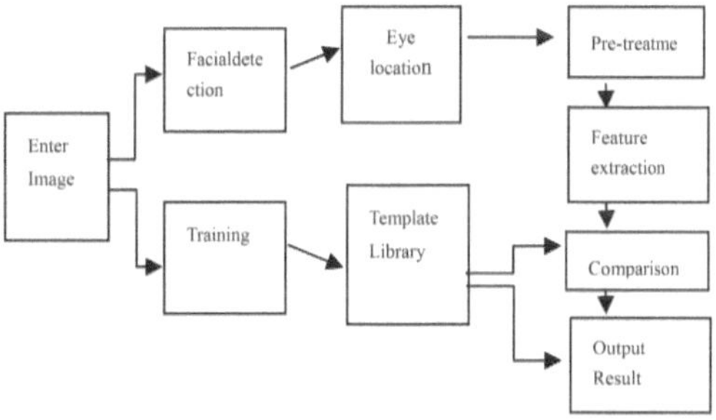

Figure 1. Face recognition technology process flow chart.

2.2 Preprocessing

Pretreatment (before processing) is an important part in the process of face recognition. Pretreatment is in image analysis; the input image feature extraction, segmentation and matching before processing. Because of different image acquisition environments for the

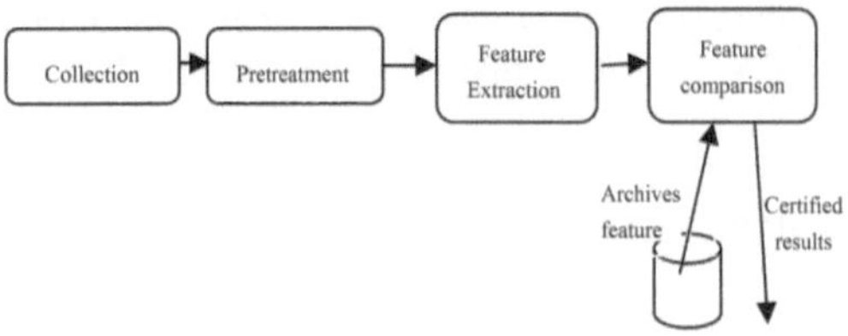

Figure 2. Flow chart of pretreatment.

input image, such as the merits of the light, shade degree, and equipment performance etc., sometimes there are shortcomings such as noise, or the contrast is not enough, as well as the distance, and the size of the focal length. Make the size of the face in the middle of the whole image and position uncertainty. In order to keep the face in face image size, location, and consistency of the face image quality, the image must be preprocessed. The main purpose of the image preprocessing is to eliminate the image giving irrelevant information, to filter out interference and noise, and to record only useful information. It must also enhance the information that can be detected and maximally simplify data, in order to improve the feature extraction, image segmentation, and matching and recognition reliability. The preprocessing flow chart is as shown.

When the image is optimized in a preprocessing stage, as far as possible, any interference of the processed image from light, the imaging system, the external environment etc. must be removed or reduced, in order to improve the quality for subsequent processing. Furthermore, in order to make different face images look as much as possible like the complete feature extraction, training and recognition skills under the same conditions are needed. The face image preprocessing mainly focuses on a facial centralizer, facial image enforcement, and image intensity normalization. A facial centralizer is needed in order to get the face image the location of the human face correct. Facial image enforcement in order to improve the quality of the face image not only makes the image clearer to see, but also makes the image more conducive to computer processing and recognition. The purpose of image intensity normalization is to get the standardization of the face image with the same size and gray value range.

2.3 *Holomorphic filter pretreatment of face recognition*

(1) The AR face database.
In the acquisition of environment camera parameters, light environment, and camera distance that was set up by the computer vision centre (Barcelona, Spain) should be strictly controlled. The database contains 126 people (70 male, 56 female), a total of 3276 colour positive images, and this database is primarily to test the face recognition experiment with reference to the condition of the facial expressions (natural, smile,

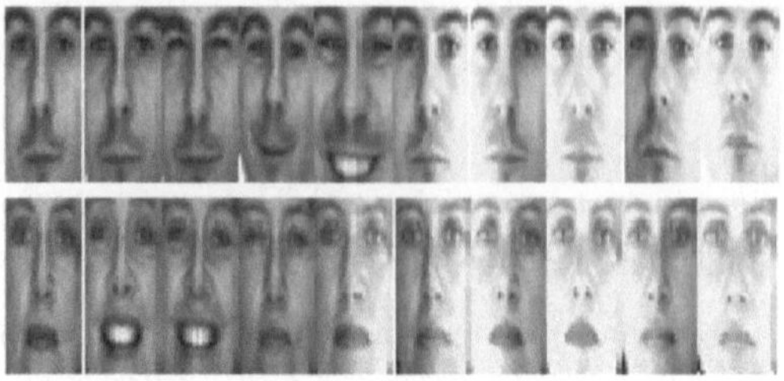

Figure 3. AR face part of the face image in the repository.

anger, cold) and illumination (left light source, light source and double side light source from the right) environmental change. Test recognition test:

(2) Linear discriminant analysis.
Linear Discriminant Analysis (LDA) is a statistical method for feature extraction. Unlike the PCA thought, LDA, by finding the optimal projection direction, making all samples on the projection of discrete degree between class and class in the ratio of the discrete degree for the biggest, so achieves the best classification of different types of samples. Therefore, in theory, LDA is more suitable than PCA for the pattern recognition problem. Classical linear discriminant analysis is used in the Fisher criterion function, so the linear discriminant analysis is also known as Fisher linear discriminant analysis (Fisher LDA).The Fisher criterion function is defined as:

$$J(W) = \arg\max_{W} \frac{|W^{T} S_b W|}{|W^{T} S_w W|} \tag{3.1}$$

Among (3.1), S_b, S_w is a training sample of the total scattering matrix between classes and the total scattering matrix in class. Formula (3.1) in the $J(W)$ Rayleigh entropy of generalized, can be solved by using the Lagrange multiplier method; the denominator is equal to a non-zero constant and then $W^T S_w W = C \neq 0$, though the laser function is defined as:

$$L(W,\lambda) = W^T S_b W - \lambda(W^T S_w W - C) \tag{3.2}$$

Type λ in the LaGrange multiplier, Formula (3.2) for W partial derivatives is:

$$\frac{\partial L(W,\lambda)}{\partial W} = S_b W - \lambda S_w W \tag{3.3}$$

Make the partial derivative is zero, obtained that $S_b W^* - \lambda S_w W^* = 0$ that is

$$S_b W^* = \lambda S_w W^* \tag{3.4}$$

W^* Isthe maximum value of J (W), When sw is non-singular, type (3.4) times S_w^{-1} on both sides, it is

$$S_w^{-1} S_b W^* = \lambda W^* \tag{3.5}$$

Table 1. The number of training samples changing the face preprocessing method recognition rate (%) of the comparison (ARdatabase).

The number of training samples	132	165	198	231	264	297	330
Histogram equalization	89.09	96.36	97.47	98.70	97.64	98.32	98.99
holomorphic filtering	90.71	96.62	97.58	98.72	97.92	98.72	99.23

Solving the Formula (3.5) for solving the general matrix characteristic value problem of $\mathbf{S}_w^{-1}\mathbf{S}_b$.

In summary, when non-singular, in mathematics, solving function (3.1) is equivalent to solving the characteristic value problem of sw and $\mathbf{S}_w^{-1}\mathbf{S}_b$. Making J (W)'s biggest transformation matrix being composed of the W maximum Eigenvalue of the former one of eigenvectors from $-\mathbf{S}_w^{-1}\mathbf{S}_b$.

Regarding the sample as a point in space, the distance between the point and the point is the most direct classification basis, and therefore the subspace method is one of the most common distance classifiers. And the AR face database adopts Distance classifier is L2 Distance (Euclidean Distance Euclidean short):

$$d(\mathbf{x},\mathbf{y})=\|\mathbf{x}-\mathbf{y}\|=\sqrt{\sum_{i=1}^{n}(x_i-y_i)^2} \tag{3.6}$$

(3) Analysis of the recognition rate.
Through in-depth study, it was found that the accuracy of face recognition and other biometric recognition is not very high. This is because the facial features are easily influenced by illumination, facial expressions, gestures, and other factors, which lead to difficulties in recognition. In carrying on the pretreatment to the image, the following table is a comparison with the recognition rate.

It can be seen from the table that the preprocess image face recognition rate has increased and the holomorphic filtering keeps a high recognition rate.

2.4 *Face detection*

Face detection is the first step in face analysis and refers to any contain image. Using a strategy to search to determine the presence of the face, if it is there, then it returns the human face position, size, and posture. For face detection, the general concern about two indexes: the face detection rate (Detection Rate) in a given image, detecting the face, and the ratio of total face. Number of false detections (False Detection): detecting the non-face treated as a human face. This indicator is very important, and some detection algorithms can even give a 100% detection rate, but the number of its false detections may be very large. An idealized perfect face detection algorithm must have a face detection rate of 100% and 0 error detection. At this stage, it may be that only a human brain has this ability.

2.5 *Eye location*

This link includes two stages which are the human eye's rough and accurate positioning. First: according to Chinese traditional three court five eyes of prior knowledge, it is certainly possible to roughly locate the general area of the eyes where the area may contain both eyebrows, hair keratin, and other interference that further reduce the computational domain. Second: the human eye's approximated area by a certain threshold is converted to a binary image, by a vertical gray-level projection and then the histogram is obtained. There is a big difference between the gray-level of the eyes and the surrounding skin. From the histogram of the peaks and troughs, it is possible to judge the eye up and down on the edge of the Y coordinate and then seethe precise position of the eye.

PERCLOS is the abbreviation of Percentage of Eyelid Closure over the pupil time, meaning the percentage of eye closure time in unit time. The principle of PERCLOS is to count the proportion of times when eyes close in a certain period of time. The judgemental standard in our system is o PERCLOS- 80. It refers to the eyelids covering the area of the pupil by more than 80%, which is when the eyes are considered to be closed. Formula 5.1 shows the metrical principle of PERCLOS. By measuring t1 to t4, it is possible to calculate the value of PERCLOS:

$$F=\frac{t_3-t_2}{t_4-t_1}*100\% \tag{5.1}$$

F represents the percentage of the eyes' closing time, which is the value of PERCLOS.

(1) Judgement of the eyes' state

Through the maximum class square error method (Otsu), in a different light it is possible to process different thresholds of binarization in the precise area of a human eye, obtaining the eyes' optimum shape when they are closing or opening respectively. Through the comparison of N consecutive frames it can be judged that in the least dark area of the pixel values it can be considered that the driver is in the closing state; other cases are in the open or half-open state.

(2) Analysis of eyestrain

The system selects the evaluation index of the recently established degree of fatigue by PERCOLS, through the time allocation of a closed frame in N consecutive frames. It is possible to analyse the state of fatigue, with an opened frame recorded with a value of "1", and a closed frame recorded with a value of "0". In this way, people can be obtained the staggered sequence about "1" and "0" after N consecutive frames, so the proportion of the value of "0" in this sequence can be used to describe the analysis of fatigue state. When the percentage is higher than a certain proportion in the experiment, it can be considered that the driver is driving in a state of fatigue.

Through the operation process in the above steps, the system can analyse whether the driver is in a state of fatigue from video streaming of the device. Once obtained, it then gives different levels of an alert alarm, to achieve the designed goal of the system.

Eyes index part: This system has been selected for twenty consecutive frames to judge the degree of fatigue. It was agreed that:

a) when the value of PERCLOS is less than 0.4, the system determines that the driver is driving in the normal driving condition without any treatment;
b) when the value of PERCLOS is greater than 0.4, the system determines that the driver is in a mild state of fatigue. The warning tone is given and if the driver wakes after this prompt, PERCLOS will sink to 0.4 or less and then the warning tone disappears
c) when the driver is still unable to stay awake after mild prompting, the value of PERCLOS will continue to increase. When it is greater than 0.7 it shows that the driver has been in a state of fatigue and the system will sound a rapid interferential noise, strongly warning the driver to stop and rest.

Here, the human face in the vertical direction is divided into four parts, and the face in the horizontal direction is divided into seven parts. An approximate area of face is selected with 1/2 area under the upper part in height, and 5/7 area width on the centre of the face. In width, in the face detection module, we have the positioning of the coordinates of the face, according to the agreement of the prior knowledge, We can do a human eye coordinate positioning on the approximate area. This ensures that whatever the detecting ratio of a face, our rectangular box can be sure to include the human eyes in a rectangular box however small the eyes are in the image, and can eliminate approximately the influence of hair, nose, and other factors. In this way, we can set up our interest in the area of the face area according to the prior knowledge, which can eliminate the interference for the following algorithm to improve the operation rate.

After determining the eye region, we need to use an adaptive thresholding operation on the human eye image processing, and then obtain further details for human eyes as a judgement of eye location. Adaptive thresholding operation plays a key role in the precise location and eye state judgement; we need to get complete information on the whole of the human eye, especially the full shape of the human eye, as a prerequisite to locate the centre of the human eye.

The following figure shows the human eyes' shape under the same light but with different thresholds, and under a different light but with the same threshold.

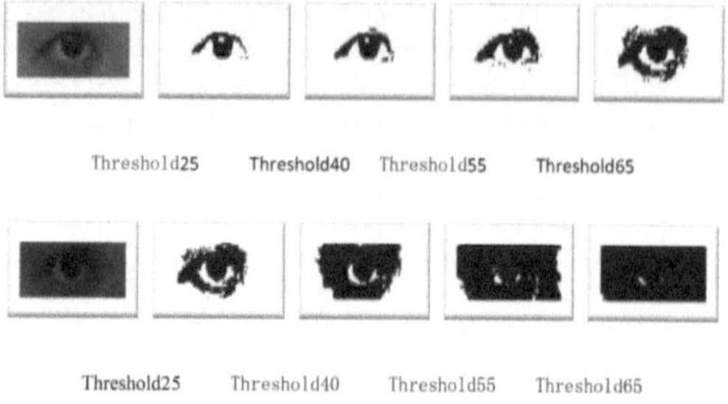

Figure 4. Threshold figure.

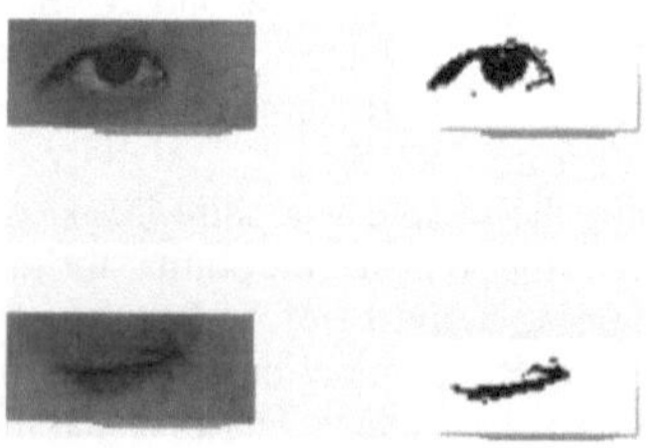

Figure 5.

Thus, the fixed threshold value cannot be suitable in all lights. What can be done is to realize the adaptive threshold, depending on different lights. The entire image in the gray scale is different; the threshold would be set to adjust automatically to get the best binary image of the eyes. Reading the relevant literature, we know that the system uses the largest class of different adaptive thresholding methods.

2.6 *Judgement of eye state*

When the judgement of eye state is correct it does not directly affect the analysis of the system and the output of the results of fatigue degree. In this system, we can judge the eye area in binary image pixels in order to judge the eye's opening or closing state changes; the eye area is defined as the number of pixels of the eye image in the precise area where the human eye is, namely the positioning area of the number of black pixels. When the eye is completely open, the blackest pixels are the largest set as Pmax. When the eyes are closed completely, they are set at the smallest of Pmin. When the eye is closed completely, the area is minimum.

After positioning the precise eye area, the system will select the driver's opening image data, which is stationary after ten consecutive frames, and the image data is in a frame with the eyes closed. The number of black pixels in each frame can then be calculated in order to obtain the average number, as shown below:

$$P_{max} = \sum_{1}^{10} P_i \tag{6.1}$$

$$P_{min} = \sum_{1}^{10} P_i \qquad (6.2)$$

After processing the video image frame, the system calculates the area Px eye area of each frame, and judges it according to the following equation:

$$S_t = \begin{cases} \text{"1"} & P_t > P_{min+\sigma} \\ \text{"0"} & P_t < P_{min+\sigma} \end{cases} \qquad (6.3)$$

In the formula, '1' represents the eyes opened, while '0' represents the eyes closed, with an allowable fluctuation value.

The analysis about the degree of fatigue.

Face detection and eye location judges the eyes' state in order to ensure the driver's fatigue state. The system needs to set up a mathematical model to predict the fatigue process, to monitor effectively the driver's fatigue.

Condition: This system uses PERCLOS' transformation as the detection basis. We did not use the P80, p70 standard as the basis to judge the closed eyes.

According to frontal analysis, from each frame image of the state of the driver's eyes opened and closed in two types, we cannot judge fatigue only according to the current state of a frame of a driver's eyes. Using the '0' and '1' to represent closed eyes and opened eyes, so the state of the driver's eyes is at '0' and '1' and is composed of a time series. At this time, PERCLOS is the proportion of '0' in one minute. This system constantly selects twenty consecutive frames for a judgement about the degree of fatigue. It is deduced that:

a. when the PERCLOS value is less than 0.4, the system judges that the driver is in normal driving condition and does not do any processing;
b. when the PERCLOS value is more than 0.4, the system judges that the driver is in a fatigued state, and gives a warning tone; the driver is alerted by the warning tone; PERCLOS in the next phase of the judgement falls to 0.4 and it therefore follows that the warning tone ceases;
c. when drivers could not keep awake after being warned, the PERCLOS' value increased continually. When it is more than 0.7, on account of the driver entering a state of fatigue, the system will give a short intermittent noise, warning the driver to stop and rest completely.

3 CONCLUSIONS

With the development of computer vision in the military field and in the field of intelligent traffic monitoring, videos of moving target detection and tracking are certain to be extensively applied and developed.

This system is different from the detection systems that we already have. Our evaluating standards in this research not only state law face, they also include

Figure 6. The final test results.

real-time eye detection, drunkenness detection, occlusion detection, and head motion detection, by OpenCV image analysis, which means that we can judge successfully the degree of a driver's abnormal behaviour. In the condition of what we have to face is not clear in state laws, but greatly improves the accuracy of the system.

This dissertation is from last year and most of the research is an application of OpenCV to realize the driver. Behaviour surveillance and recognition technology, based on machine vision, has successfully developed the final test results, as shown in Figure 7-1. Experimental results indicate that the system is robust, accuracy is good, and stable performance is also very reliable. Reaching the expected goal in the stage of experiment, the system can automatically analyse the driver's head pose, eye movements, head movements, and facial features of information by the OpenCV platform to determine the driver's mental state, giving corresponding early warnings. The experiments' results were significant and reached an ideal state of reliability.

4 FUNDING

This work was supported by the Open Funding of Guangdong Provincial Key Laboratory of Intelligent Transportation System (no.201401003), and the Fundamental Science and Technology Program of Ministry of Public Security (no. 2013GABJC013).

REFERENCES

Ana S N, Yoob J, Choi S. Manifoldrespecting discriminant nonnegative matrix factorization [J] Pattern Recognition Letters, 2011, 32(6):832–837.

Chen G, Han B. Improve kmeans clustering for audio data by exploring a reasonable Sampilingrate[C]//2010 Seventh International Conference on Fuzzy Systems and KnowledgeDiscovery (fskD). Ye-nta.Shandong: [s.n.], 2010:1639–1642.

Ergun Gumus, NiyaziKilic, Ahmet Sertbas, Osman N. Ucan Evaluation of face recognition techniques using PCA, wavelets and SVM [J]. Expel System switch Application, 2010. (37):6404–6408.

G. QueHec, M. Lamard, G. Cazuguel, B. Coherer, C. Ro-ux. Wavelet optimization for con tentbased image retrieval in medical databases [J]. Medical Imag Analysis, 2010, (14):227–241.

Hazlm Kemal Ekenel, Johannes Stall Kamp, Rainer Stiefelhagen. A videobased door monitoring system using local appearance-based face models [J]. Computer Vision and Image UnderStanding, 2010, (114):596–608.

Karantasis K I, Accelerating data clustering on GPU based clusters under shared memory abstraction[C]//2010 IEEE International Conference on ClusterComputingWorkshops and posters(C LUSTER WORKSHOPS). [S.L.]: IEEE Press, 2010:1–5.

Karlsson C.New directions in regional economic development [M]. London; [s.n.], 2009.

Tan X Y, Trigg's B. Enhanced local texture feature sets Far face recognition under difficult lighting conditions [J]. IEEE Transactions on Image Processing, 2010.19 (6):1635–1650.

Xu KS. Revealing social networks of spam mars through spectral clustering[C]//IEEE InternationalConferenceon Communications. [S.1.]: IEEE Press, 2009:1–6.

Zhao Li, Xindong Wu, Hong Peng Nonnegative Matrix Factorization on Orthogonal Subspace [J]. Pattern Recognition letters, 2010, (31):905–911.

Zhi R C, Flierl M, Ruan Q, et a1. Graph preserving Sparce nonnegative matrix factorizationn with application to facial expression recognition [J]. Systems, Man and Cybernetics, Part B: Cybernetics, IEE-E Transactions on, 2011, 41(1):38–52.

Future Mechatronics and Automation – Yang (Ed.)
 ISBN 978-1-138-02648-3

Real-time fault detection and diagnosis of ASCS in AMT heavy-duty vehicles

Y.N. Zhao, H.O. Liu, W.S. Zhang & H.Y. Chen
National Key Laboratory of Vehicle Transmission, Beijing Institute of Technology, Beijing, China

ABSTRACT: Taking the Automated Shift Control System (ASCS) in a heavy-duty vehicle equipped with an Automated Manual Transmission (AMT) as the re-search platform, an on-board real-time fault detection and diagnosis system based on a lower computer and a handheld fault diagnosis tester is designed. The system consists of three components: a dedicated hand-held fault diagnosis tester, the fault diagnosis subroutine embedded in Transmission Control Unit (TCU), and the off-line data analysis software on PC. The bench test is carried out through artificially setting failure points and the test results show the accuracy of diagnostic procedures.

1 INTRODUTION

Automated Shift Control System (ASCS) is an important component of Automated Manual Transmission (AMT), whose working reliability and stability directly determine the performance and operation quality of AMT. In addition, the complex electronic control system of ASCS makes high demands on maintenance technicians. How to make the fault self-diagnosis accurately and send the corresponding fault codes and important driving data timely to assist maintenance have great significance in improving system reliability, driving safety of the vehicle and AMT product promotion (Peng Jianxin et al. 2012).

Firstly, the overall program of ASCS fault detection and diagnosis system is determined by analyzing the AMT control system structure and working principle. Then, the handheld fault diagnosis tester based on microcontroller XC164CS of Infineon Company and the real-time diagnosis algorithm for main faults modes are designed. Finally, the correctness of the system is verified by bench test (Zeng Ai 2008).

2 ASCS AND ITS OVERALL PROGRAM OF FAULT DIAGNOSIS

2.1 *ASCS composition and principle*

The AMT studied herein, achieves automatic manipulation on the gear shifting and clutch engaging /disengaging by the use of an electrically controlled hydraulic system. Figure 1 shows the composition of ASCS. The system comprises Transmission Control Unit (TCU), handle type gear selector, auto clutch mechanism assembly, shifting mechanism assembly, hydraulic supply system and relevant cables, oil and gas pipelines.

2.1.1 *Transmission Control Unit (TCU)*

TCU is the core of ASCS and responsible for collecting all kinds of input signals. Through real-time computing and processing as well as communication bus, TCU drives actuators in order to achieve the automatic shift control and fulfill diagnostic functions at the same time. Assume that TCU is working properly in the fault diagnosis process.

2.1.2 *Sensors*

(1) Sensors of switching signal. These signals are from gear selector, brake pedal, parking brake and auxiliary gearbox (the structure of transmission consists of a main gearbox and an auxiliary one, so the shifting process involves control of both gearboxes);
(2) Sensors of frequency signal: transmission's input shaft speed sensor N1, transmission's output shaft speed sensor N2;
(3) Sensors of analog signal: clutch displacement sensor Lc, gear-selecting cylinder stroke sensor Ty, gear-shifting cylinder stroke sensor Tx, etc.

2.1.3 *Actuators and their solenoid valves*

Actuators include clutch cylinder, gear-selecting cylinder, gear-shifting cylinder, and auxiliary gearbox cylinder, etc.

Additionally, the system also uses several signals from the other subsystems of the vehicle, such as engine speed signal NE, the throttle opening signal, etc.

Specific working principle can be found in the reference Wang Hongliang et al. (2009).

2.2 *The main modes and causes of ASCS faults*

According to system mechanisms, the main modes of ASCS faults can be divided into the clutch mechanism fault and gear shifting control mechanism fault.

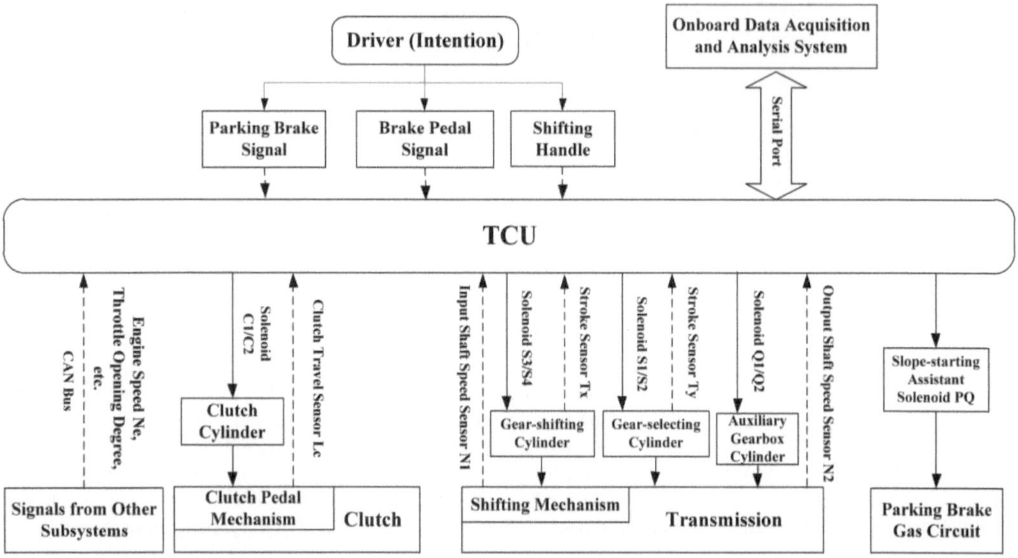

Figure 1. Schematic diagram of ASCS composition.

Take the clutch mechanism for example. The solenoid valves C1 and C2 are electrified under the control command of TCU, thereby driving the clutch cylinder to disengage the clutch. In this case, the value of the clutch displacement sensor Lc as the clutch actuator feedback signal will increase. The clutch will remain at the current location if only C2 is electrified. The clutch will engage and the value of LC will decrease in the meanwhile if both C1 and C2 are powered off. After the clutch is fully engaged, the value of LC reaches the minimum.

When the clutch mechanism fails, its main fault modes include clutch's failure to disengage (or disengaging is not complete) or engage. The main causes of the faults include: ① hydraulic system fault resulting in too low oil pressure; ② clutch actuator fault, such as solenoid valve blockage, short/open circuit of cables, or cylinder jamming resulting in cylinder rod's failure to move; ③ clutch displacement sensor LC fault.

2.3 *Overall program of ASCS fault diagnosis system*

The ASCS fault diagnosis system for AMT heavy-duty vehicles uses the method of combined diagnosis containing online fault diagnosis and handheld fault diagnosis tester, namely, ASCS's electronic control unit TCU detects and analyses the relevant information and data while taking control of vehicle starting and shifting, and when a fault is found, the fault will be stored in the EEPROM in the form of Diagnostic Trouble Code (DTC) which is sent out via serial communication. Read DTC from serial port of the system by using a handheld fault diagnosis tester, and synthesize other related driving data to further confirm the fault cause.

ASCS fault diagnosis system is divided into two levels:

(1) fault self-detection and self-diagnosis in the process of vehicle travelling
(2) fault analysis in the process of vehicle maintenance

The fault self-detection and self-diagnosis in the process of vehicle travelling refer to fault diagnostic subroutine that is written in C language and embedded into the main control flow of TCU. And the subroutine adopts the diagnostic methods based on signal detection, analytical redundancy, and ASCS control logic, etc. These methods are cross used in the program, achieving diagnostic functions for most important parts of the ASCS without changing the hardware of TCU.

The fault analysis in the process of vehicle maintenance is that during the process of vehicle maintenance and debugging, read DTC with a handheld fault diagnosis tester sent by TUC via serial communication, and monitor changes of data in the acquisition interface of tester, in order to make more precise fault analysis and judgment.

The handheld fault diagnosis tester will store the data. During vehicle maintenance and debugging in the plant, maintenance technicians can communicate with serial port of PC via tester and export the data stored in the tester memory to the PC, in the condition that the upper computer is available.

The data acquisition and analysis software (Xi Junqiang et al. 2001) (programmed in Visual Basic 6.0) installed on the PC can plot important offline data curves, in addition to achieving all related functions of the handheld fault diagnosis tester. The curve plotting function is very helpful to analyze the cause of fault.

In summary, the overall program of ASCS fault diagnosis system is shown in Figure 2 and all components use serial communication.

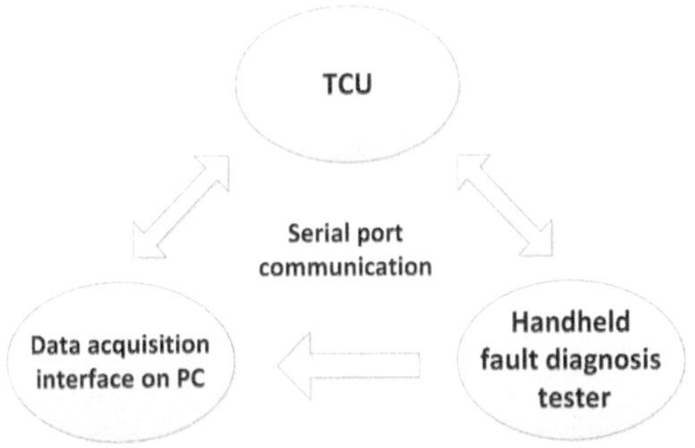

Figure 2. The overall program of ASCS fault diagnosis system.

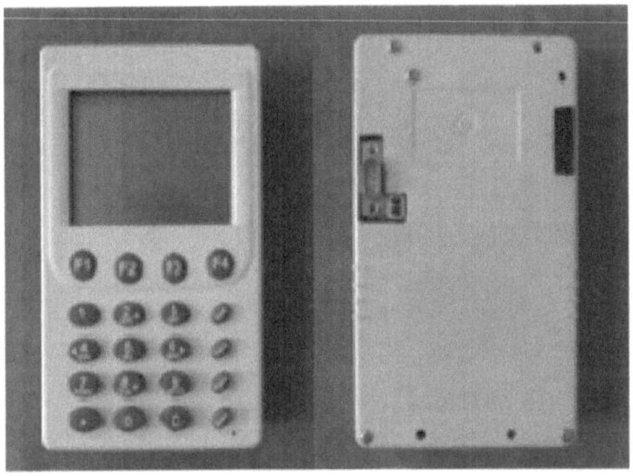

Figure 3. The handheld tester.

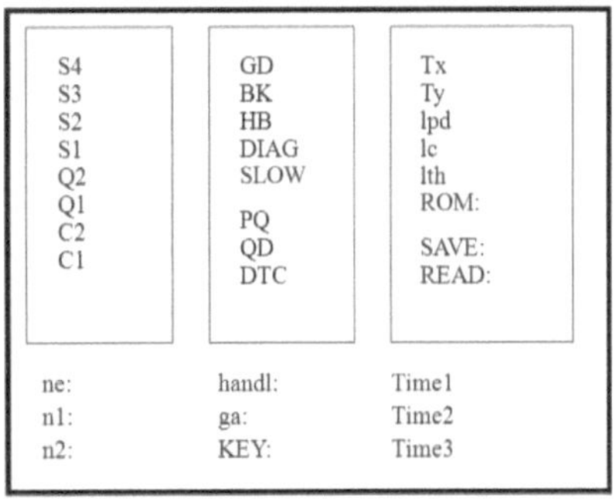

Figure 4. LCD format of fault diagnosis tester.
Note: The symbols have the same meaning as Figure 1.

Table 1. List of part of diagnostic trouble codes.

DTC	Code Meaning	Flag
P0700	Transmission actuator fault	t_act_error
P0715	Transmission output speed sensor circuit fault	n2_error
P0750	Shifting solenoid valve S1 fault	s1_error
P0805	Clutch position sensor fault	lc_error
P0850	Neutral position fault	p_n_g_error
P0900	Clutch actuator fault	c_a_error
P1701	Clutch disengaging fault	c_d_error
P1711	Reverse cables of solenoid valve S1, S2	s1s2_r_error

3 DESIGN OF HANDHELD FAULT DIAGNOSIS TESTER

The appearance of handheld fault diagnosis tester is shown in Figure 3. The equipment is designed to mainly achieve the following four functions:

(1) serial communication: communicate with TCU and PC via RS232 interface.
(2) liquid crystal display: the tester receives important data from TCU and display the working status of TCU in text mode on the LCD screen. LCD format is shown in Figure 4.The data frame format from TCU to tester is: length 18 bytes, sent once every 10ms, communication baud rate 19200bps.
(3) data storing / exporting
(4) additionally, the memory can be erased by keyboard operation.

4 DESIGN OF ASCS FAULT DIAGNOSIS PROGRAMS

4.1 *Settings of diagnosis trouble codes*

Diagnostic trouble codes (DTC) of the system are defined according to the standard of SAE J2012 (SAE 2002), which can be divided into two categories: device fault code and symptom fault code. Settings of part of DTC are shown in Table 1.

4.2 *Design of online self-diagnostic programs*

The design of online self-diagnostic programs will be explained with the following example of diagnostic program based on signal detection.

The diagnostic algorithm based on the signal limits detection is used and the timing cycle is 200 ms. Compare the values of clutch displacement sensor Lc, gear-shifting cylinder stroke sensor Tx, gear-selecting cylinder stroke sensor Ty, input shaft speed sensor N1, and output shaft speed sensor N2 with their respective maximum and minimum values. Its corresponding sensor will be considered to have a fault if a certain comparative value exceeds the normal range. If the same fault accumulates over five times, the corresponding DTC will be sent. The diagnostic program flowchart based on signal detection is shown in Figure 5.

5 BENCH TEST OF ASCS FAULT DIAGNOSIS SYSTEM

5.1 *Description of the test bench*

The test bench is schematically shown in Figure 6. The motor output shaft is connected with the transmission output shaft, reversely driving the transmission input

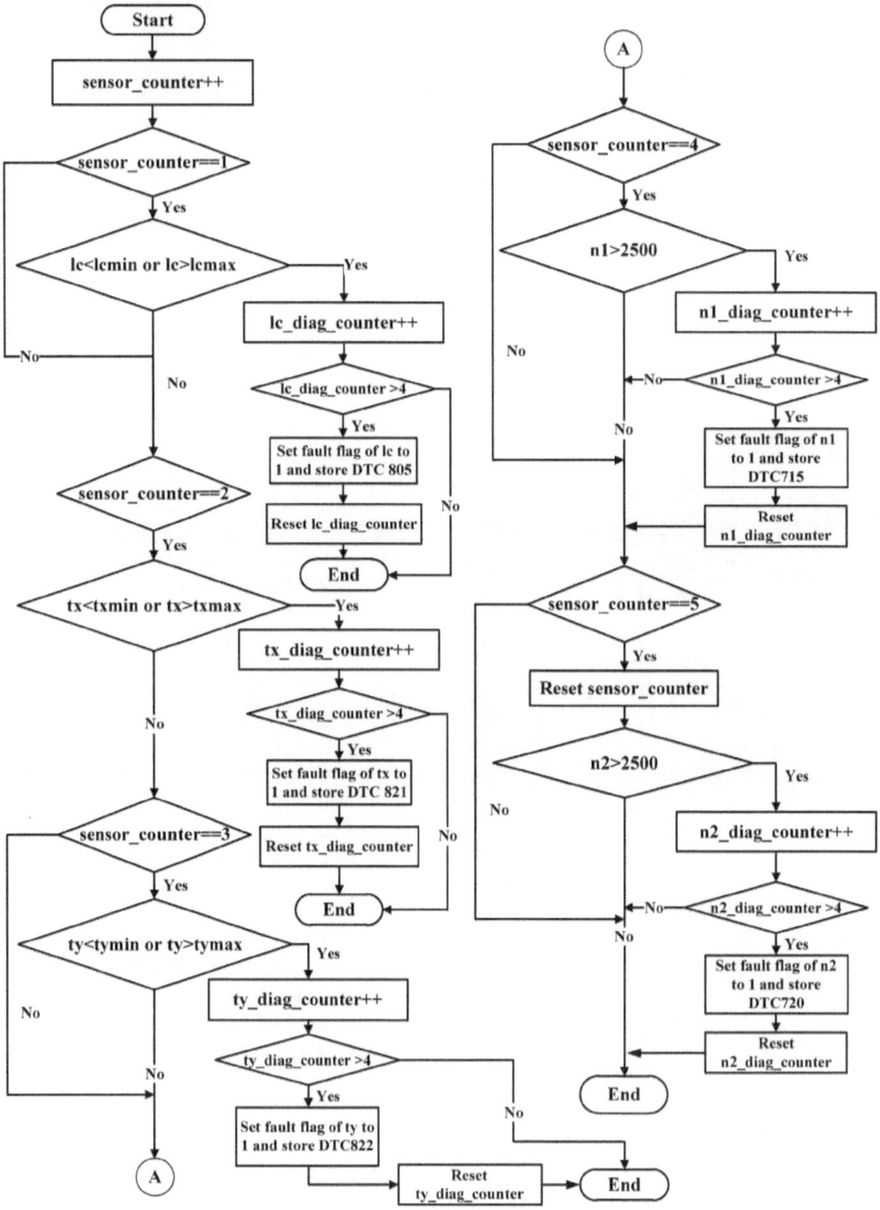

Figure 5. Diagnostic program flowchart based on signal detection.

shaft to rotate. Figure 7 shows the AMT transmission used in the test.

5.2 *Test and analysis of its results*

In order to test the fault diagnosis function, bench test was conducted by artificially setting fault. With the focus on the fault diagnosis of clutch displacement sensor, the analysis of test results is as follows.

Rotate the shaft of clutch displacement sensor Lc to make the signal value exceed its calibration value (the minimum of Lc is 500, and the max is 810 in the present system), and check whether the corresponding DTC 805 is sent. The test methods for gear-shifting stroke sensor Tx , gear-selecting stroke sensor Ty are similar like this. Figure 8 shows the data displayed on the tester when sensor Lc fails. After the data stored in the tester is exported to PC, data analysis software can be used to plot data diagram, as shown in Figure 9.

6 CONCLUSIONS

On the basis of overall analysis of fault type and fault cause of the system, a design of fault diagnosis system

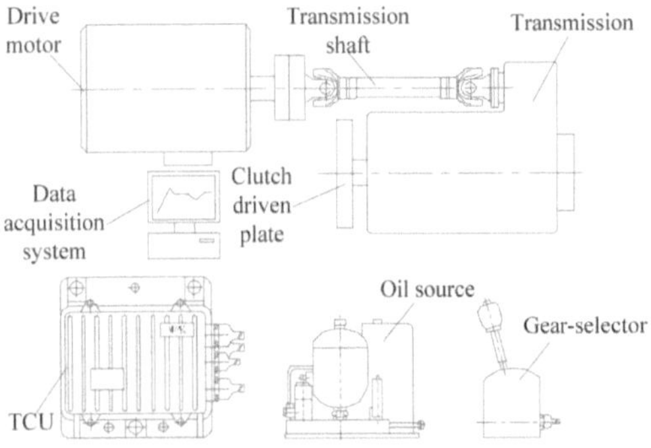

Figure 6. Schematic layout of test bench.

Figure 7. AMT transmission used in the test.

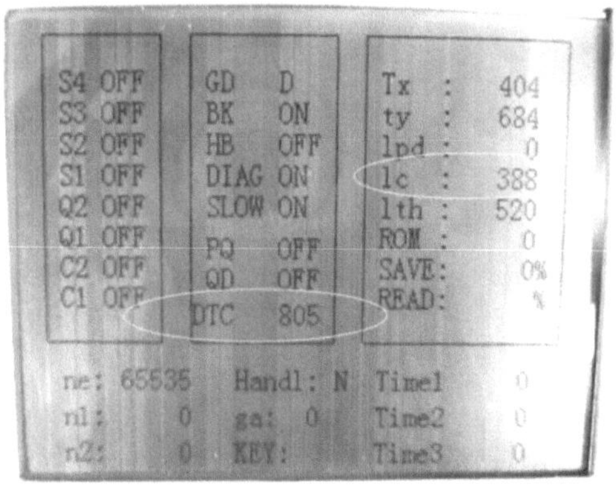

Figure 8. Display on the tester when sensor Lc fails.

for the vehicle is proposed. Through the coordination and cooperation of diagnostic functions between diagnostic system's lower computer (TCU) and upper computer (handheld fault diagnosis tester, VB data

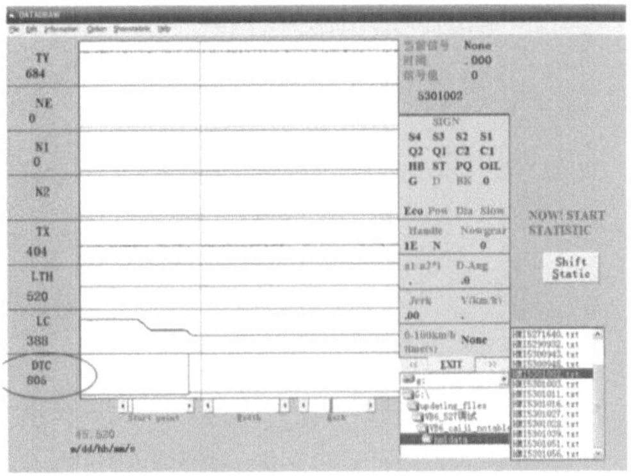

Figure 9. Data plotting on PC.

analysis software on the PC), the faults of most of ASCS parts can be diagnosed, which has important implications for driving safety and vehicle reliability of AMT system.

REFERENCES

Peng Jianxin, Liu Haiou, Hu Yuhui, et al. 2012. Research on fault detection and diagnosis strategy based on AMT system behavior. 2012 IEEE International Conference on Vehicular Electronics and Safety. IEEE, 186–190.

SAE. 2002. Diagnostic Trouble Code Definitions. J2012(4): 31~38.

Wang Hongliang, Liu Haiou, Chen Huiyan. 2009. Automatic shift control systems (ASCS) in off-road vehicles. Journal of Beijing University of Technology, 29(3): 214–218.

Xi Junqiang, Chen Huiyan, Ding Huarong, Dou jiange. 2001. On-board data acquisition system for automatic shift system of vehicle. Journal of Beijing Institute of Technology (English Edition), 10(4): 436–442.

Zeng Ai. 2008. Design of a Fault Diagnostic System for AMT in a Heavy-duty Vehicle. Beijing: Beijing Institute of Technology.

Future Mechatronics and Automation – Yang (Ed.)
© 2015 Taylor & Francis Group, London, ISBN 978-1-138-02648-3

WSN node localization technology research based on improved PSO

PingYing Ren
Zhengzhou University, Zhengzhou, China

LeRui Chen
Zhengzhou Railway Vocational and Technical College, Zhengzhou, China

JinSheng Kong
Zhengzhou University, Zhengzhou, China

ABSTRACT: With the wide application of wireless sensor networks, the sensor node localization problem has become one of the problems giving most cause for concern. In view of the general positioning technology defects and limitations, based on the TOA of the maximum likelihood estimation method, an improved particle swarm optimization algorithm is introduced, which improves the positioning accuracy. Through the simulation, this paper analyses the influence of ranging error, the number of unknown nodes, and node communication radius to the position, respectively The experimental results show that, under the same test conditions, the improved algorithm can achieve more accurate positioning and enhance the ability of PSO to jump out of the local optimum and the maximum probability search for the global optimal solution. At the same time it effectively avoids the shortcomings of the standard particle swarm optimization algorithm, which is easy to premature and fall into local optimum, and maximum probability find the global optimum solution.

Keywords: Node localization; improved PSO; maximum likelihood estimation; global optimal

1 INTRODUCTION

As information technology advances, wireless sensor networks have achieved rapid development. In wireless sensor networks, sensor network node localization is used to obtain additional information about a sensor node of a monitored area by obtaining an absolute or relative position in the plane or spatial information. Therefore, the positioning of nodes in wireless sensor networks has become one of the key technologies. Typically, the positioning algorithm is divided into two categories based on distance and without distance. The ranging localization algorithm based on the hardware requirements is relatively high, and the accuracy is relatively high. The non-ranging localization algorithm is without additional measuring equipment, but measuring is more rough and using field has its limitations [1].

For the features of ranging algorithms for positioning and the insufficient of the particle swarm algorithm, an improved particle swarm algorithm and the TOA maximum likelihood estimation method are combined. The new positioning algorithm measures first the distance between nodes using the TOA method, and then uses the maximum likelihood estimation method to estimate the unknown nodal coordinates. In addition, the coordinates are precisely positioned through the application of the adaptive mutation particle swarm optimization algorithm. This method can solve the shortcomings of the traditional PSO, which is easy too premature and local optimum; the positioning is more accurate and faster. The simulation results' analysis shows that this method is better than the standard PSO and basic maximum likelihood estimation method at obtaining coordinates that are more precise.

2 NODE LOCALIZATION

Node localization is, in two-dimensional (or three-dimensional) space, the distance information from one node to another of the three or more (three) beacon nodes that are known, thereby determining the position coordinates of the node [2]. The coordinates of unknown nodes are usually estimated using the least squares method. The multilateral measurement method is also known as the "maximum likelihood estimation method." It is shown in Figure 1. Assuming the distance to be measured from an unknown node u to the n ($n \geq 3$) anchor node, where the position coordinates of the anchor node i are (x_i, y_i) and the distance

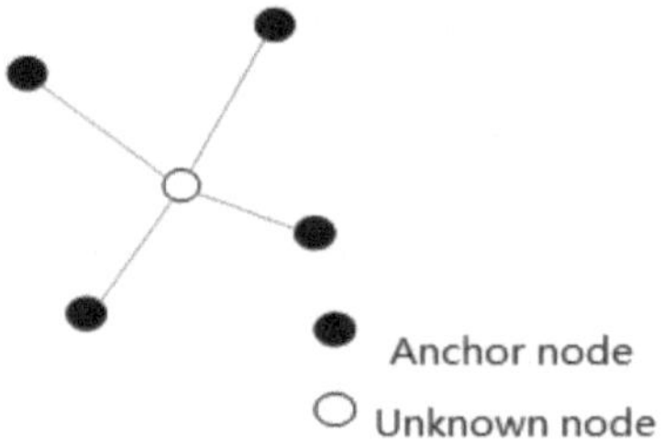

Figure 1. Schematic multilateral measurement.

to the unknown node u is $d_i(i = 1, 2, \ldots i, \ldots, n)$, you can get Equation (1):

$$\begin{cases} (x_1 - x^{est})^2 + (y_1 - y^{est})^2 = d_1^2 \\ (x_2 - x^{est})^2 + (y_2 - y^{est})^2 = d_2^2 \\ \cdot \\ \cdot \\ \cdot \\ (x_n - x^{est})^2 + (y_n - y^{est})^2 = d_n^2 \end{cases} \tag{1}$$

(x^{est}, y^{est}) is the coordinate of the unknown node u. Linearization of the problem is that the pre-(n-1) equations of Equation (1) minus the last equation, we have:

$$\begin{cases} x_1^2 - x_n^2 - 2x^{est}(x_1 - x_n) + y_1^2 - y_n^2 - 2y^{est}(y_1 - y_n) = d_1^2 - d_n^2 \\ x_2^2 - x_n^2 - 2x^{est}(x_2 - x_n) + y_2^2 - y_n^2 - 2y^{est}(y_2 - y_n) = d_2^2 - d_n^2 \\ \cdot \\ \cdot \\ \cdot \\ x_{n-1}^2 - x_n^2 - 2x^{est}(x_{n-1} - x_n) + y_{n-1}^2 - y_n^2 - 2y^{est}(y_{n-1} - y_n) = d_{n-1}^2 - d_n^2 \end{cases} \tag{2}$$

Equation (2) is converted to the form $AX^{est} = b$ where:

$$A = \begin{bmatrix} 2(x_1 - x_n) & 2(y_1 - y_n) \\ \cdot & \cdot \\ \cdot & \cdot \\ \cdot & \cdot \\ 2(x_{n-1} - x_n) & 2(y_{n-1} - y_n) \end{bmatrix}$$

$$b = \begin{bmatrix} x_1^2 - x_n^2 + y_1^2 - y_n^2 + d_n^2 - d_1^2 \\ \cdot \\ \cdot \\ \cdot \\ x_{n-1}^2 - x_n^2 + y_{n-1}^2 - y_n^2 + d_n^2 - d_{n-1}^2 \end{bmatrix}$$

The solution $\hat{X} = (A^T A)^{-1} A^T b$ of the above equation is obtained using the least squares method, that is, the estimated value of the unknown node coordinates.

For some reasons present during the ranging must make d_i exist certain measured errors, with the result that the obtained position coordinates of the unknown node are inaccurate. Assuming that the position coordinate error of the unknown node is:

$$E = \sum_{i=1}^{n} \left(\sqrt{(x^{est} - x_i)^2 + (y^{est} - y_i)^2} - d_i \right) \tag{3}$$

Due to the imprecise distance d_i, we set a threshold ε here. Assuming that the obtained position coordinates (x^{est}, y^{est}) can make $|E| \leq \varepsilon$, then ε smaller, more close to the real coordinate values of the unknown nodes.

3 ADAPTIVE PARTICLE SWARM OPTIMIZATION ALGORITHM WITH MUTATION

For the shortcomings of traditional PSO prematurity and local convergence, we use the Adaptive Particle Swarm Optimization algorithm with Mutation (APSOwM) [3] which combines an adaptive mutation strategy and particle swarm algorithm. The idea of this method is as follows: During the process of PSO, the APSOwM based on the change rate of the optimal fitness, adaptively modifies the inertia weight value, and sensitively handles the global and local search capability of the algorithm. When the average particle distance of the particle swarm is smaller than the limit value, or the global optimal solution does not change significantly in a number of iterations, some particles of the particle swarm are processed with variation to improve the diversity of PSO. Particle swarm can continue to evolve and there is greater probability to pursue the global optimal solution.

4 THE STANDARD PARTICLE SWARM OPTIMIZATION

In the PSO algorithm, each individual is called a "particle," representing a potential solution. Assuming, in the D-dimensional measurement space, each particle is a node within the measuring space, let m particles constitute a group; m is also called the "population size." Each individual particle can be estimated the fitness of its own position by the rules, and remember the best position found currently, which is called "local optimum value pbest." In addition, the best location found by all the particles of the population is known as the "global optimum gbest." These two variables makes optimal particle moving in these directions is close to some extent [4].

Assuming that $Z_i = (z_{i1}, z_{i2}, \ldots, z_{id}, \ldots, z_{iD})$ is the D-dimensional position vector coordinate of the particle $i(i = 1, 2, \ldots, m)$, the adapted value is obtained by the setting fitness function. The merits of locating the position of the particle can be judged. $V_i = (v_{i1}, v_{i2}, \ldots, v_{id}, \ldots, v_{iD})$ is the flight speed of the particle i, that is, the distance travelled by the particle. So far, $P_i = (p_{i1}, p_{i2}, \ldots, p_{id}, \ldots, p_{iD})$ is the optimal position of the particle searched, and $P_g = (p_{g1}, p_{g2}, \ldots, p_{gd}, \ldots, p_{gD})$ is the optimal position of the swarm.

In each iteration, the particle velocity and position information according to Equation (4) and Equation (5) is updated:

$$v_{id}^{k+1} = \omega v_{id}^{k} + c_1 r_1 (p_{id} - z_{id}^{k}) + c_2 r_2 (p_{gd} - z_{id}^{k}) \tag{4}$$

$$z_{id}^{k+1} = z_{id}^{k} + v_{id}^{k+1} \tag{5}$$

Among them, k is the number of iterations, r_1 and r_2 are the random numbers, and $r_1, r_2 \in [01]$ are the two parameters used to keep the diversity of population. c_1 and c_2 are the learning factor, namely the acceleration factor. $v_{id} \in [v_{\min} v_{\max}]$, when speed exceeds this range, v_{id} take boundary values. Likewise, $z_{id} \in [z_{\min} z_{\max}]$. ω is the inertia weight factor, $0.1 \leq \omega \leq 0.8$, its role is to weigh the local optimum capability and global optimization capability [5]. The function form of ω is:

$$\omega = \omega_{\max} - \frac{\omega_{\max} - \omega_{\min}}{K} \times k \tag{6}$$

Among them, $\omega_{\max}$ is the initial weight, $\omega_{\min}$ is the ultimate weight, k is the maximum number of iterations, and k is the current iteration.

Since the actual working environment of the paper considered is a three-dimensional space (D = 3), the fitness function is set by the Equation (3):

$$F(x_j^{est}, y_j^{est}, z_j^{est}) = \sum_{i=1}^{20} \left(\sqrt{(x_j^{est} - x_i)^2 + (y_j^{est} - y_i)^2 + (z_j^{est} - z_i)^2} - d_i \right) \tag{7}$$

5 THE ADAPTIVE PARTICLE SWARM OPTIMIZATION ALGORITHM WITH MUTATION

Since the particle swarm algorithm does iterative optimization by way of finding the optimal particle of the particle swarm, during the iterative process, if a particle finds the current optimal position, the other particles will move closer to it. However, if the particle finds the optimum position is a local optimum position, it will lead the swarm into a local optimum, and the algorithm will converge prematurely. After analysis of the PSO optimization process, the particles are gathered into a range, so that the loss of the diversity of particle swarm characteristics results in premature particle swarm.

The reference [6] uses the average particle distance of particle swarm to represent the diversity of particle swarm. The average distance d (k) is defined as follows:

$$d(k) = \frac{1}{m \cdot L} \sum_{j=1}^{m} \sqrt{\sum_{d=1}^{D} (Z_{jd}^{est} - \overline{P}_d)} \tag{8}$$

Among them, L is the maximum length of the search space diagonal and $\overline{P}_d$ is the mean of d-dimensional coordinates of all particles. The average distance shows the dispersion degree of the distribution of each particle in the particle swarm. The average particle distance is smaller, particle swarm more aggregation.

Define the change rate of the optimal fitness g:

$$g = \frac{|F(k) - F(k-5)|}{|F(k-5)|} \tag{9}$$

Among them, F (k) is the optimal fitness value of k generation in the particle swarm; g represents the relative change rate of the optimal fitness value for successive 5 generations iterating in the swarm. Set w according to g, which is shown in expression (10):

$$\omega = \begin{cases} \alpha_1 + \frac{r}{2}, g \geq 0.05 \\ \alpha_2 + \frac{r}{2}, g < 0.05 \end{cases} \tag{10}$$

Among them, $r \in [01]$, $\alpha_1 > \alpha_2$. Based on experience to take $\alpha_1 = 0.6$, $\alpha_2 = 0.2$. When $g \geq 0.05$, it is said that the fluctuation of the particle swarm optimal fitness value is relatively strong in an iterative process, and the swarm has yet found the direction gathering, which situation $\omega = \alpha_1 + \frac{r}{2}$ conducive to convergence of the algorithm. When $g < 0.05$, it is said that the particle swarm optimal fitness value tends to a certain value in an iterative process, and the swarm is into the development period, which situation $\omega = \alpha_2 + \frac{r}{2}$ conducive to seek a more accurate solution.

6 SIMULATION STUDY

In order to verify the positioning effect from combining the adaptive mutation particle swarm optimization algorithm and the maximum likelihood estimation method based TOA, this paper uses the Matlab 7.0 simulation. Set the simulation area in three-dimensional space $100\,m \times 100\,m \times 100\,m$. Set the particle swarm at various parameters: $c_1 = c_2 = 1.518$, K = 200, repeat the calculation 100 times, the maximum step length is $V_{\max} = \frac{\text{regional}}{10} = \frac{}{10} = 10\,m$, and $\omega = 0.76$. Set the positioning space with randomly assigned 20 particles.

The wireless sensor network node positioning mechanism and algorithm performance will directly affect its usefulness. Moreover, positioning accuracy is the most important evaluation of positioning technology. The evaluation generally uses the proportion of error value and node communication radius to represent, which is expressed as (11):

$$E = \frac{\sum_{j=1}^{n} \sqrt{(x_j^{real} - x_j^{est})^2 + (y_j^{real} - y_j^{est})^2 + (z_j^{real} - z_j^{est})^2}}{n \times r} \tag{11}$$

Among them, $(x_j^{real}, y_j^{real}, z_j^{real})$, is the real coordinate position of the unknown node, n is the number of unknown node, and r is a wireless sensor network communication radius [7].

The experiment in the simulation environment sets 100 sensor nodes, which are randomly distributed in the simulation area, and the number of anchor nodes is 20. The number of unknown nodes is 80.

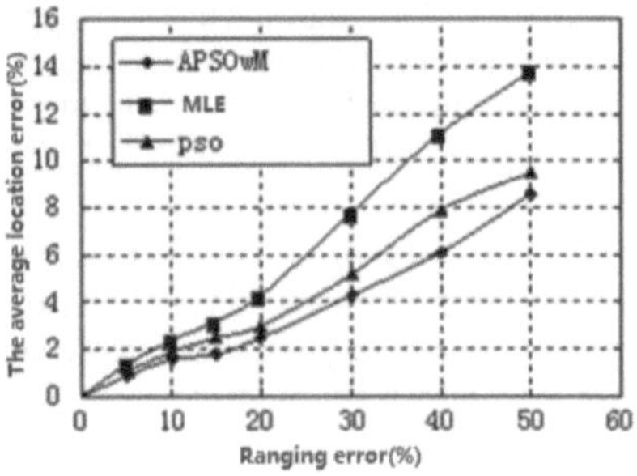

Figure 2. The graph showing the average location error changes with ranging error.

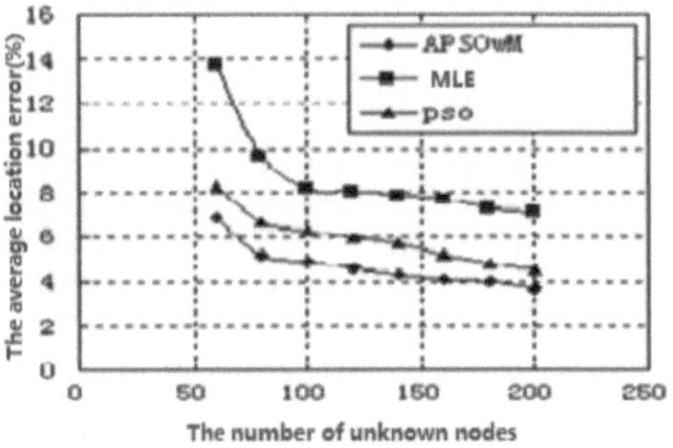

Figure 3. The graph showing the average location error changes with the number of unknown nodes.

7 THE INFLUENCE OF RANGING ERROR ON THE NODE LOCALIZATION

By Equation (3) we know that the size of ranging error will affect the accuracy of positioning, thus affecting the size of the positioning error. Set the number of unknown nodes as 80, the number of anchor nodes as 20, and the node communication radius at 30m. The graph showing average location error changes with ranging error for the three algorithms is shown in Figure 2.

Analysis is derived from Figure 2. The effect of particle swarm optimization positioning is better than the maximum likelihood estimation method, and the adaptive particle swarm optimization algorithm with mutation is better than the standard particle swarm algorithm. With the increase of ranging error, the increase in the magnitude of the average location error of the adaptive particle swarm optimization algorithm with mutation is smaller than that of the other two methods. Obviously, the adaptive mutation particle swarm optimization algorithm has better anti-errors.

8 THE INFLUENCE OF THE NUMBER OF UNKNOWN NODES ON THE NODE LOCALIZATION

Equation (11) shows that the number of unknown nodes will affect the average location error. Set the node communication radius at 30 m with a ranging error of 10%. The graph where the average location error changes with the number of unknown nodes for the three algorithms is shown in Figure 3.

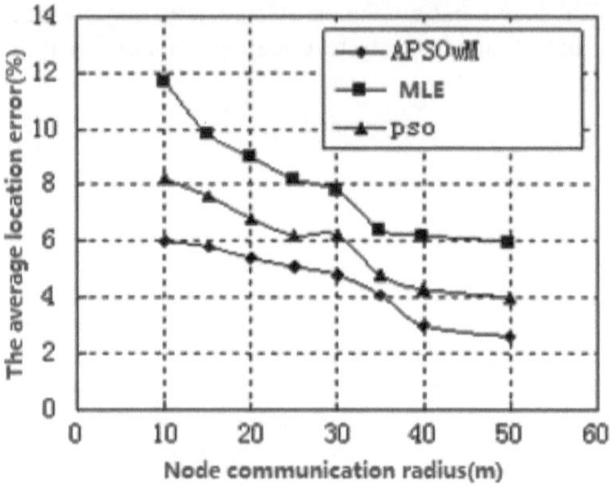

Figure 4. The graph showing the average location error changes with the node communication radius.

Derived from the analysis of Figure 3, in the case of the number of unknown nodes changes, the average location error of APSOwM is slightly lower than that of PSO. However, it is far less than the maximum likelihood estimation method. Obviously, in this case, the adaptive mutation particle swarm optimization has better convergence and better positioning accuracy.

9 THE INFLUENCE OF THE NODE COMMUNICATION RADIUS ON THE NODE LOCALIZATION

Equation (11) shows that the node communication radius, *r*, also affects the positioning error. Therefore, in the case of the communication range changes, set the number of the unknown node as 80, and the ranging error is 10%. The graph where the average location error changes with the node communication radius for the three algorithms is shown in Figure 4.

Derived from the analysis of Figure 4, in the APSOwM algorithm, the influence the node communication radius makes on the positioning accuracy is less than for standard particle swarm optimization. The positioning errors of the two algorithms are better than that of the maximum likelihood estimation, especially in the case of the node communication radius small, the average positioning error of the APSOwM algorithm minimizes. It has more obvious advantages; obviously, the APSOwM algorithm searches for stronger, more precise positioning.

10 CONCLUSION

In wireless sensor networks, the most critical aspect is the positioning of nodes. Based on analysing the characteristics of various location technologies, the adaptive mutation PSO maximum likelihood estimation positioning algorithm based on TOA ranging was proposed, and programming and simulation was performed by Matlab 2012. The average location error comparison for the three different methods was based on three conditions: different ranging errors, a different number of unknown nodes, and different communication radii. The results show that, in each case, the positioning accuracy of the APSOwM algorithm is better than the maximum likelihood estimation algorithm, and more accurate than standard particle swarm optimization positioning [8].

11 ABOUT THE AUTHORS

Pingying REN, female, 1988-, Kaifeng in Henan, graduate of the Electrical Engineering Department of Zhengzhou University. The main research direction: Network Intelligence Optimization Control.

Lerui CHEN, male, 1985-, Nanyang in Henan, Assistant of the Electrical Engineering Department of Zhengzhou Railway Vocational and Technical College. The main research direction: intelligent information processing

Jinsheng KONG, male, 1963-, Nanjing in Jiangsu, professor of Zhengzhou University. The main research direction: intelligent information processing and control of complex systems.

ACKNOWLEDGMENT

This paper is generated in the context of the project, which is funded under the Education Department of Henan Province Science and Technology Key Project Support Programme (contract to NO 13A413451)

REFERENCES

[1] LI Ti-hong, FENG Shu-qian. Indoor Wireless Sensor Network Difference Localization Algorithm Research [J]. Computer Simulation, 2010, 27(7):102–104.

[2] Nasipuri A and Li K. A Directionality Based Location Discovery Scheme for Wireless Sensor Networks [J]. Prco. First ACM Int'l Workshop Wireless Sensor Networks and Application. pp. 105–111, Sept. 2002.

[3] Jeffrey Hightower, Gaetano Borriello. Location systems for ubiquitous computing [J]. IEEE Computer, August 2001, 34(1):57–66.

[4] Liu Zhi-kun, Liu Zhong, Tang Xiao-ming. Node self-localization algorithm based on modified particle swarm optimization [J]. Journal of Central South University (Science and Technology), 2012, 43(4):1371–1376.

[5] Yang Chun-hua, Gu Li-shan, Gui Wei-hua. Particle Swarm Optimization Algorithm with Adaptive Mutation [J]. Computer Engineering, 2008, 34(16):188–190.

[6] Parham H. Namin, Mohammad A. Tinati. Node Localition Using Particle Swarm Optimization [J]. IEEE Networks, 2011:288–293.

[7] Zhu Shi-juan, Zhu Qing-bao. Research on the Divisional Particle Swarm Optimization Algorithm for Solving Function Optimization Problems [D]. Nanjing Normal University, 2012.

[8] Yang Dan-tong, He Jin-sheng, Bai Hong-tao. Constraint Particle Swarm Optimization Algorithm for Wireless Sensor Networks Localization [J]. Computer Science, 2011, 38(7): 46–50.

Future Mechatronics and Automation – Yang (Ed.)
© 2015 Taylor & Francis Group, London, ISBN 978-1-138-02648-3

An indoor control system based on LED visible light

Wang Yang Yu, Zi Yang Chen, Yan Zhe Zhao & Cheng Yang Hu
Department of Communication Engineering, Jilin University, Changchun, China

ABSTRACT: In this paper, by analysing the application background of Visible Light Communication (VLC), we put forward a new scheme of intelligent control through transmitting control instructions with an indoor LED lamp. We introduced the composition, working principle, and realized the process of the system. A Microprogram Control Unit (MCU) helps the system to carry on auxiliary control. Using serial debugging software, the receiving control code is displayed on a computer screen in order to verify the feasibility of the scheme. Finally, the physical simulation and test shows that the visible light control system information transmission rate is 9600(bit/s). The transmission error rate is 2×10^{-5}. This paper provides a new idea and reference for indoor home furnishing control systems.

1 INTRODUCTION

Using LED lights as communication base stations to transmit information has become one of the hot research topics in the field of wireless optical communication at home and abroad. White LEDs with high brightness, a low power consumption, and having the advantages of a long service life, a small size, and offering green environmental protection, are regarded as the fourth generation of light source for energy saving and environmental protection. Not only that, white LEDs also have the advantages of high response sensitivity and good modulation characteristics. By using the above excellent properties of LED in a lamp, the signal is modulated to LED visible light for transmission, thereby achieving a new type of wireless optical communication technology, Visible Light Communication (VLC). VLC technology has the advantages of being eye-safe. It is not necessary to apply for a radio frequency card and it has no electromagnetic interference. It provides a new data access method for optical communication etc. It has been widely concerned. Recently, the research on visible light communication system is just as effective in China. Alongside in-depth research concerning LED light source modulation characteristics, the indoor wireless channel, and visible light modulation technology, the advantages of visible light communication technology have been gradually revealed: the LED visible light for indoor wireless channel controls can greatly reduce the cable control cost. The control terminal can transmit the signal through the network, and the new technology can cooperate with the Internet closely, so it becomes a very attractive prospect. A visible light communication system is very suitable for application in exhibition guides, intelligent traffic systems, hospitals, laboratory information transmission, etc.

This paper uses a white LED to design an indoor visible light control system. In order to establish the indoor control system with the white LED as a delivery centre, an MCU as a simple control command send interface, can control each device once it has a PIN detector. In this paper, the LED visible light short-range transmission system that we designed has high reliability, through the different operation of people in the control interface; we use the serial port to send different code element information. The PIN detector has MCU decoding recognition, to control the appropriate equipment to work. The use of the indoor wireless transmission control mode gives full play to the advantages of LED visible light illumination and communication: it reduces the time of setting up the communication link, it reduces the cost of laying the cable control, it reduces the effects on the environment of electromagnetic radiation, it is fast and convenient, it has high reliability, a low energy consumption, and provides a new idea for the realization of 'green control'.

2 THE SCHEME AND PRINCIPLE

2.1 *The indoor control model based on VLC*

Firstly, we established the indoor intelligent control model, which takes LED visible light as the transfer station; the model structure is the point to surface. The structure is as follows:

The operation of the man-machine interface; this part is the main variety of control interface. The control information that the operator wants to send is transmitted to the LED lamp and spread out in the form of optical wireless communication. This part can be an intelligent terminal, such as an intelligent mobile

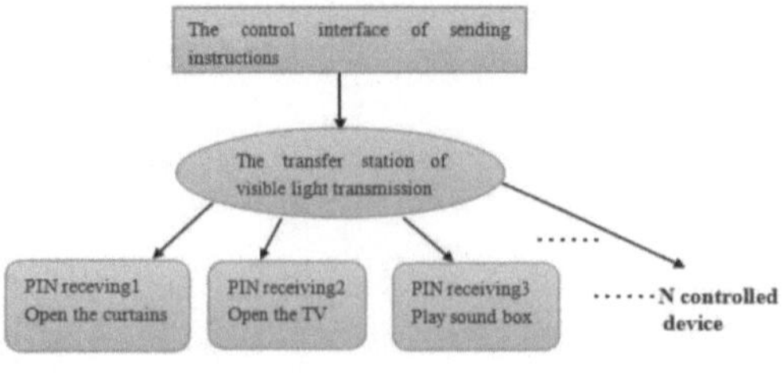

Figure 1. Control network model.

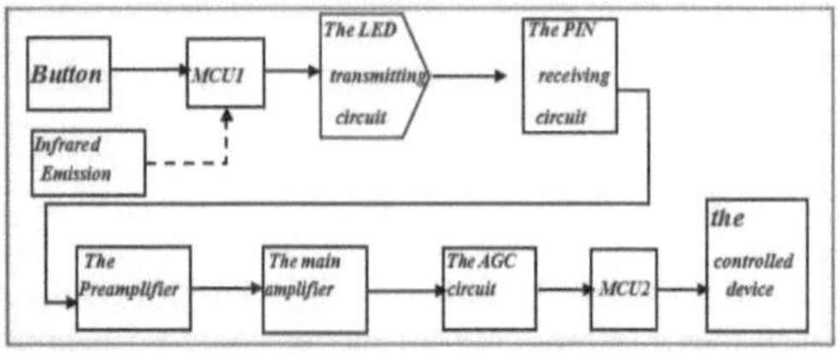

Figure 2. The experimental system.

phone, PC, etc. These terminals can send instructions to the transfer station of the LED lamp. It can also be a simple manual operation interface, such as the keyboard, where each key can transmit the corresponding control information to the transmitting station.

Information sending part: The base station is composed of a LED lamp; the transmitter circuit can convert command information to the signal that can be transmitted in the light carrier. After the white LED drive circuit, the signal will be modulated to the illuminating lamp. The LED lamp sends the signal to each controlled device in the form of an optical pulse. The LED lighting covers the whole of the receiving space; wherever there is light, it can receive information. The optical wireless broadcast control network can combine the indoor equipment to become a part of the intelligent system.

The receiving and the controlled device: The PIN light detection circuit of the controlled device receives the light signal, and then converts the optical signal into an electrical signal. The receiver circuit converts the signal after amplification, it shapes the processing, and the shaped control instruction code signal which is transmitted to the receiving end (MCU2), according to the received instruction, can identify and determine whether to carry out the operation to complete the whole control action. The controlled device can be indoor curtains, power switches, TV, audio equipment, etc. Therefore, we achieve the goal that a control centre can control various devices.

2.2 *The working principle of the system*

The scheme route is shown in Figure 2:

The information sending part consists of the control centre and LED lamp; the command centre consists of the MCU and peripheral equipment, such as the keyboard, a GSM module, an infrared remote control, etc. When a button is pressed, the MCU generates a control signal; keys can also be the portable infrared remote control. The generated control signal is sent by the MCU asynchronous serial port TxD in the form of a TTL level signal, and transmitted to the drive transmitting circuit of the LED, making the signal modulate to light. The receiver circuit system consists of a PIN photoelectric detection circuit, a preamplifier, a main amplifier circuit, an automatic gain control amplifier circuit, and an MCU. The PIN photoelectric detection circuit receives the optical pulse signal and converts it into an electrical signal and then weak signals travel through the amplifying circuit. After that, through AGC amplification circuit to make the amplitude of the output signal stable, and then converts the signal into the TTL signal, next, giving it into the MCU asynchronous serial port RXD. After receiving the control signal, MCU recognizes the signal and determine whether to control the device to work on the receiving end. The MCU uses a STC89C52RC transmission between two MCU asynchronous serial port control signals; the signal is transmitted to the controlled equipment in the form of light in the air channel.

2.3 *The LED drive transmitting portion*

The wireless channel part is mainly composed of a visible light transmitting part and the PIN receiving part of the system. For the LED drive transmitting part it was decided to use digital modulation; digital modulation LED is also known as the light intensity modulation. Digital modulation technique is to use PCM to control the LED turn-on and turn off pulse, and driving circuit plays a major role in opening and closing. The nonlinear of the LED P-I characteristics has a small influence on digital modulation, so the circuit switching rate, also the data modulation rate is the foremost issue of the digital modulation to consider. The digital modulation principle is shown in Figure 3. The input signal is at high level, the output pulse of the LED digital modulation circuit and photoelectric receiving circuit. The modulation circuit is designed in this paper and the circuit as shown in the following figure. The above is the modulation circuit, below is the receiving circuit. The digital modulation circuit is equivalent to a control switch, and provides tens to hundreds of current (MA) to the LED in the Q2 saturation conduction, no light in the Q2 is off, circuits composed of Q1 improves the driving ability of the MCU. The TTL control signal can control the LED fast light off to send light signals. The main task of the light receiving end is to recover the optical carrier carrying the information transmitted via a wireless channel with minimum noise and distortion. Therefore, the output characteristics of the light receiving end reflect the properties of the visible light control system. The photoelectric receiving circuit consists of a photoelectric detector and preamplifier circuit. The PIN photoelectric detector selected was the Hamamatsu Corporation

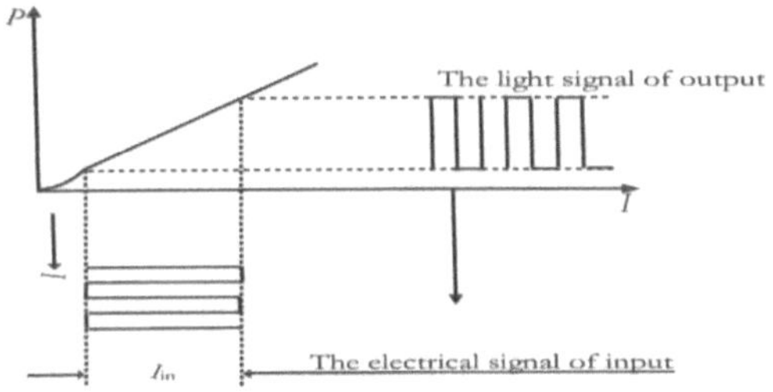

Figure 3. The principle of LED digital modulation.

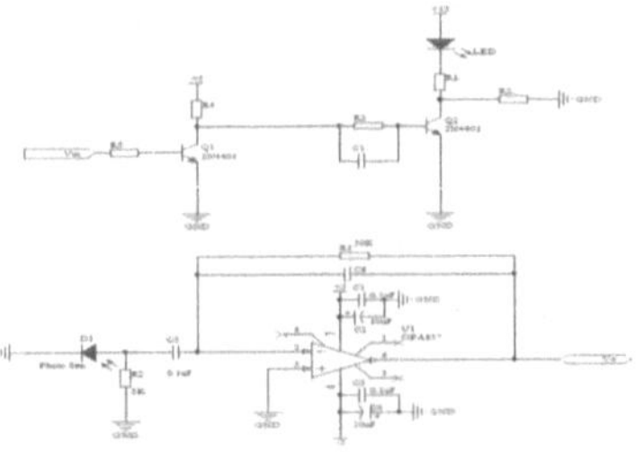

Figure 4. Digital modulation circuit and photoelectric receiving circuit.

S6968; the blue light response characteristics are sensitive in visible light, the response frequency is 50M, and the addition of a blue light filter before the receiver, can reduce the receiver noise interference. Selection of high input impedance preamplifier, low noise, high response speed of the op amp OPA847.

2.4 *Preamplifier*

The PIN photodiode changes the optical signal into the electrical signal. Generally, it needs to be amplified through a multiple-stage amplifier and then outputted by the circuit terminal. We know from a Vries equation, that the noise factor f of the multiple-stage amplifier depends approximately on the first order noise factor, and receiver's minimal noise factor depends largely on the preamplifier. Therefore, the preamplifier should have characteristics such as low noise and high gain, and a certain amount of bandwidth. According to the $E_n \sim I_n$ noise model of amplifier, the internal noise source of the receiving end is converted into the equivalent noise of the preamplifier input. To calculate the noise factor F is:

$$F = (E_n^2 + I_n^2 + 4KTR_s\Delta f) / 4KTR_s\Delta f \quad (1)$$

In Equation 1, R_s is the equivalent resistance of power, E_n and I_n is the equivalent input noise voltage and current, K is the Boltzmann constant, T is the absolute temperature, Δf and is the noise bandwidth. The derivative of Equation (1) is $df/dR_s = 0$, so the optimum source resistance is $R_{s0} = E_n/I_n$. Therefore, in the adjacent area of the optimum source resistance, F can take a smaller value; this is the basis for the correct choice of the source resistance in order to achieve the best noise performance. The photoelectric detector can be equivalent to a signal source, which has a high source resistance, and it needs to be matched with a high impedance load resistance to obtain the maximum gain. A transimpedance amplifier can meet the direct coupling requirements of $R_{s0} \approx R_s$. Making up the feedback resistor between the input and output and forming parallel voltage negative feedback. It can make the circuit gain stability, dynamic range is increasing, band broadening and does not need to balance the circuit, the circuit is shown in Figure 3. The preamplifier is designed by the integrated operational amplifier, OPA657. It can produce an output voltage which is big enough to overcome the effect of noise from the subsequent circuit, and its wide band can match the signal rate, thus acquiring better noise and interference between codes.

Table 1. Control code.

KEY	CODE	KEY	CODE
Key 1	0X00	Key 9	0X08
Key 2	0X01	Key 10	0X09
Key 3	0X02	Key 11	0X0A
Key 4	0X03	Key 12	0X0B
Key 5	0X04	Key 13	0X0C
Kye 6	0X05	Key 14	0X0D
Key 7	0X06	Key 15	0X0E
Key 8	0X07	Key 16	0X0F

3 SYSTEM TESTING

The human-machine operation interface of the experimental system is the keyboard. When a button is pressed, the MCU serial TxD outputs an 8-bit control code, the de baud rate is 9600bit/s. The control codes launch out through the LED in the form of light. After receiving the control code, the receiver MCU identifies and judges. When the MCU needs to perform the operation, MCU controls the equipment. A controlled device in this experimental system is the sound module, MCU controls the voice module to broadcast pre-recorded sound information. Different control codes can control the different recording to broadcast.

We used a matrix keyboard to make the MCU send the control information; the keys of the matrix keyboard are coded as follows: once the button is pressed, the MCU1 transmits the code:

We tested the transmission performance of the information transmission system. The MCU1 outputs a TTL square wave signal at 9600 de baud rate, the signal is modulated to the LED light. We used an oscilloscope to observe the received signal and the sent signals, and signals were compared as shown below. The results showed that we could receive the sent signal without distortion. When the control signal was transmitted, we used the serial debugging assistant to test the bit error rate at 9600 de baud rate. We used a serial debugging

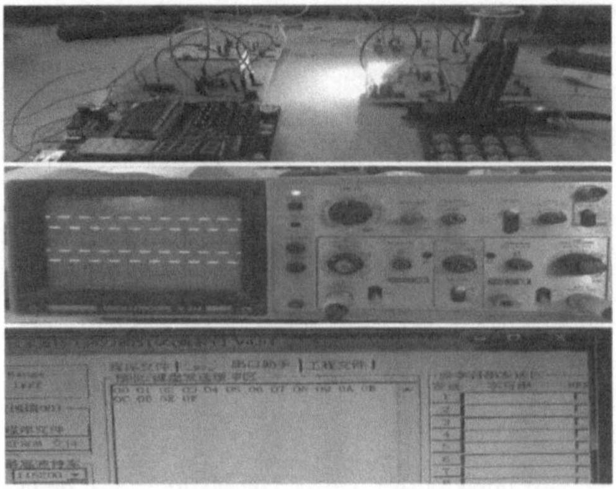

Figure 5. The real circuit and test waveforms, and receiving data.

assistant to send a string of long random characters to the asynchronous serial port RXD of MCU1; the MCU1 sent received information to the LED lamp through the asynchronous serial port TxD. The information was transmitted to the asynchronous serial port of MCU2 through the indoor air channel. Then MCU2 sent the received signal to the PC to display the sent characters. The symbol of the sent information was compared with the symbol of the received information. We counted the number of wrong symbols, denoted as "a." If the number of the sent symbols is "b," we can calculate the bit error rate. The bit error rate is written as P, and $P = a/b$. After many tests, we can calculate that $P = .2 \times 10^{-5}$.

We tested the effect of the specific control code transmission at 9600 baud by using serial debugging software, and the method can be divided into several steps. First, we pressed 16 buttons on the sending end, so that this side of the MCU will transmit the corresponding control code to the air channel. Next, the control code travels across the air channel and reaches the MCU, which is at the end of the air channel. Finally, the receiving end of the MCU uses its TxD serial to sending the control code to the PC, which can display this code by the serial debugging assistant. The result is shown below, and we easily find that the serial debugging software can receive the 16 different control codes corresponding to 16 different keys without any mistakes.

4 CONCLUSIONS

By analysing the function of this system, it is not surprising to find that the data rate and the transmission error rate can certainly match the whole control system. The method, which uses an array of LED lamps, can widen the control range, and furthermore, the number of controlled devices can increase significantly if a better MCU is chosen. This indoor control system can easily adapt to a variety of indoor environments through software and hardware optimization, which can greatly enhance the portability of the system in indoor environments.

Visible light communication is still in the initial stages of exploration at home and abroad, and the method, which uses visible light to control, is still in the research stage, but its application prospects are very attractive. In this study, we used the coding system, and quickly verified the feasibility of the indoor control system based on LED visible light, thanks to the existing technology and equipment conditions. Meanwhile, we transmitted the real-time control code of the LED to serial debugging software in a computer to demonstrate that the control system is under our control, so it provides us with a new way of indoor intelligent control management. With the development of technology, an indoor LED wireless control system will not only be limited to this, but it may overturn the initial definition in the value of indoor lighting. LED lights will not only provide the function of road lighting, but may also exist as the access point in entire indoor communication systems, providing new direction and impetus to the realization of indoor intelligent control systems.

REFERENCES

Hoa Le Minh, Dominic O'Brien, Grahame Faulkner et al. 100-Mb/s NRZ Visible Light Communications Using a Postequalized White LED [J]. IEEE PHOTONICS TECHNOLOGY LETTERS, 2009, 21(15): 1063~1065.

Iang Hyung-Joon, et al. PWM-based PPM format for dinmming control in visible light communication system [C]. CSNDP, 2012:1~5.

Kaiyun Cui, Gang Chen, Zhengyuan Xu, and Richard D. Roberts. Line-of-sight Visible Light Communication System Design and Demonstration [C]. CSNDSP 2010, 2010: OWC-21.

Taubman D. High performance scalable image compression with EBCOT [J]. IEEE Trans. Image Processing, 2009, 9 [7]: 1151~1170.

Tronghop D, HWang J, Jung S, et al. Modeling and analysis of the wireless channel formed by LED angle in visible light communication [C]. Information Networking (ICON), 2012 International Conference on IEEE, 2012: 354~357.

M. Zhang, Y. Zhang, Y. Yuan, J. Zhang, Mathematic models for a ray tracing method and its applications in wireless optical communications [J]. Optical Express, 2010, 18(17): 18431~18437. Chemical Industry Press.

Zukauska, A., Michael S. Shur, Remis Gaska. Introduction to solid state lighting [M]. Beijing: Chemical Industry Press, 2006.

Future Mechatronics and Automation – Yang (Ed.)
© 2015 Taylor & Francis Group, London, ISBN 978-1-138-02648-3

Experimental research on ultrasonic separation of two-dimensional normal mode

Chenhui Hua
Chengdu College of University of Electronic Science and Technology of China, Baiye Road No1 High-tech Zone Chengdu Sichuan Province, China
School of Mechatronics Engineering of University of Electronic Science and Technology of China, Sichuan Province, China

Jiexiong Ding
School of Mechatronics Engineering of University of Electronic Science and Technology of China, Sichuan Province, China

ABSTRACT: In this paper, an ultrasonic separator based on two-dimensional normal mode has been microfabricated from layers of silicon and Pyrex. The "swallow tail" structure within the fluid chamber effectively avoids three dimensional normal modes to be excited and helps to improve separation efficiency in two dimensions.

The matching experimental platform is set up for the separator. Experiment results show that the separator can realize the separation of particles (yeast is chosen here) suspending in liquid under two-dimensional normal mode. In order to improve the separation efficiency, experimental temperature and suspension liquid concentration are analyzed experimentally and their influence rules are summarized The results show that the optimal experimental temperature is around 8 for the separator, and mass suspension liquid concentration around 2% is most suitable for particles separation. This is significant for further studying on MEMS ultrasonic separating methods based on two-dimensional normal modes.

1 INTRODUCTION

Ultrasonic standing waves can be used to generate radiation forces on particles suspending in fluid. The force has a number of potential applications in microfluidic devices, such as separation [1, 2, 3], positioning [4, 5], washing [6] and assaying [7, 8, 9] of bio-functionalized particles or cells. Most of the previously reported miro-machined systems are based on one-dimensional ultrasonic fields.

In recent years, ultrasonic separation or manipulation involving two dimensions has become a popular branch. Most fluid chambers have a square ($5 \times 5\,mm^2$) [10] or nearly square ($375 \times 110\,\mu m^2$) [11] cross-section within the presented devices. They are not appropriate for separation. Hill [12] describes a two dimensional MEMS separator within which the fluid chamber has a width to height ratio of approximately 30:1 (width equals 5 mm). The structure will result that the clean and turbid streams mixed together at the outlets because perfect laminar flow near the outlets would be destroyed. And three dimensional normal modes cannot be avoided effectively.

The ultrasonic separator based on two-dimensional normal mode described in this paper can effectively avoid the problems mentioned above. The character and validity of the resonant structure have been previously discussed and proved [13, 14]. In order to increase the separation efficiency, experimental investigation is carried out to analyze those factors that would influence separation.

2 EXPERIMENTAL METHODS

2.1 *Separator*

The device consists of a $30 \times 8 \times 0.2\,mm$ chamber. As shown in figure 1, from bottom to top, the separator has four layers: piezoelectric transducer (PZT), matching layer (silicon), fluid layer, reflection layer (Pyrex glass). The fabricating process has been proposed in the paper [15]. Two-dimensional normal modes on the plane of X and Y can be excited by the acoustic transducer which is fixed on the matching layer, and particles are forced into several bands coming out from the right outlets on Z direction.

The "swallow tail" structure in the fluid chamber has great advantages comparing to devices mentioned in the instruction. This unique structure near the inlet and outlets effectively cuts down the reflection of acoustic wave in Z-direction, and then the acoustic resonance in Z-direction can be avoided. Simultaneously the entrance-exit structure reduces the energy loss. It

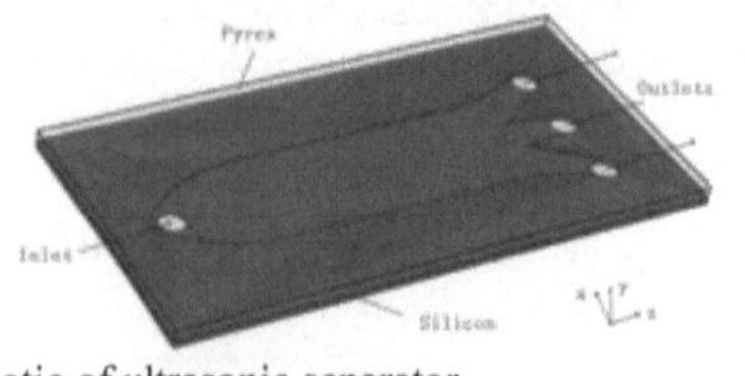

(a) Schematic of ultrasonic separator

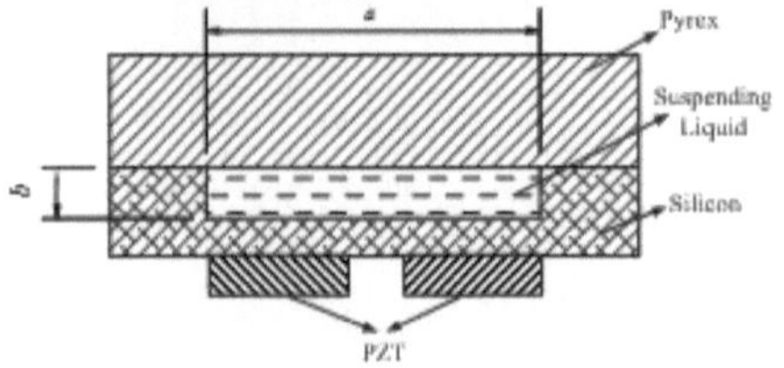

(b) Cross-section of the separator

Figure 1. Ultrasonic resonance separator based on two-dimensional normal mode.

is due to that viscous fluid will produce energy loss in the process flow, one unit of gravity the energy loss h_f can be expressed as [16],

$$h_f = \lambda \frac{lv^2}{2dg} \tag{1}$$

Where l is the flow length, and d is pipe diameter, λ is the loss coefficient. The entrance-exit structure reduces the flow length l, enlarges the pipe diameter d when the fluid comes in, thus leading to reduce the energy loss h_f. It is helpful to improve the separation efficiency. Besides the wedge shape training wall near the outlets helps to divide the clean and turbid streams more smoothly.

2.2 *Experimental platform*

The corresponding experimental platform is established, as shown in figure 2. A microscope camera is used to catch the pictures of particles convergence in the experimental processing. The support arm of camera is designed with four crew adjusters, one is for fastening, one for allocation, and the other two are for fine tunings that helps the camera catch pictures at the best position. Signal generator here is Agilent 33250A, the output frequency range is from 0.001 Hz to 80 MHz, and its resolution is 1 μHz.

The ultrasonic separation system is detailed schematically as shown in figure 3. The separator bed is designed to match with the structural characteristics of the separator, and made of aluminum. Within the separator bed a ductwork is designed with one inlet and three outlets to fit for the separator. O sealed rings are used to connect inlet and outlets of the separator with the bed to avoid fluid spilling. And four pipe connectors are designed to connect the pipes with the ductwork within the separator bed. Fixing devices are

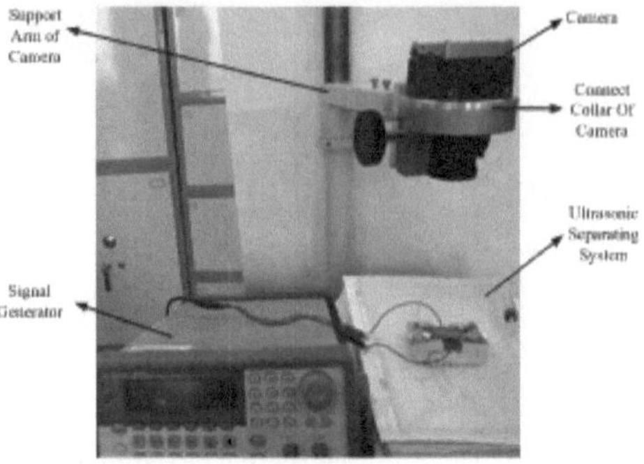

Figure 2. Experimental platform.

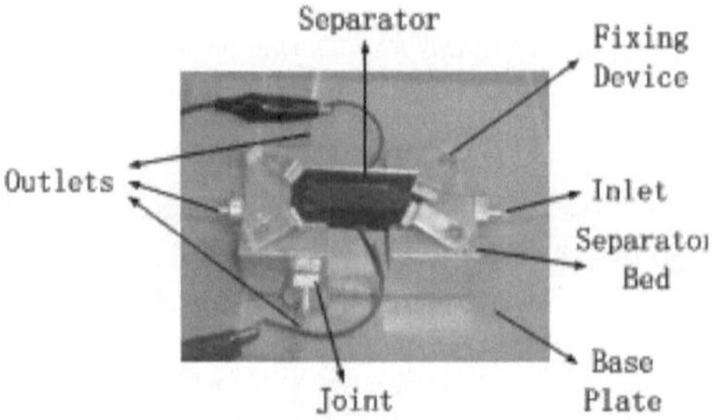

Figure 3. Ultrasonic separation system.

composed of pleated sheets and screws in order to fix the separator on the separator bed. Yeast particles are chosen as suspended load of the suspension fluid in the experiments, because it is easier to observe separation effectiveness as the suspension fluid of yeast shows yellow. And a white base plate is used to hold the ultrasonic separation system, and also helps to identify the separation effectiveness easily.

3 RESULTS AND DISCUSSION

3.1 *Experimental separation performance*

In this section, the separator is proved effectively. According to the "swallow tail" structure of the separator as shown in figure 1, there are two ideal cases for particles separation. When (1, 1) normal mode is excited, the fluid with particles aggregation comes out from the middle outlet, and the fluid without particles comes from the other two outlets on sides. When (2, 1) normal mode is excited, the case becomes inverse. Figure 4 shows that fluid comes from two side outlets are more turbid than from the middle outlet. Then it can be firstly judged that probably (2, 1) normal mode is excited in the experiment. However, the repeatability of experiment like this is not always good. To improve the separation effectiveness some factors are analyzed and discussed in the following.

3.2 *Suspension liquid concentration*

Suspension liquid concentration may influence separation effectiveness. Figure 5 shows experimental

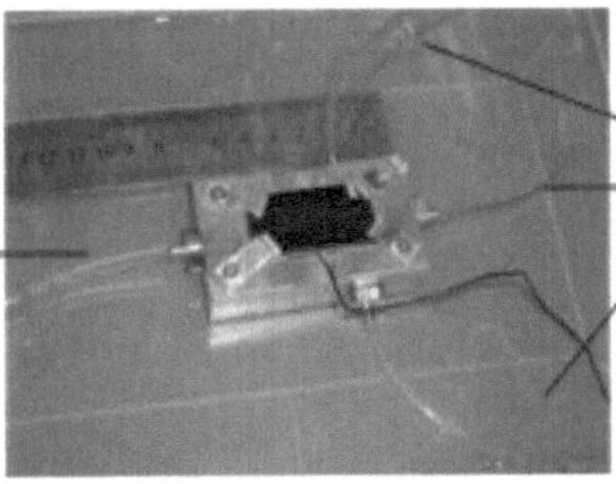

Figure 4. Experimental picture when (2, 1) normal mode is excited.

(a) 1%

(b) 2%

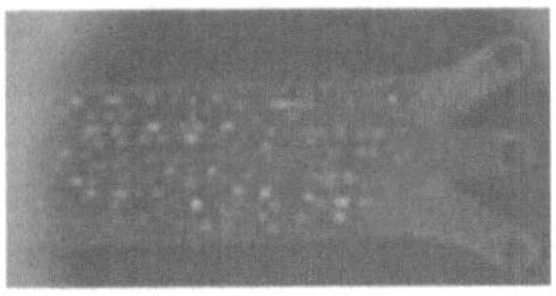
(c) 4%

Figure 5. Comparison experiments of different suspension liquid concentration.

phenomena of three comparison experiments with same experimental conditions except mass concentrations. These experiments are excited with resonance frequency of (2, 1) normal mode. One has mass concentration of 1%, one of 2%, and one of 4% respectively.

The secondary acoustic radiation force becomes significant when the particles are very close to each other. Allowing the superposition of rigid-sphere and compressibility contributions, between two identical compressible spheres in a plane standing wave Weiser [17] obtained for the secondary radiation force

$$F_B(x) = 4\pi a^6 \left\{ \frac{(\rho-\rho^*)^2(3\cos^2\alpha - 1)}{6\beta^* d^4} v_{(x)}^2 - \frac{\omega^2\rho^*(\beta-\beta^*)^2}{9d^2} p_{(x)}^2 \right\} \quad (2)$$

Where d is the center-to-center distance of the spheres, β and β^* are the compressibility of the particles and the host fluid, respectively. α is the angle between the centerline of the particles and the propagation direction of the incident sound wave; $v_{(x)}$ and $p_{(x)}$ are the velocity and pressure, respectively, of the unperturbed incident field at the position of the particles. ρ and ρ^* are the density of the particles and the host fluid, respectively.

Equation (3) explains that, the smaller the particles distance d, the stronger the secondary radiation force $F_B(x)$. The effect of secondary radiation force increases the velocity and density of the particles clusters. Gupta [18] has applied the secondary radiation force on collection of suspended particles, and Grossner [19] on the sediment of suspended particles.

The secondary radiation force $F_B(x)$ increases with increasing suspension liquid concentration. It is easily to find that in figure 5, the higher concentration, the more obvious particles aggregation phenomenon. However, mass concentration of 4% in figure 5 (c) is not helpful for separation; many clusters of particles may sediment beforehand in the cavity or plug up the outlets.

Contrast to figure 5 (b), figure 5(a) is influenced by secondary forces much less. Figure 5(b) shows that, particles aggregation distributes on the two sides of the cavity as supposed to excite (2, 1) normal mode. The result of comparison experiments tells that, mass suspension liquid concentration of 2% is more suitable than that of 1% and 4% for particles separation. So the suitable range of suspension liquid concentration may lie around 2%.

3.3 *Experimental temperature*

According to the function expression C = F (L, T) mentioned above, experimental temperature T will also influence the ultrasonic velocity C. If the suspension liquid concentration L is certain, the ultrasonic velocity C will vary with the temperature, and then the resonant frequency of the resonator varies.

Three comparison experiments have been carried out based on the same experimental conditions except different temperatures, as shown in figure 6. Mass suspension liquid concentration of 2% is chosen. Three experiments are driven by calculated resonant frequency. But the actual resonant frequency varies with the temperature. Therefore, it is hard to make the resonance in the fluid chamber as supposed. Through the comparison experiments, it is indicated that about 8°C is the right temperature for this resonator.

4 CONCLUSION

The advantages of the unique structure of the ultrasonic separator based on two-dimensional normal mode in this paper are embodied in several aspects. Firstly, this unique structure near the inlet and outlets can effectively cut down the reflection of acoustic wave in Z-direction, and avoid three dimensional normal modes to be excited. According to the experimental images (in figure 5, 6), there are no bands of particles formed in Z-direction. Secondly, the "swallow tail"

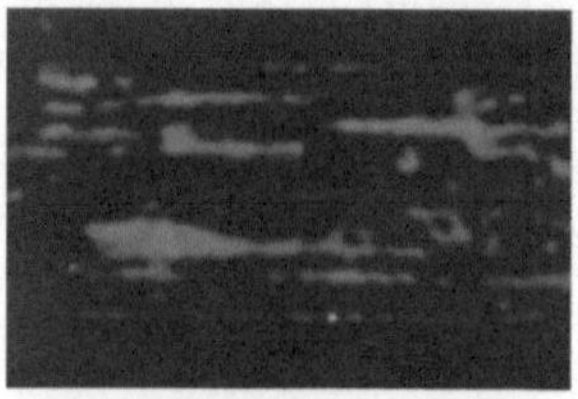

(a) about 8°

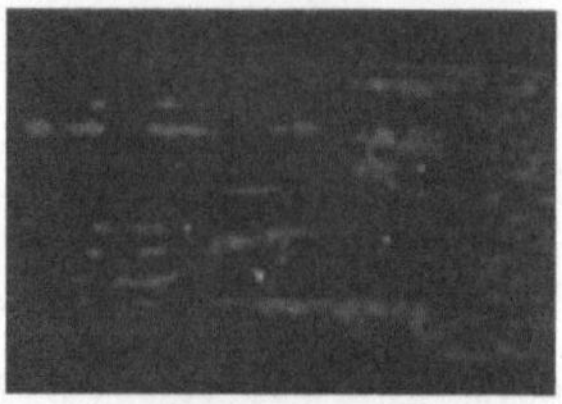

(b) about 15°

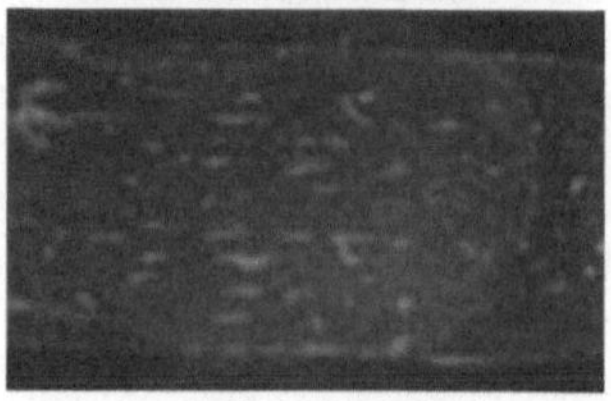

(c) about25°

Figure 6. Comparison experiments in different experimental temperatures.

structure of the outlets and the training wall of the inlet in the fluid chamber promises the laminar flow stable, simultaneously reduces the energy loss and improves the separation efficiency. Thirdly, the fluid chamber has a large width height ratio of 40:1. On the basis of the guarantee process, the throughput of a single separator is successfully increased

To improve the separation efficiency and the repeatability of experiments, experimental temperature, suspension liquid concentration have been analyzed and discussed. The results show that the optimal experimental temperature is around 8° for the separator, and mass suspension liquid concentration around 2% is most suitable for particles separation.

Some other factors may also influence the separation efficiency, such as the driving methods, structure dimensions of the separator, and physical characteristics of the suspended particles and so on. However, the related theory and experiments still need further study.

ACKNOWLEDGMENT

This paper is generated in the context of the project which is funded under the National Natural Science Foundation of China (contract to NO 50675031).

REFERENCES

Adams J D, Ebbesen C L, Rune, Barnkob A, Yang H J, Soh H T, Bruus H 2012 High-throughput, temperature-controlled microchannel acoustophoresis device made with rapid prototyping *J. Micromech. Microeng* 22 1-8

Adrian Neild, Stefano Oberti, Gerald Radziwill, et al. 2007 Simultaneous Positioning of Cells into Two-Dimensional Arrays Using Ultrasound *Biotechnology and Bioengineering* 97(5) 1335–1339

Chenhui H, Jiexiong D, Luyuan F 2012 Research on Driving the MEMS Ultrasonic Separator based on two-dimensional Normal Mode, *Proceedings of 2012 IEEE International Conference on Mechatronics and Automation* 1479–1484

Groschl M. 1998 Ultrasonic separation of suspended particles *Part I. Fundamentals Acustica* 84(3) 432–447

M.T. Grossner, A.E. penrod, J.M. belovich. 2007 Lesser known ultrasound-assisted heterogeneous sample-preparation procedures. TrAC Trends in Analytical Chemistry, 154–162

S. Gupta, D.L. Feke. 1997, Acoustically driven collection of suspended particles within porous media. Ultrasonics. 35(2): 131–139

Haake A, Neild A, Radziwill G and Dual J 2005 Positioning, displacement, and localization of cells using ultrasonic forces *Biotech. Bioeng* 92 8–14

Harris N.R., Hill M., Beeby S., etal 2003 A silicon microfluidic ultrasonic separator *Sensors and Actuators* B 950 425–434

Hawkes J J, Barber RW, Emerson D R and Coakley WT 2004 Continuous cell washing and mixing driven by an ultrasound standing wave within a microfluidic channel *Lab Chip* 4 446–52

HChenhui H; Jiexiong D; Yuxiang W 2009 Research on Normal Vibration Modes and Lateral Standing Wave in MEMS Ultrasonic Separator *The 2009 IEEE International Conference on Mechatronics & Automation* 318–322

Laurell T, Petersson F and Nilsson A 2007 Chip integrated strategies for acoustic separation and manipulation of cells and particles Chem. *Soc. Rev.* 36 492–506

Lilliehorn T, Nilsson M, Simu U, Johansson S, Almqvist M, Nilsson J and Laurell T 2005 Dynamic arraying of microbeads for bioassays in microfluidic channels *Sensors Actuators* B 106 851–858

Manneberg O, Hultström J, Hertz H M, Brismar H and Wiklund M 2006 Elementary manipulation functions for gentle and long-term handling of cells in microchannels by ultrasonic standing waves *Proc. 10th Ann. Eur. Conf. on Micro & Nanoscale Technologies for the Biosciences* (NanoTech Montreux, Switzerland)

Peter Glynne-Jones, christine E. M. démoré, Congwei 2012 Array-Controlled Ultrasonic Manipulation of Particles in Planar Acoustic Resonator *IEEE TransacTIons on UlTrasonIcs, FErroElEcTrIcs, and FrEqUEncy conTrol*, 59(6) 1258–1266

Townsend R.J., Hill M., Harris N. 2008 Performance of a quarter-wavelength particle concentrator. *Ultrasonics* 48 515–520

Wang songling 2007, Fluid mechanics *Beijing china power press*

Wiklund M, Günther C, Jäger M, Fuhr G and Hertz H M 2006 Ultrasonic standing wave manipulation technology integrated into a dielectrophoretic chip *Lab Chip* 6 1537–44

M.A.H.Wweiser, R.E.Apfel 1984, Interparticle forces on red cells in a standing wave field, Acustica 56 114–119

Xiao Hanli, Ding Jiexiong, Hua Chenhui, et al 2010, Research on the Fabrication of MEMS Ultrasonic Separators Based on SOI Materials, *Semiconductor Technology Vol. 35 No. 12* 1145–1173

Future Mechatronics and Automation – Yang (Ed.)
© 2015 Taylor & Francis Group, London, ISBN 978-1-138-02648-3

Adaptive fuzzy PID control for the quadrotor

Duo Qi, Jin-fu Feng, Yong-li Li & Jian Yang
Air Force Engineering University, Xi'an China

Fei-fei Xu
Unit 93320, Qiqihar China

Kun Ning
Unit 93383, Mudanjiang China

ABSTRACT: The quadrotor is an under-actuated system, which is nonlinear, multivariate, and highly coupled. It is easily interfered by external factors, such as airflow. During flying, achieving stable and efficient control is very difficult. In this paper, a simplified mathematical model was put forward, and an adaptive fuzzy PID controller was designed. The conventional PID control and adaptive fuzzy PID control are compared. It can be obtained from tests and simulations that besides the advantages of mature technologies and being easy to implement, the adaptive fuzzy PID control is superior in accuracy and robustness.

1 INTRODUCTION

The quadrotor is a new unmanned air vehicle (UAV) that possesses the following characters: vertical take-off and landing (VTOL), multi-rotor and so on. Compared with fixed wing vehicles, the quadrotor can take off in a narrow space, achieve to hover, fly upside and down. Because of the high maneuverability, it fits in a wide application range involving extreme low-altitude detection, interference, surveillance, and aerial photography in rigorous environment, and has a promising application whatever in military and civil field[1-2].

The quadrotor is an under-actuated system, which is nonlinear, multivariate, and highly coupled [3]. It is easily interfered by external factors, such as airflow. During flying, achieving stable and efficient control is very difficult. ZHOU Quan[4] studied attitude stabilization control of quadrotor with PID controller, and test results proved its feasibility. However, the experiment was carried out under the ideal condition without considering external influences. Bai Yongqiang[5] presented a robust compensation controller by integrating a robust attitude controller with an PD position controller, which made the quadrotor realize indoor hovering. WANG Lu[6] designed a sliding mode controller of the under-actuated subsystem by defining a generalized sliding surface, and it had stronger robustness than conventional PID method. Meanwhile, sliding mode control brought the whole system buffeting.

In this paper, a simplified mathematical model is put forward firstly, and then an adaptive fuzzy PID controller is designed. Compared with conventional PID control, in which the parameters are constant, it finishes parameter setting with fuzzy control method on line, and improves the control effect. Finally, this controller is verified by an indoor test platform and MATLAB/SIMULINK simulation.

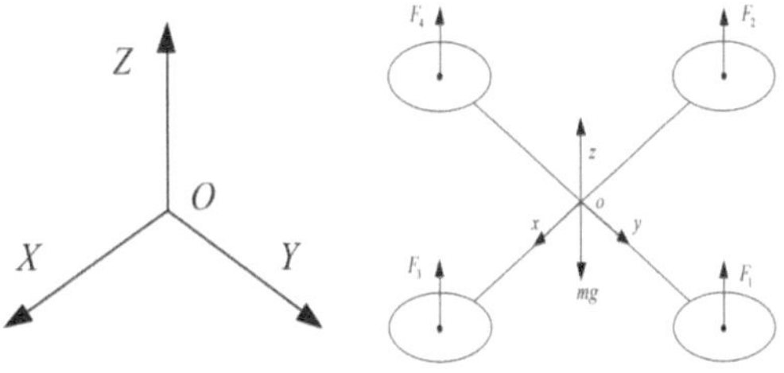

Figure 1. The ground coordinate system and the body coordinate system.

2 MATHEMATICAL MODEL OF THE QUADROTOR

A mathematical model is the precondition to design and apply control method. This section presents a simple model of the quadrotor.

As shown in Fig.1, the quadrotor possesses a simple and symmetrical structure with four rotors and can be controlled by the rotational speed of them. Define the ground coordinate system(X,Y,Z)and body coordinate system(x,y,z), and the movements in the body coordinate system can be conversed to axes of the ground coordinate via transformation matrices as follows.

φ, θ, ψ is roll angle, pitch angle and yaw angle respectively.

$$R(x,\phi)=\begin{bmatrix}1 & 0 & 0\\ 0 & \cos\phi & \sin\phi\\ 0 & -\sin\phi & \cos\phi\end{bmatrix} \tag{1}$$

$$R(y,\theta)=\begin{bmatrix}\cos\theta & 0 & -\sin\theta\\ 0 & 1 & 0\\ \sin\theta & 0 & \cos\theta\end{bmatrix} \tag{2}$$

$$R(z,\psi)=\begin{bmatrix}\cos\psi & \sin\psi & 0\\ -\sin\psi & \cos\psi & 0\\ 0 & 0 & 1\end{bmatrix} \tag{3}$$

Define $R(\phi,\theta,\psi)$ as the transformation matrix from body coordinate system to ground coordinate system, and

$$R(\phi,\theta,\psi)=R(z,\psi)\cdot R(y,\theta)\cdot R(x,\phi)=$$

$$\begin{bmatrix}\cos\psi\cos\theta & \sin\psi\cos\phi+\cos\psi\sin\theta\sin\phi & \sin\psi\sin\phi-\cos\psi\sin\theta\cos\phi\\ -\sin\psi\cos\theta & \cos\psi\cos\phi+\sin\psi\sin\theta\sin\phi & \cos\psi\sin\phi-\sin\psi\sin\theta\cos\phi\\ \sin\theta & -\cos\theta\sin\phi & \cos\theta\cos\phi\end{bmatrix} \tag{4}$$

Assume that 4 propeller shafts are parallel the vertical axis, and in the body coordinate system, force of the quadrotor is:

$$F=\begin{bmatrix}F_x\\ F_y\\ F_z\end{bmatrix}=\begin{bmatrix}0\\ 0\\ \sum_{i=1}^{4}F_i\end{bmatrix} \tag{5}$$

Taking $R(\phi,\theta,\psi)$ into consideration, force of the quadrotor in the ground coordinate system is:

$$\begin{bmatrix}F_x\\ F_y\\ F_z\end{bmatrix}=R(\phi,\theta,\psi)\cdot F=(\sum_{i=1}^{4}F_i)\begin{bmatrix}\sin\psi\sin\phi-\cos\psi\sin\theta\cos\phi\\ \cos\psi\sin\phi-\sin\psi\sin\theta\cos\phi\\ \cos\theta\cos\phi\end{bmatrix} \tag{6}$$

According to Newton's Second Law, motion equations of line displacement along x, y, z axes in the ground coordinate system are:

$$\begin{cases}\ddot{x}=(F_x-K_1\cdot\dot{x})/m\\ \ddot{y}=(F_y-K_2\cdot\dot{y})/m\\ \ddot{z}=(F_z-mg-K_3\cdot\dot{z})/m\end{cases} \tag{7}$$

where, K_i is aerodynamic drag coefficient, which can be ignored under the condition of low speed. Motion equations of angle displacement based on Euler equation are:

$$\begin{cases}\ddot{\phi}=l(F_4-F_2-K_4\dot{\phi})/I_x\\ \ddot{\theta}=l(F_3-F_1-K_5\dot{\theta})/I_y\\ \ddot{\psi}=(M_1-M_2+M_3-M_4-K_6\dot{\psi})/I_s\\ \quad=(F_1-F_2+F_3-F_4)/I_s'\end{cases} \tag{8}$$

Further, the final mathematical model of the quadrotor is

$$\begin{cases}\ddot{x}=(F_x-K_1\cdot\dot{x})/m\\ \ddot{y}=(F_y-K_2\cdot\dot{y})/m\\ \ddot{z}=(F_z-mg-K_3\cdot\dot{z})/m\\ \ddot{\phi}=l(F_4-F_2-K_4\dot{\phi})/I_x\\ \ddot{\theta}=l(F_3-F_1-K_5\dot{\theta})/I_y\\ \ddot{\psi}=(M_1-M_2+M_3-M_4-K_6\dot{\psi})/I_s\\ =(F_1-F_2+F_3-F_4)/I_s'\end{cases} \tag{9}$$

where, I is moment of inertia for the certain shaft, and I_s' is scale factor.

Define the following inputs, which are the controlled variables as:

$$\begin{cases}u(1)=F_1+F_2+F_3+F_4\\ u(2)=F_4-F_2\\ u(3)=F_3-F_1\\ u(4)=F_1-F_2+F_3-F_4\end{cases} \tag{10}$$

In conclusion, motion equations of the system are given by:

$$\begin{cases}\ddot{x}=(u(1)(\sin\psi\sin\phi-\cos\psi\sin\theta\cos\phi)-K_1\cdot\dot{x})/m\\ \ddot{y}=(u(1)(\cos\psi\sin\phi-\sin\psi\sin\theta\cos\phi)-K_2\cdot\dot{y})/m\\ \ddot{z}=(u(1)\cos\theta\cos\phi-K_3\cdot\dot{z})/m-g\\ \ddot{\phi}=(u(2)-K_4\cdot\dot{\phi})l/I_x\\ \ddot{\theta}=(u(3)-K_5\cdot\dot{\theta})l/I_y\\ \ddot{\psi}=(u(4)-K_6\cdot\dot{\psi})/I_s'\end{cases} \tag{11}$$

Because of the low speed, $K_i=0$, $i=1\sim6$, and the final motion model can be simplified as:

$$\begin{cases}\ddot{x}=(u(1)(\sin\psi\sin\phi-\cos\psi\sin\theta\cos\phi))/m\\ \ddot{y}=(u(1)(\cos\psi\sin\phi-\sin\psi\sin\theta\cos\phi))/m\\ \ddot{z}=(u(1)\cos\theta\cos\phi)/m-g\\ \ddot{\phi}=u(2)l/I_x\\ \ddot{\theta}=u(3)l/I_y\\ \ddot{\psi}=u(4)/I_s'\end{cases} \tag{12}$$

3 DESIGN OF ADAPTIVE FUZZY PID CONTROLLER

Parameter setting is the key of PID control. Adaptive fuzzy PID, belonging to intelligent areas, is applying basic theories and methods of fuzzy math to denote conditional rules and operations with fuzzy sets. It stores these rules and knowledge in rule-base, and then automatically adjusts the parameters according to relative fuzzy inference. Basic architecture of adaptive fuzzy PID controller is as Fig. 2. Where, rin is input, yout is output, e denotes error, and ec denotes the rate of error change. Design of adaptive fuzzy PID controller is made up of two components: fuzzy logic controller and PID controller.

A two-dimension fuzzy controller is applied in this paper, where fuzzy controller changes parameters k_p, k_i and k_d according to the variance of e and ec measured on line.

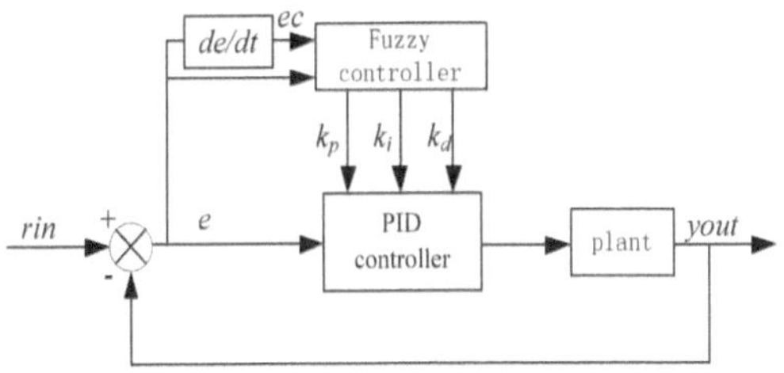

Figure 2. Basic architecture of adaptive fuzzy PID controller.

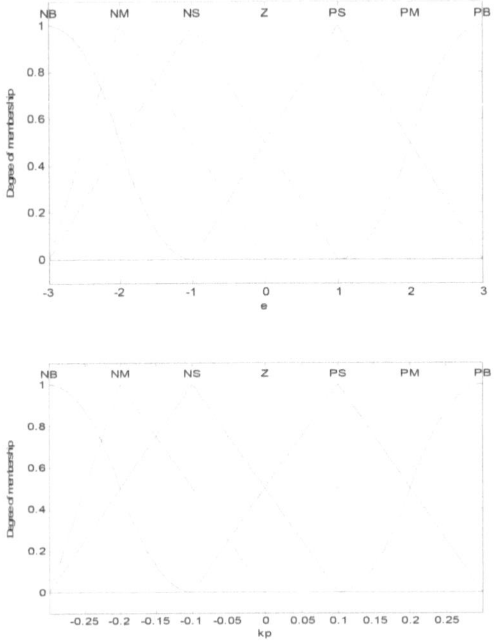

Figure 3. Membership function of e and kp.

The universe of discourse of *e* and *ec* range from −3 to 3, classified into 7 grades: NB(negative big), NM(negative medium), NS(negative small), O(zero), PS(positive small), PM(positive medium) and PB(positive big). k_p, k_i and k_d are in the range of −0.3 to 0.3, −0.06 to 0.06, and −3 to 3 respectively, and they are classified the same as *e* and *ec*. Successful implementation of a fuzzy logic controller depends on the heuristic rule-base from which control actions are derived, so the main of the fuzzy logic controller is the rule-base. Rule-base in this paper is basically the same as what widely used.

Membership functions have a great impact on the outputs of fuzzy control: the steeper the functions are, the higher sensitivity system will be. On the contrary, the system will become steadier. All the types of the membership functions of inputs and outputs are triangle membership function. Membership functions of *e* and k_p are shown in Fig. 3.

As known from Fig.2 above, PID controller is made up of proportional, integral and derivative control. Use fuzzy modules of MATLAB/SIMULINK to build PID controller, and its architecture is illustrated in Fig.4.

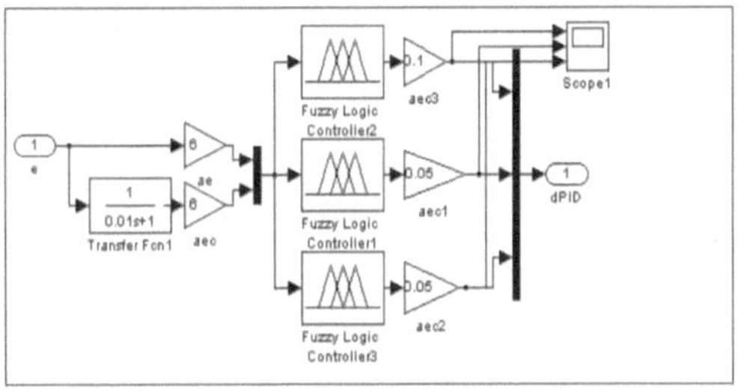

Figure 4. Architecture of PID controller.

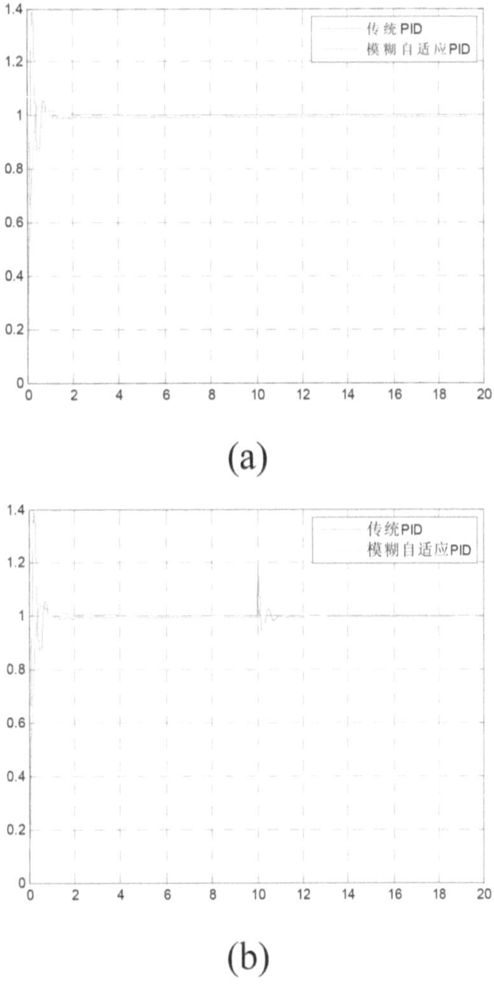

Figure 5. Architecture of PID controller.

4 TEST AND ANALYSIS OF CONTROL

According to the built controller model, test the control effect with data of the quadrotor. In this section, an adaptive fuzzy PID controller is compared

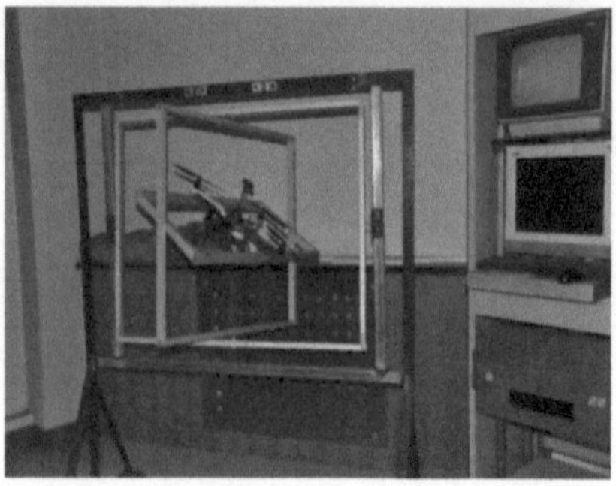

Figure 6. 6DOF test platform.

with conventional PID controller, based on MATLAB/SIMULINK and Fuzzy Logic toolboxes. Input is unit step signal, and simulation result is shown in Fig.5(a), response after adding interferences at 10s is shown in Fig.5(b).

For inspection and monitoring data in time, meanwhile, to ensure the operator's security and reduce the risk brought by outdoor uncertainties, such as wind, buildings, a six degrees of freedom (6DOF) test platform is designed, which is shown in Fig.6. The test results basically agree well with simulation results.

Test and simulation results show that compared with conventional PID control, adaptive fuzzy PID control has higher adjustment accuracy. And in a relatively rigorous environment (adding interferences), adaptive fuzzy PID controller can change the parameters on line to reduce oscillation to make the system steady. However, there is so little difference between adjustment time, and this is because the quadrotor is an underactuated system, which is nonlinear, multivariate, and highly coupled. Reasoning process is complicated and running time becomes longer.

5 CONCLUSION

Aiming at the shortages of conventional PID control, an adaptive fuzzy PID controller is designed in this paper. It can be obtained from test and simulation that adaptive fuzzy PID controller has not only advantages of mature technology and being easy to implement, but also is superior in accuracy and robustness.

REFERENCES

Li Guodong, Song Zili, Wu Hua & Liu Changan. 2013. Design of stability augmentation hybrid controller for a quadrotor unmanned air vehicle *Journal of Harbin Institute of Technology* 45(1):86–90.

Nie Bowen, Ma Hongxu, Wang Jian & Wang Jianwei. 2010. Study on actualities and critical technologies of micro/mini quadrotor *Electronics Optics & Control* 17(9):46–52.

Wang Daobo. 2008. Nonlinear flight control strategy for an under actuated quad-rotor aerial robot *Proceedings of 2008 IE International Conference on Networking Sensing and Control* 938–942

Zhou Quan, Huang Xianghua & Zhu Lihua. 2009. Experiment study on attitude stabilization control of quadrotor micro-aircraft. *Transducer and Microsystem Technologies* 28(4):72–79.

Bai Yongqiang, Liu Hao, Shi Zongying & Zhong Yisheng. 2012. Robust flight Control of quadrotor unmanned air vehicles. *ROBOT* 34(35):519–524.

Wang Lu, Li Guangchun, Wang Zhaolong & Jiao Bin. 2012. Sliding mode control of an underactuated quadrotor UAV *Journal of Harbin Engineering University* 33(9): 1248–1253.

Future Mechatronics and Automation – Yang (Ed.)
 ISBN 978-1-138-02648-3

Design and implementation of cloud computing platform for mechatronics manufacturing

Tiantian Liu
Cloud Computing Center, Chinese Academy of Sciences, Dongguan, China

Qiang Yue
Guangdong Electronics Industry Institute, Dongguan, China

Tongkai Ji
Cloud Computing Center, Chinese Academy of Sciences, Dongguan, China

Xiaoqing Wu
Guangdong Electronics Industry Institute, Dongguan, China

ABSTRACT: Cloud computing has good prospects in the field of mechatronics manufacturing industry, because it dramatically reduces the construction cost, brings data integration capability, and has efficient and flexible working models. In this paper, a case study of Dongguan mechatronics manufacturing cloud computing platform is introduced. By descripting the design and implementation of this platform, we hope this efficient, low-carbon, and low-cost construction model will be promoted in more mechatronics manufacturing fields.

1 INTRODUCTION

Manufacturers are seeking new technologies and methods in order to shorten product development cycle and improve innovation capacity, especially for design stage of mechatronics products. Currently, making use of advanced CAD and CAE tools to build a unified product definition model throughout various stages of a product life cycle is a solution for many manufacturing enterprises. However, the current advanced CAD/CAE product price is quite expensive. Furthermore, most CAD/CAE products are for different applications, and have different data structures, resulting in Isolated Islands of automation, which is not conducive to the exchange of information and sharing of design data between designers and developers. Therefore, this paper considers the common demand of product design and performance optimization for mechatronics manufacturers and builds a manufacturing public computing platform combining cloud computing technology. This mechatronics manufacturing cloud computing platform provides various services to manufacturing enterprises enabling them to make full use of the platform resources, such as computing capacity, storage, networking, software and design data, with very low cost.

The paper is organized as follows. We first propose the architecture of the mechatronics manufacturing cloud computing platform in Section 2. Section 3, 4 and 5 describe IaaS, PaaS, and SaaS layers, respectively. In Section 6, the test result is presented. Finally, we conclude our paper in section 7.

2 ARCHITECTURE

First, the platform builds infrastructure resource service platform. On this basis, related mechatronics manufacturing middleware and applications are integrated, supporting management of massive data, exchanging of mechatronics manufacturing resources and collaborative manufacturing. The platform hierarchy architecture is shown in Figure 1. It mainly includes several layers: the host environment layer, the virtual resource layer (IaaS), the development and usage layers (PaaS), and the CAD / CAE applications and support software layer (SaaS).

(1) The host environment layer: is the physical infrastructure layer of the whole platform, including servers, storage servers, and network devices. These hardware scatters in different places, such as in the existing manufacturing enterprises' data centers or the third party data centers.
(2) The virtual resource layer (IaaS): achieves the virtualization and management of infrastructures and offers infrastructure as a service (IaaS) *(Hwang 2011)*. This layer provides hypervisor and core functions, such as user management, data management, systems management, authentication and

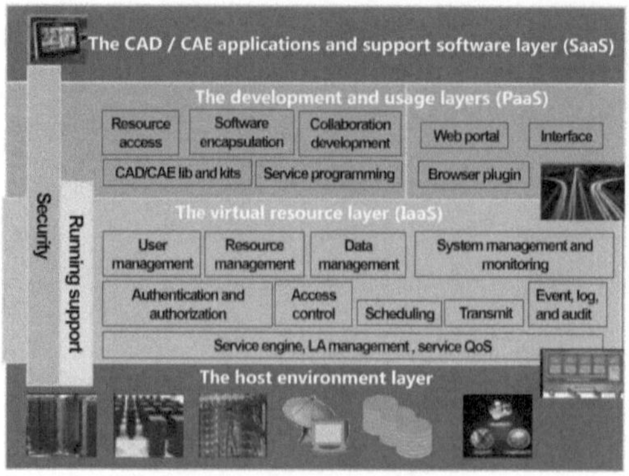

Figure 1. Architecture of the mechatronics manufacturing cloud computing platform.

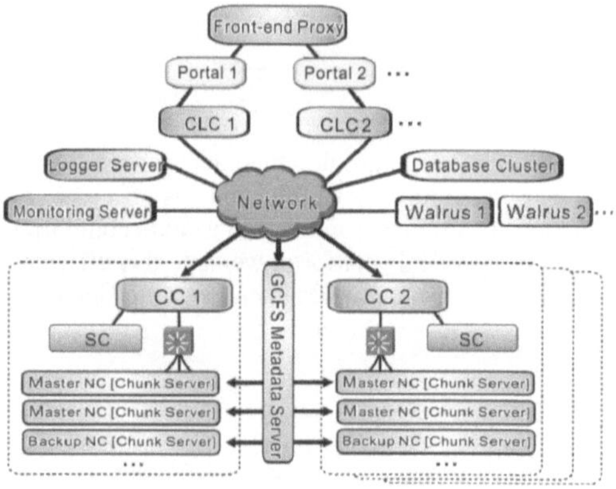

Figure 2. Modules of G-Cloud OS.

authorization, scheduling, transaction processing and other common virtual resource managements. The distributed file system is also utilized in this level to efficient store the users' data.

(3) The development and usage layers (PaaS): through the core system calls, system libraries, running environment supports and development kits, this layer provides developers some implementation-independent user interface, supporting for the development and operation of other cloud manufacturing components.

(3) The CAD / CAE applications and support software layer (SaaS): deploys a variety of manufacturing design applications, such as the portal system, browser plugs and graphical interface, to support user applications.

3 IAAS: CLOUD OS AND DISTRIBUTED FS

In this layer, we use the G-Cloud cloud operating system (G-Cloud OS) to support the virtualization and unified management of large-scale IT resources, including computing resources, storage resources, network resource and data resources, and system software, such as operating systems, databases and application servers, providing the uniform basic services for the upper layers and uses, such as flexible computing services and distributed cloud storage capabilities. G-Cloud OS is a management platform which consists of virtual machine management, system management, image management, monitoring, user management, certificate management. It also provides CLI system management tools, including tools for querying and making images. For system security, it achieves secure Web communications, secure communications among components and virtual machines, and user authentication, etc.

G-Cloud OS is built with a SOA architecture based on REST/SOAP web service technology. The system consists of a number of modules as shown in Figure 2. The main work process of the various modules is: first, a user takes an operation on his operation proxy, for example, he creates a virtual machine. The portals distribute the resource request using a load balancing strategy to the CLCs, who make resources scheduling and send the resource instruction to CCs. Finally virtual machine is created in one of NCs, and feedback is sent back to the user. All the modules can be deployed in a redundant configuration to avoid single point of failure and achieve load balance.

GCFS is a high-performance, low-cost, highly reliable distributed file system, based on low-priced generic server and hard drivers. GCFS uses a distributed storage architecture like GFS (Google file system, *Ghemawat 2003*) that supports a storage cluster of 4-500 servers. It provides a unified user access interface, supporting standard POSIX interface standards. GCFS can avoid the shortcomings such as low utilization rate, poor hardware compatibility, complex management and extending difficulties when using SAN, NAS and other traditional shared storage. GCFS is effectively integrated in G-Cloud OS as shown in Figure 2. The components of GCFS are shown in Figure 3.

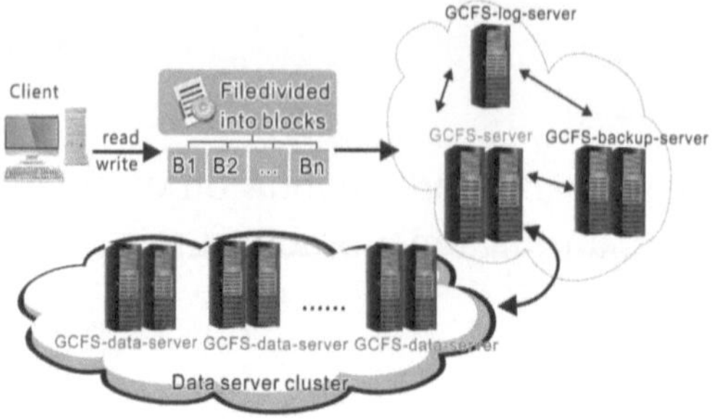

Figure 3. Components of GCFS.

Master metadata server (GCFS-server) is responsible for managing the whole system. Backup metadata server (GCFS-backup-server) is an optional component which has a real-time synchronization with the GCFS-server so that when GCFS-server fails, it can automatically take over GCFS-server's work. Log Server (GCFS-log-server) is also an optional component which is responsible for recording the change log of GCFS-server and GCFS-backup-server and taking over their work when they fail. Data storage server (GCFS-data-server) is where user data is stored. When storing a file, GCFS first divides the file into several blocks, then stores these blocks among GCFS-data-servers. The blocks are stored with several copies to avoid hardware failures, while the number of copies can be specified manually. A great number of data servers create a virtualized disk space for users. When clients successfully mount GCFS system, they can use this virtualize disk just as using NFS.

4 PAAS: SOFTWARE ENCAPSULATION AND DATA EXCHANGE AND CONVERSION

Currently, there exist many different types of CAD/CAE software with large differences causing the difficulty of integrated usage. PaaS layer designs a unified interface model to encapsulate different types of CAD/CAE software by hiding the calling heterogeneity between different CAD/CAE software and achieving consistency of user access. We uses the Portlet-model-based encapsulation strategy (*Linwood 2004*) and related software systems in this layer. First, the various stages of product design task are decomposed and allocated as different nodes in the design chain, and the basic framework model is established. For specific encapsulation requirements, a multi-level encapsulation template set is built, and by instantiating the template, the web encapsulation of traditional CAD/CAE software is implemented. Then, through the resources-tasks perceived modules in the design nodes and the Portlet in CAD/CAE software, resources and tasks can be identify automatically. A web service library is hosted in front of software to implement the real-time output of the above different types of design data. XML and product model data exchange standards are combined to achieve the information transmission and interaction between Portlets in nodes. Queuing services of CAD/CAE software is solved by using intelligent algorithms.

The system framework is built as shown in Figure 4. Software to be encapsulated is integrated in this framework by hosting at the Portlet server, which is physically implemented as a Java-enabled Web server. The web portal Portlet is consisted of two parts: the web service software on the web server developed by "Applet-servlet" programming model, and the CAD/CAE software resource web page developed by "JavaScript & HTML". Here, web services include remote call services, design task queue management services, visualization input/output services and design collaboration services.

Different CAD/CAE software has different data structure and characteristics model, which is not conducive to exchanges and cooperation between enterprises or departments. Currently, there are two ways for

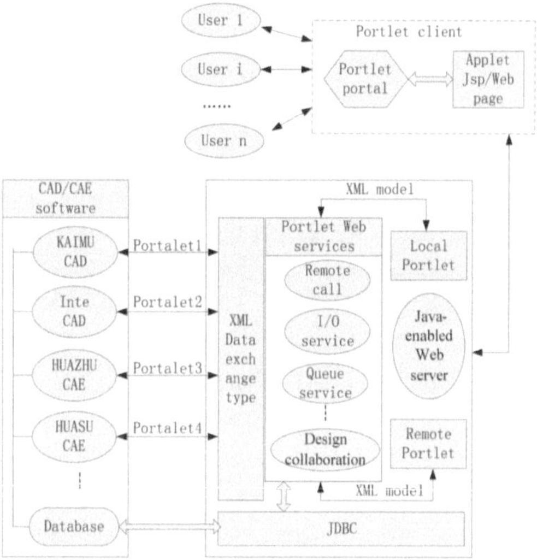

Figure 4. Framework of Portlet-model-based encapsulation.

data exchange and conversion between heterogeneous CAD/CAE: geometry-based data conversion and characteristic parametric conversion (*Szykma 2001*). The geometry-based data conversion method introduces a large amount of data, limiting its utilization. Therefore, characteristic parametric conversion method is commonly used between heterogeneous CAD/CAE. We develop the data conversion software based on characteristic parametric conversion method to achieve a two-way data conversion between different CAD/CAE software. It fully converts the design model's information, such as geometry, feature tree, dimensions and constraints, from the source system which builds this model to the target systems. And in the target systems, the characteristic-based editing can be done on the converted model. Therefore, the communication and collaboration difficulties are solved.

5 SAAS: COLLABORATION SERVICES AND DATA MANAGEMENT

The main industry application in the cloud platform is for integration and coordination. First, we have conducted research about the manufacturing characteristics of local manufacturing enterprises. The mold business process model is built. Then, the service encapsulation system and coordination system are established. The overall framework of coordination system is shown in Figure 5. It consists of two parts: one is a 3D remote collaboration system, and the other is a product data management system.

The manufacturing platform addresses a wide range of user needs. First is the 3D virtual image data processing. Using neutral display model, OBJ format, which uses triangular facets to approximately represent parts, and using Java and OpenGL programming technology, we implement a 3D scene in

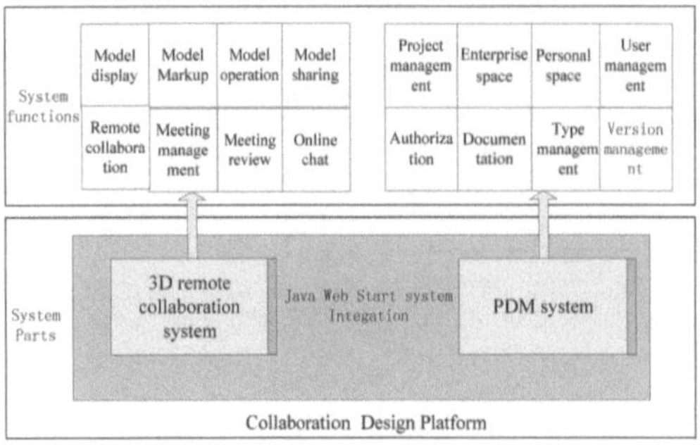

Figure 5. Collaboration design platform.

the application. Several 3D image operating techniques are achieved, including methods of image remarking, image rotation, zooming and movement. Second is the virtualized remote desktop. Using Java and JFace programming techniques, message communication mechanism of virtualized remote desktop is implemented, including message formats, message representation, message packaging and parsing, lightweight network transmission. Finally, 3D remote collaboration systems and enterprise registration space are implemented by combining 3D image display technology and virtualized remote desktop technology. Mechatronics mold enterprise registered space is designed and built, providing private workspace for each designer to manage, share, review and save design documents. Designers security assurance mechanism and version control management are established for the shared documents, so that designers can easily use the design resources in cloud and assess models using remote collaboration system.

Product data management system (PDM) is a software framework (or data platform) based on a distributed network, master-slave structure, graphical user interface and database management technology (*Kropsu-Vehkapera 2009*). PDM keeps overall management for users, tools, equipment, product data as well as data generation process in the concurrent engineering. The main modules are shown in Figure 6.

6 EVALUATION

The platform supports more than 100 servers and has a computing capacity of more than 10 trillion times. The storage capacity is greater than 1PB. The platform supports data conversion between more than three CAD/CAE software, such as AutoCAD, KAIMU CAD, Inte CAD, HUAZHU CAE, HUASU CAE. The GCFS resource monitoring interface is shown below.

The platform can support more than 500 users to work online at the same time. The data storage reliability and availability are evaluated by pressure test. The pressure testing tool is jmeter (version 2.9), and the script generation tool is badboy (version 2.0). Multiple threads are set up to simulate the behaviors of user

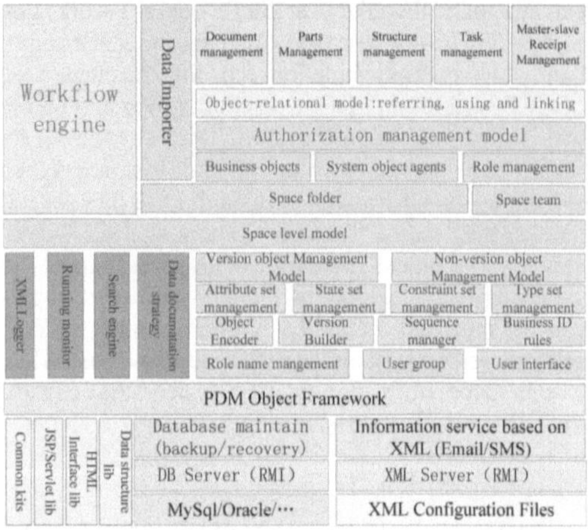

Figure 6. PDM System.

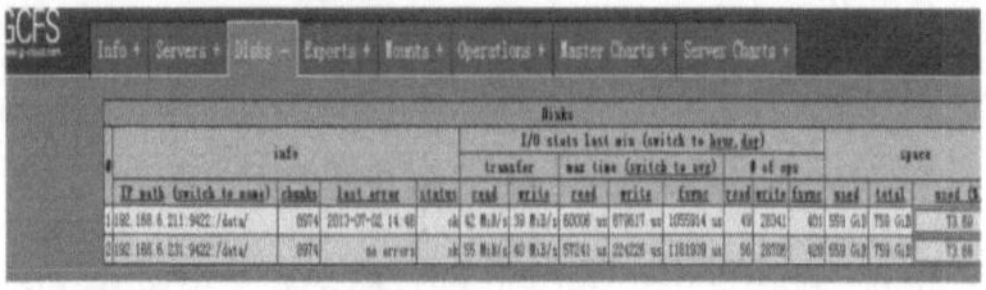

login, reviewing and other 3D operating. The concurrency is set to be 500. The following table shows the test results. Results show that the response speed meets performance objectives. The success rate is greater than 99.9%.

Label	# Samples	Average	Median	90% Line	Min	Max	Error %	Throughput	KB/s
http://192.168.6.205	500	973	742	1627	6	2745	0.00%	89.2/s	12.1
Total	500	973	742	1627	6	2745	0.00%	89.2/s	12.1

7 CONCLUSIONS

In this paper, Dongguan mechatronics manufacturing cloud computing platform is introduced. It provides various services, such as computing capacity, storage, networking, software, design data, and collaboration with very low cost. We hope this efficient construction model will be promoted in more places.

ACKNOWLEDGEMENTS

The work was supported by Guangdong province high-tech zone development guide special Project-Key Technology Research (2011B010700043) and the introduction of innovative R&D team program of Guangdong Province (201001D0104726115).

REFERENCES

H. Kropsu-Vehkapera, H. Haapasalo, J. Harkonen, and R. Silvola. 2009. Product data management practices in high tech companies. Industrial Management & Data Systems, 109(6): 758–774.

Jeff Linwood, David Minter. 2004. Building Portals with the Java Portlet API, Berkeley, CA, USA: APress.
Kai Hwang, Geoffrey C. Fox, Jack J. Dongarra, Morgan Kaufmann. 2011. Distributed and Cloud Computing: From Parallel Processing to the Internet of Things. Burlington, MA, USA: Morgan Kaufmann Publisher.
Sanjay Ghemawat, Howard Gobioff, and Shun-Tak Leung. 2003. The Google File System. SOSP'03:29–43.
S. Szykma, S. J. Fenves, W. Keirouz, et al. 2001. A foundation for interoperability in next-generation product development systems. Computer-Aided Design, 2001, 33(7): 545–559.

Future Mechatronics and Automation – Yang (Ed.)
 ISBN 978-1-138-02648-3

A fuzzy comprehensive assessment model and application of traffic grade on an emergency in a city

Fu Wang, Jian Gao & Zhi-li Xiong
Transportation Research Center, Wuhan Institute of Technology, Wuhan, China

Yi Jiang
Department of Civil Engineering, Wuhan Polytechnic University, Wuhan, China

ABSTRACT: It is necessary to found a grade for traffic in an emergency for fast, efficient, and ordered traffic evacuation. Thus, different traffic grades should be correspond to different management departments and management measures. Because many factors of a disaster are numbered according to difficulty, the chief factors were selected such as influencing scope and persons, road network, traffic management as an assessment index, it was found that the fuzzy comprehensive assessment model of traffic gradein an emergency, and the aggregations of assessment index and weight were set by the method of specialist, and the maximum degree of membership was selected as the assessment standard. Finally, the fire of Han-zheng Street in Wuhan was regarded as an example with which to apply and certificate the model.

Keywords: Traffic engineering; traffic grade; fuzzy comprehensive assessment; emergency

1 INTRODUCTION

It is necessary to evaluate a grade for traffic in a traffic accident emergency in a city. By evaluating the grade, the different level of traffic emergency can match different levels of emergency management and emergency management measures; the evacuation and rescue can be implemented more efficiently, thereby reducing casualties and property losses. Early accidents tend to be graded after their rating and, generally, we gather statistics for the casualties, property damage, and economic losses after the accidents, and according to the size of the loss, match it to its classification. We do so only after the event summary, which does not provide guidance and reference for managers to deal with the accident disaster. If the emergency departments quickly assess the traffic emergency level of events and start the corresponding plan after the disastrous accident, it will no doubt help the disaster relief. Accident disasters, however,are affected by many factors and most of the influencing factors and the influence degree of accidents is difficult to quantify accurately; the fuzzy comprehensive evaluation method was very suitablefor solving such problems[1]. The disaster accident handling was selected, (which is the main influence factor, such as number, scope, and the road network and traffic management). as the evaluation index, the fuzzy comprehensive evaluation model was used to judge the character of the traffic accident disaster emergency grade, and the traffic organization plan was put forward according to the different level of emergency, thus providing a traffic accident disaster rescue and evacuation plan.

2 THE DETERMINATION OF THE TRAFFIC EMERGENCY LEVEL EVALUATION INDEX

Determining the judging indexes of the traffic emergency level is the basis of the fuzzy comprehensive evaluation. The selection of evaluation indicators will directly affect the accuracy of the comprehensive evaluation[2]. To reach accurate traffic emergency grades of evaluation indicators, on the basis of references and experts, and in combination with the traffic characteristics of urban disaster accidents, the disastrous accident's sphere of influence, the numbers of impact, the density of the road network, and the traffic management level are the four indicators chosen for the fuzzy evaluation factors.

2.1 *The factor of the disastrous accident's sphere of influence*

The accident's scope and the harm degree are particularly important. The scope of the impact of various disastrous accidents have relatively large gaps. In general, the scope of the disaster for the affected area is closely related to its dangers, and its sphere of influence on the implementation of emergency traffic measures has a profound impact. By consulting experts and assessing the size of the accident, the scope

Table 1. The hierarchical division of the range of factors for disastrous accidents.

symbol	The influence of emergencies
a_1	Affect the whole city
a_2	Affect an administrative area or CBD of the city
a_3	Affect a large community
a_4	Affect a residential area

Table 2. Hierarchical division of factors of the number of disastrous accidents.

symbols	The number of people impacted by emergencies
b_1	More than 100,000 people
b_2	30,000 to 100,000 people
b_3	1,000 to 30,000 people
b_4	1,000 people or less

of the disaster will be divided into four levels as shown in Table 1[3].

2.2 *The factors of the number of disastrous accidents*

The occurrence of accidents will inevitably affect people's activities. How much influence the number of accidents has, plays an important role in judging the level of disasters; the greater the number of people affected, the greater the pressure on the evacuation of the traffic organization. Personal safety is more relevant when considering the most important measures in disasters. Therefore, the affected number of accidents is regarded as an important factor. By consulting experts and according to the level of the factors influencing the scope demarcation in proportion to the number of how many people are affected, the accident disaster influence can be divided into four levels as shown in Table 2.

2.3 *The factor of the accident area road network density*

A road network is important for social and economic development. It is also the basis of a traffic emergency evacuation. The road network density is a basic index for reflecting the nature of the road network. The size of the road network density plays a decisive role in the speed and efficiency of a traffic emergency evacuation.

With reference to the evaluation of urban road network density levels in the urban road traffic management evaluation system(2008)[4], to meet the needs of the research content, the accident disaster area of the city road network density is divided into four levels by consulting experts as shown in Table 3.

Table 3. Hierarchical division of factors of regional road network density.

symbol	Road network density
c_1	6 km/km^2 or more
c_2	5~6 km/km^2
c_3	4~5 km/km^2
c_4	4 km/km^2 or less

Table 4. Hierarchical division for factors of area traffic control range%.

symbol	d_1	d_2	d_3	d_4
megalopolis	[75, 100]	[65, 75)	[55, 65)	[0, 55)
Class A city	[70, 100]	[60, 70)	[50, 60)	[0, 50)
Class B city	[65, 100]	[55, 65)	[45, 55)	[0, 45)
Class C,D city	[60, 100]	[50, 60)	[40, 50)	[0, 40)

2.4 *The factor of the accident area traffic control level*

The urban road traffic management evaluation system (2008) chooses the index of city road control level evaluation as the proper path control rate, and it also chooses the index as the accident factor of regional traffic control. Traffic management ability is directly related to the case of an emergency traffic channel capacity, and it also has an important impact on traffic accident disaster emergency rescue. With reference to the evaluation index system of the urban road traffic management evaluation system(2008) for proper control rate level, by consulting experts, the city disaster accident area traffic control is divided into four levels as shown in Table 4.

3 THE JUDGEMENT OF TRAFFIC EMERGENCY GRADES

Evaluation sets are a collection of the evaluation results judged by judges, it is represent by letters V. The national emergency pre-plans for sudden public events[5]divide various emergencies into four grades: extremely large accidents, major accidents, accidents, and general accidents. Correspondingly, the city traffic accident disaster emergency level is divided into four grades: level I (particular), level II (major), level III (large), and level IV (general).

Therefore, the traffic accident disaster emergency evaluation set V= {level I, level II, level III, and level IV}.

4 THE DETERMINATION OF A SINGLE FACTOR FUZZY EVALUATION MATRIX

Through consultation with experts, the experts showed their influence on an influence factor rating for the accident[6] by voting. For example, for the range of

Table 5. Evaluation set of scope affected by disasters.

	Evaluation set			
	Level I	Level II	Level III	Level IV
a_1	1(a_{11})	0(a_{12})	0(a_{13})	0(a_{14})
a_2	0.3(a_{21})	0.6(a_{22})	0.1(a_{23})	0(a_{24})
a_3	0 (a_{31})	0.25(a_{32})	0.65(a_{33})	0.1(a_{34})
a_4	0 (a_{41})	0.1(a_{42})	0.3(a_{43})	0.6(a_{44})

Table 6. Evaluation set of members affected by disasters.

	Evaluation set			
sequence	Level I	Level II	Level III	Level IV
b_1	1(b_{11})	0(b_{12})	0(b_{13})	0(b_{14})
b_2	0.3(b_{21})	0.6(b_{22})	0.1(b_{23})	0(b_{24})
b_3	0.1(b_{31})	0.4(b_{32})	0.5(b_{33})	0(b_{34})
b_4	0(b_{41})	0.2(b_{42})	0.3(b_{43})	0.5(b_{44})

Table 7. Evaluation set of factors of regional road network density.

	Evaluation set			
sequence	Level I	Level II	Level III	Level IV
C_1	0.05 (C_{11})	0.1 (C_{12})	0.2 C_{13}	0.65 C_{14}
C_2	0.15 (C_{21})	0.2 (C_{22})	0.35 C_{23}	0.3 C_{24}
C_3	0.25 (C_{31})	0.4 (C_{32})	0.3 C_{33}	0.05 C_{34}
C_4	0.6 (C_{41})	0.3 (C_{42})	0.1 C_{43}	0 C_{44}

Table 8. Evaluation set of factors influenced by accident disaster area traffic control range.

	Evaluation set			
sequence	Level I	Level II	Level III	Level IV
d_1	0.02 (d_{11})	0.13 (d_{12})	0.3 (d_{13})	0.55(d_{14})
d_2	0.1(d_{21})	0.2 (d_{22})	0.45(d_{23})	0.25(d_{24})
d_3	0.3(d_{31})	0.5(d_{32})	0.15(d_{33})	0.05(d_{34})
d_4	0.6(d_{41})	0.3(d_{42})	0.1(d_{43})	0(d_{44})

factors affecting the second level a_2 (an administrative area or CBD of a city), 30% of the experts considered it as level I, 60% of the experts believed it to be level ¢ò, only 10% of the experts believed it to be level II, and no experts believed that it was level ¢ô. Then, a factor of accidents affecting the scope of the review set was obtained as shown in Table 5. Similarly, other evaluations of each factor can be drawn from Tables 6, 7, and 8.

Among them, a_{ij} (I = 1, 2, 3, 4; j = 1, 2, 3, 4) said the level of membership for level i to the degree of evaluation j.

For the four evaluation indexes a, b, c, and d, according to the specific situation of a disaster in combination with the evaluation tables shown above, the single factor fuzzy evaluation matrix is obtained, that is, the influence scope of a disaster is a_a, the influence number of a disaster is b_b, its regional road network density is c_c, its scope of area traffic control is d_d. From this, we can get the single factor fuzzy evaluation matrix of the disaster:

$$R=\begin{vmatrix} a_{a1} & a_{a2} & a_{a3} & a_{a4} \\ b_{b1} & b_{b2} & b_{b3} & b_{b4} \\ c_{c1} & c_{c2} & c_{c3} & c_{c4} \\ d_{d1} & d_{d2} & d_{d3} & d_{d4} \end{vmatrix}$$

5 THE WEIGHT SETS OF TRAFFIC LEVEL OF EMERGENCY

In the factors set, various factors to the importance of the evaluation object value are different. So the factors a, b, c, and d should be given the corresponding right to a_i(i = 1,2,3,4), the right weight collection called factor right weights set A[7]. Normally, the right weight of each a_i(i = 1,2,3,4) should satisfy a sex and not negative, namely:

$$\sum_{i=1}^{4} a_i = 1,\ a_i \geq 0 \quad (i=1, 2, 3, 4).$$

In the fuzzy comprehensive evaluation, the weight is an important numerical level that reflects some sense; it is also a balanced role of the different degree of differences between evaluation factors[8]. Only by weighted comprehensive, can the intrinsic relationship between different evaluation factors be revealed, and make evaluation results conform closer to the actual situation. For the influence factors that cannot be quantitatively described, the experienced experts and engineers will determine the weights according to the actual situation. In the evaluation of traffic accident disaster emergency grades, the greater the influence factor's influence on the traffic emergency, the bigger its weight will be.

Through the expert questionnaire and statistical analysis, we will get the city traffic accident disaster emergency grades of weighting sets: A = (0.25 0.45 0.2 0.1), Experts say the evaluation of the influence numbers of a disaster to the grades is one of the most important.

6 THE FUZZY COMPREHENSIVE EVALUATION OF THE TRAFFIC EMERGENCY LEVELS

Generally speaking, when carries on the fuzzy comprehensive evaluation, if the importance of the evaluation indicators, which is also the weight concentrated weight , are the same, at this time as long as we put the judgment matrix column elements together, then we can get the judged "score" respectively in each element[9]. If the weights of the evaluation indexes are different, we need to implement the fuzzy matrix multiplication operation. The judging indexes of urban traffic accident disaster emergency levels are different to the importance of the level of evaluation, so this needs to be operated[10]:

$B = A \times R$ (1)

where $b_j = \bigvee_{i=1}^{4} (a_i \wedge r_{ij}), j=1, 2, 3, 4.$

After matrix B comes through the normalized processing, we get the v_j of judge set V which corresponds to the biggest b_{jmax} as the final results of the evaluation of the traffic emergency level.

7 CASE ANALYSIS

Hanzheng Street is located in the bustling area of Hankou in Wuhan. Nowadays, Wuhan has East Hanzheng Street and West Hanzheng Street. Individual manager has reached 13,200, the daily average cargo throughput is more than 400 tons, and the daily average number of visitors to the markets is around 160,000. In recent years, due to the commercial/ residential mix,old electrical wiring, and the connection and safety problems, firesfrequently happen. In December 2005, a fire killed four people; On January 3, February 2 and February 5, 2009, Han-zheng Street took place three fire. fire on February 5 killed one person, more than 10 people were rushed to hospital, hundreds of people were relocated, 580 million yuan of direct economic were lost.

In the Hanzheng Street fire, for example, we used the fuzzy comprehensive evaluation method to determine traffic emergency levels.

After investigation, it was found that the Hanzheng Street district road network density is about 5 km/km^2, the traffic control range is about 55%, affecting about 25,000 people. According to the traffic emergency level comprehensive evaluation model, its sphere of influence factors is a_2, the number of factors affecting it is b_3, the road network density factor is c_2, the traffic control range factor is d_3, Thus, a single factor fuzzy evaluation matrix was draw:

$$R = \begin{vmatrix} 0.3 & 0.6 & 0.1 & 0 \\ 0.1 & 0.4 & 0.5 & 0 \\ 0.15 & 0.2 & 0.35 & 0.3 \\ 0.3 & 0.5 & 0.15 & 0.05 \end{vmatrix}$$

By putting the single factor fuzzy evaluation matrix into Equation (1), we get:

$$B = (0.25 \quad 0.45 \quad 0.2 \quad 0.1) \times \begin{vmatrix} 0.3 & 0.6 & 0.1 & 0 \\ 0.1 & 0.4 & 0.5 & 0 \\ 0.15 & 0.2 & 0.35 & 0.3 \\ 0.3 & 0.5 & 0.15 & 0.05 \end{vmatrix} = (0.18 \quad 0.42 \quad 0.335 \quad 0.065)$$

The maximum value of the matrix B is 0.42; we can then determine the maximum degree of utilization of fire emergency traffic level II in principle.

8 EPILOGUE

Through access to information and by seeking expert advice, the scope, the number, density of road network and traffic management and control were selected as the evaluation indexes.

By experts, the right set of evaluation index of reviews and grade evaluation weight set were determined, and the maximum degree of membership fuzzy comprehensive evaluation method was regarded as the standards.

The model has been put into practical application, and the applicability has been tested.

REFERENCES

[1] Long,X.Q. & Tan,Y.L. 2011.Urban traffic congestion evaluation based on fuzzy comprehensive evaluation. *Transport Standardization*.11:114–116.

[2] Zhang,H. & Yan,X.P.,etc.2011. Relationship between driver safety cognition and vehicle based on fuzzy comprehensive evaluation. *Journal of Jiangsu University(Natural Science Edition)*, 32(5):511–515.

[3] Lu,H.P. 2006.*Transportation planning theories and method*. Beijing: Tsinghua University press.

[4] *Urban road traffic management evaluation system(2008 edition)*. Beijing: Ministry of Public Security, the Ministry of Housing and Urban-Rural Development, 2008.

[5] The overall national public emergency contingency plans, Beijing: State Council, 2006.

[6] Wang,T. & Chen,J.2011. Urban traffic road safety evaluation based on fuzzy mathematic method. *Journal of Transport Information and Safety*, 29(4):99–103.

[7] Ricardo&Pinto, etc. 2011.Dynamic routing combined to forecast the behavior of traffic in the city of Sao Paulo using neuro fuzzy network. *Computer Technology and Application*, 2(1):36–41.

[8] Mei,Z.Y.& Xiang,Y.Q.,etc. 2008.Optimization model of parking guidance coordinated with traffic flow control based on fuzzy algorithm. *China Journal of Highway and transport*, 21(2):84–88.

[9] Ling,J.M. & Guan,S.F.,etc. 2008.Muti-hierarchy Fuzzy Decision-making model for PPM treatment selection of highway asphalt pavement. *Journal of Highway and Transportation Research and Development*, 25 (6):25–34.

[10] Zhao, L.J. 2008.Research on the evaluation of environmental logistics based on fuzzy comprehensive evaluation. *Journal of Wuhan University of Technology*, 30(6):174–1.

Future Mechatronics and Automation – Yang (Ed.)
© 2015 Taylor & Francis Group, London, ISBN 978-1-138-02648-3

The transplant process of Linux2.6.20 on the development board of K9iAT91RM9200

B.H. Jiang & J. Mei
China Three Gorges University, Yi Chang, HuBei, China

ABSTRACT: Research and development on the embedded system has become a new hot spot. Owing to its advantage, a kind of OS in designing the embedded application, Linux becomes the first selection by a few of companies. As a high capability and low cost RISC microprocessor, ARM has already become the most comprehensive embedded processor. This paper introduces the feature of the ARM processor and the Linux OS. Combined with the K9, a designation panel with the 32 bit AT91RM9200 processor, the complete process is analyzed including hardware architecture, development environment and tools. In particular, the crossing complied environment and the transplantation of Linux are emphatically analyzed.

Keywords: Embedded system; ARM; Linux

1 INTRODUCTION

At present, in the embedded system, embedded processor based on ARM microkernel has become the mainstream. With the widely application of the ARM technology, the establishment of the ARM architecture oriented embedded operating system has become the focus of research. Now more embedded operating systems such as VxWork, Windows CE, Palm OS, Linux and so on, have already emerged[1]. Facing so many embedded operating systems, developers still choose Linux because of its free and open source code which can let anyone modify and transplant it to their target platforms. ARM Linux supports ARM710, ARM720T, ARM920T, Strong ARM, etc. All these ARM processors have memory management unit (MMU), for without the memory management unit of the CPU, it generally uses *μCLinux* as its operating system[2]. This article chooses the microprocessor K9iAT91RM9200 based on 32 bit ARM920T processor. Based on this, a Linux embedded operating system is constructed and it is embedded into the system of 32 bit ARM kernel.

2 HARDWARE

ARM processor is a kind of 16/32 bit of high performance, low cost, low power consumption embedded RISC microprocessor. It is designed by the ARM company and empowered to each semiconductor manufacturers. It provides comprehensive technical support for the development of a complete system. Almost all major semiconductor manufacturers in the world produce general chips based on ARM architecture, or use the technology related ARM in their special chips. At present, ARM processor has occupied more than 75% market on the 32 bit RISC embedded products. ARM microprocessor core is suitable for various areas and is rapidly becoming a portable communication equipment, handhold computing, multimedia digital consumption and embedded solutions which solve the RISC batch production standards in the market. ARM processor is also very popular at home, so this article chooses the ARM processor.

ARM processor has a series of kernel structure used in different areas. In this paper, K9i development board of 32 bit AT91RM9200 processor is used. AT91RM9200 include a high-speed on-chip SRAM workplace and a low latency external bus interface (EBI) to complete the seamless connection between external memory and internal

Memory required by the application. EBI has controller of synchronous DRAM, Burst Flash and static memory, and a special circuit is designed to connect with Smart Media, Compact Flash and NAND Flash. Advanced interrupt controller (AIC) has im- proved the interrupt handling performance of ARM920T processor by multi-vector, prioritization of interrupt source and shortening the interrupt processing transfer time. Peripheral data controller (PDC) provides DMA channels for all of the serial peripheral, makes data transmission successfully between it and external memory or internal memory without the processor, which reduces the cost of processor during continuous data stream transferring. In addition, double pointer PDC controller greatly simplifies the AT91RM9200 buffer links. Parallel I/O (PIO) controller as a general data I/O multiplexing peripheral input/output port, adapts the device configuration to maximum extent. Each line of the port contains an input change interrupt, the ability of opening Drain and the pull-up resistor of the programmable.

3 LINUX KERNEL OF OPERATING SYSTEM

Usually, the operating system is made up of kernel and some system service program (command interpretation, library files, links and compiling the program, etc.). The kernel is the soul of the operating system and it provides some ports in virtual machine for user process to make the user process run in parallel, and occupy system resources in public and without mutual interference.

Generally speaking, Linux operating system is constituted by four main parts, namely, kernel, shell, file structure and utility tools. The kernel of Linux operating system uses single block structure which has dynamical load and unload modules[3].System using the dynamical load and unload function of kernel module can easily add a new component or uninstall components which is no longer needed in the kernel. The dynamic loading of kernel module keeps the size of the kernel image minimum and has the maximum flexibility, at the same time to test the new kernel code without recompiling the kernel and rebooting. Thus the users can construct their own kernel according to their need of system. The publicity of source code provides the possibility of transforming their kernel, especially to the operating system of the special requirements[4].

4 THE DEVELOPMENT ENVIRONMENT AND DEVELOPMENT TOOLS BASED ON LINUX

In order to achieve the application development based on Linux, it is necessary to establish a complete Linux development environment. The embedded system is usually constrained by its resources. Therefore, writing software directly on the hardware platform of the embedded system is difficult and sometimes even impossible. At present, the general solution is to write on the general computer program and then generating the binary code format that can run on the target platform through the cross compiling and then downloading to run on a specific location on the target platform. Needing the support of cross development environment is a striking feature of embedded application software. Cross development environment refers to compiling, link and debugging of embedded application software environment. It is different from running environment of embedded application software, and it generally uses the host machine/target model.

4.1 *The cross compiling and link*

After completion of embedded software coding, compiling and link are needed to generate the executable code. Because of the development process mostly using general-purpose computer of Intel x86 series

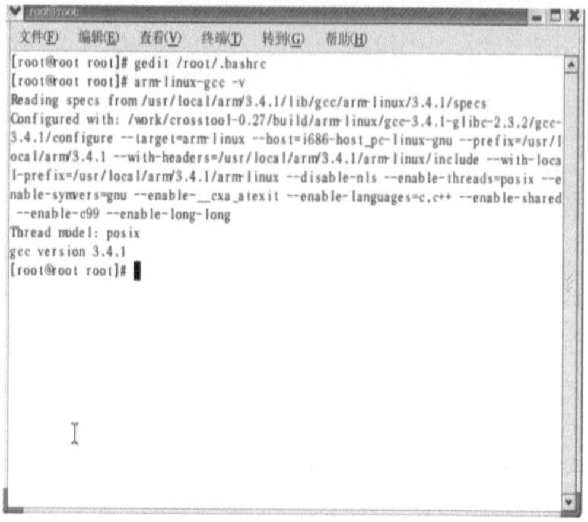

Figure 1. Compiling information.

CPU, while the processor chip of target environment is ARM9 microprocessor, embedded systems do not have enough memory or storage resources to compile executable code which requires to build a good cross development environment and then carries out cross compile and link.

The cross compiler and cross linker are able to run on the host and to generate compiler and linker of binary code to run on the target machine. For example, in GCC cross development based on ARM architecture, arm-linux-gcc is the cross compiler and the arm-linux-ld is the cross linker. Embedded systems in the process of links usually require the use of smaller function libraries to produce smaller executable code. So function libraries with special processing are used in the practice. For embedded Linux system, C function library glibc and mathematical function library libm have become stronger in the function and bigger in the volume .They have been difficult to meet the actual need. So we need their elaboration version such as uClibc, uClibm and newlib, etc[5].

4.2 *Cross compiling environment*

The GNU tool is issued in the form of the source code. For different architectures, GNU provides corresponding development packages. Download a cross compiler tool chain online: arm-linux-gcc-3.4.1.tar.bz2.Login to Linux as the root user and then enter the directory of /usr/local/arm and execute the following command:

tarjxvf arm-linux-gcc-3.4.1.tar.bz2 -C /

Modify the PATH variable and add /usr/local/arm/3.4.1/bin. For example, export PATH=$PATH:/usr/loca /arm/3.4.1/bin. Log back in Redhat (don't need to restart the machine, to start->logout), these settings take effect. At the command line, enter the arm-linux-gcc–v, as is shown in figure 1 indicating that cross compiling environment has been successfully installed.

```
static void __init get_mac_address(struct net_device *dev)
{
static char def_mac[]={0x00,0x15,0xf2,0x9d,0x66,0x80};
        /* Check Specific-Address 1 */
        if (unpack_mac_address(dev, at91_emac_read(AT91_EMAC_SA1H), at91_emac_read
(AT91_EMAC_SA1L)))
                return;
        /* Check Specific-Address 2 */
        if (unpack_mac_address(dev, at91_emac_read(AT91_EMAC_SA2H), at91_emac_read
(AT91_EMAC_SA2L)))
                return;
        /* Check Specific-Address 3 */
        if (unpack_mac_address(dev, at91_emac_read(AT91_EMAC_SA3H), at91_emac_read
(AT91_EMAC_SA3L)))
                return;
        /* Check Specific-Address 4 */
        if (unpack_mac_address(dev, at91_emac_read(AT91_EMAC_SA4H), at91_emac_read
(AT91_EMAC_SA4L)))
                return;
```

```
                return;
memcpy(dev->dev_addr,&def_mac,6);return;

        printk(KERN_ERR "at91_ether: Your bootloader did not configure a MAC address.
\n");
}


/*
 * Program the hardware MAC address from dev->dev_addr.
 */
static void update_mac_address(struct net_device *dev)
{
        at91_emac_write(AT91_EMAC_SA1L, (dev->dev_addr[3] << 24) | (dev->dev_addr
[2] << 16) | (dev->dev_addr[1] << 8) | (dev->dev_addr[0]));
        at91_emac_write(AT91_EMAC_SA1H, (dev->dev_addr[5] << 8) | (dev->dev_addr
[4]));

        at91_emac_write(AT91_EMAC_SA2L, 0);
        at91_emac_write(AT91_EMAC_SA2H, 0);
}
```

Figure 2.

5 THE COMPILATION OF EMBEDDED LINUX KERNEL

After the above work is finished, an embedded platform is built successfully and begins to compile kernel.

5.1 *The related source code*

(1) Unpack the source code and strike AT91 patch # cd /work
    # tar jxvf linux-2.6.20.tar.bz2
    # cd /work/linux-2.6.20
    # zcat ../2.6.20-at91.patch.gz | patch -p1
    # bzcat ../linux-2.6.20-at91-exp.diff.bz2| patch -p1

```
root@root:/work/linux-2.6.20
文件(F)  编辑(E)  查看(V)  终端(T)  转到(G)  帮助(H)
  SYSMAP  .tmp_System.map
  MODPOST vmlinux
  OBJCOPY arch/arm/boot/Image
  Kernel: arch/arm/boot/Image is ready
  AS      arch/arm/boot/compressed/head.o
  GZIP    arch/arm/boot/compressed/piggy.gz
  AS      arch/arm/boot/compressed/piggy.o
  CC      arch/arm/boot/compressed/misc.o
  AS      arch/arm/boot/compressed/head-at91rm9200.o
  LD      arch/arm/boot/compressed/vmlinux
  OBJCOPY arch/arm/boot/zImage
  Kernel: arch/arm/boot/zImage is ready
[root@root linux-2.6.20]# make uImage
  CHK     include/linux/version.h
make[1]: `include/asm-arm/mach-types.h' is up to date.
  CHK     include/linux/utsrelease.h
  CHK     include/linux/compile.h
  MODPOST vmlinux
  Kernel: arch/arm/boot/Image is ready
  Kernel: arch/arm/boot/zImage is ready
  UIMAGE  arch/arm/boot/uImage
"mkimage" command not found - U-Boot images will not be built
  Image arch/arm/boot/uImage is ready
[root@root linux-2.6.20]#
```

Figure 3.

(2) The modification of source code, as is shown in figure 2.

5.2 *Configuring the kernel*

(1) Modify Makefile
    ARCH ?= $(SUBARCH)
    CROSS_COMPILE ?=
    is instead of
    ARCH ?= arm
    CROSS_COMPILE ?=arm-linux-
(2) Configuring the kernel
    # cd /work/linux-2.6.20
    # make at91rm9200dk_defconfig
    # makemenuconfig

5.3 *Compile the kernel*

(1) Compile zImage
    # cd /work/linux-2.6.20
    # makezImage
(2) Compile uImage
    # cd /work/linux-2.6.20
    # makeuImage
    An error will appear as is shown in figure 3.

The solution is: mkimage of k9uboot/tools directory file is copied to bin directory of cross compiling environment. Run the command again: # makeuImage
    # cpvmlinux /usr/local/arm/2.95.3/bin

(3) Under the compilation of tools /usr/local/arm/3.4.1/bin

./arm-linux-objcopy -O binary -S vmlinuxlinux.bin;

gzip-v9 linux.bin;

cp linux.bin.gz /usr/local/arm/k9s/k9uboot/tools;

(4) Under the directory of /usr/local/arm/k9uboot/tools

./mkimage -A arm -Olinux -C gzip -a 0x20008000 -e 0x20008000 -d linux.bin.gz uImage;

The generated image file uImage is copied to the shares folder k9s.

6 CONCLUSIONS

On the basis of analyzing the characteristics of ARM and Linux, the construction process of embedded Linux development platform is introduced. The Linux kernel compiling is analyzed in detail and finally using a simple example to illustrate the application of adding process. So we can based on this explore the various drivers and applications.

REFERENCES

[1] Ma Zhongmei, Li Shanping. Embedded system tutorial of ARM&Linux[M].Beijing, Beijing university of aeronautics and astronautics press, 2004.

[2] Wu Minghui. Embedded system development and application based on ARM [M]. Beijing, People's posts and telecommunications press, 2004.

[3] Xia Weiwei, Shen Lianfeng. The key technology analysis of embedded system and the development and application[J]. The BBS, 2003(3):5–9.

[4] Ying Haiyan. The transplant of embedded Linux operating system based on ARM [J]. Modern intelligence, 2005(5):155–156.

[5] Yuan Taisheng, Zhang Suqin. The discuss of Linux transplant issues of the embedded environment[J]. Computer application research,2003(11):61–63.

Future Mechatronics and Automation – Yang (Ed.)
© 2015 Taylor & Francis Group, London, ISBN 978-1-138-02648-3

Eliminating bridge offset voltage for AMR sensors

YongJun Wang
Guilin University of Aerospace Technology, Guilin, P.R. China

ABSTRACT: The offset voltage of Anisotropic Magneto-Resistive (AMR) sensors is one of the important factors which affect the sensor system output. This document elaborates upon the property of AMR sensors bridge offset and its influence on magnetic detection. Several compensation methods for the bridge offset have then been analysed. According to the AMR sensor characteristics, the set-reset type digital subtraction method to eliminate the offset voltage was put forward. The compensation method by testing on the electronic compass is successfully demonstrated.

Keywords: Anisotropic Magneto-Resistive, offset voltage, geomagnetism detecting, compensate, digital subtraction method

1 INTRODUCTION

Anisotropic magneto-resistance sensors are often used as the sensing element in geomagnetic detection embedded systems, such as electronic compasses, etc. An AMR sensor is generally in the form of Wheatstone bridge, which is a bridge of four components. These bridge sensors can provide many features, but when they cannot detect any sensor stimulus signals, the bridge output voltage is called an "offset voltage." This offset voltage may cause reduced performance of the sensor system if it is not compensated for. If that occurs, the performance of the sensor system may be degraded. With regard to traditional bridge offset voltage compensation methods, the Shunt Resistance Method, the Amplifier-bias Nulling, and the Switching Feedback Method are used to eliminate bridge offset voltage that can be found in [1] and [2]. This document elaborates the property of AMR sensors' bridge bias, it analyses the impact on sensor circuits, and it compensates for the offset voltage to obtain a better effect.

2 AMR SENSOR BRIDGES AND OFFSET VOLTAGE

AMR sensors are designed in Wheatstone bridge structures. The principle of the bridge is to use two potentiometer elements (or "half- bridges") difference as the output of each components in zero or sensors usually have no incentive which has the same reactance. When in the absence of any incentive, each half-bridge voltage should be the output of half of the total bridge power supply voltage, so the Wheatstone bridge output voltage becomes zero. A typical AMR Wheatstone bridge circuit structure is shown in [3].

When using an AMR sensor, the bridge offset voltage is mainly caused by the precise resistance value of each sensor element. Although Honeywell have used semiconductor manufacturing equipment in the manufacturing process, and precision matching of AMR components can be obtained, the permalloy (NiFe) thin film deposition to the flat screen printing control region of tolerance will lead to some resistance mismatch. Even with several hundreds or thousands of ohm resistance elements, tracing the resistance of the error could generate considerable bridge offset voltage, as described in [3].

In a compassing solution for an automotive system, an AMR sensor, HMC 1052, has been used. If the HMC 1052 is used to measure an earth's field vector component of 625 mGauss, when the Vb voltage is 5V, the sensor bridge output is about 0.50 mV, which is a relatively small voltage compared to the offset voltage. Consequently, the sensor bridge's output voltage sensitivity has to be compared with the bridge offset voltage for the degree desired to the undesired output signals. Therefore, compensation for the bridges' offset voltage of each sensor can play an important role in improving the measurement accuracy of the system.

3 IMPACTS ON MEASUREMENTS

Generally, the bridges' offset voltages are ignored in Wheatstone bridge output voltage computations, so the output voltage is the multiplicand of the sensitivity multiplied by the bridge voltage and by the sensitive axis field strength, which is shown by the equation below:

$$V_{out} = Vb * K * H \tag{1}$$

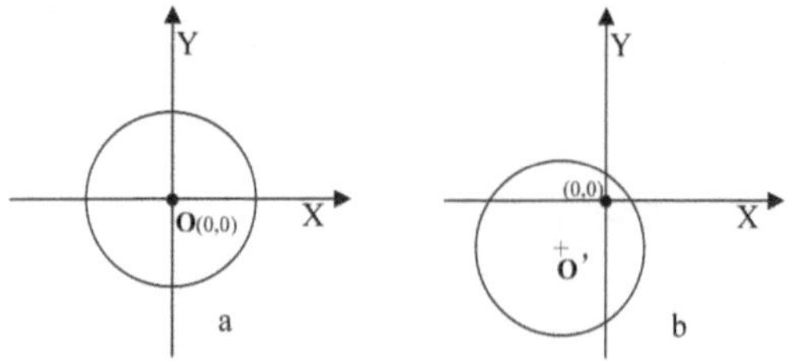

Figure 1. Non-offset effect and mix offset effect.

The value for bridge sensitivity (K) is strictly for the field applied in the sensitive direction. For bias effects, the offset voltage (Voff) is added in this equation and affects the output voltage:.

$$V_{out} = V_{off} + Vb * K * H \tag{2}$$

With Voff dependent on the feature geometry of each sensor bridge, there is a reasonable consistency of Voff for each sensor design.

Once identified through data collection, a compensation algorithm may be employed to correct the bridge offset voltage. For the compassing application, any sensors measured the earth magnetic field values at the bridge is about 625 mgauss, therefore, the sensitivity of HMC1051 specification 0.8 to 1.25 mV/V/Gauss, any bridge of geomagnetic field on possible incentive for +/−3.9 mV. When the +/−3.9 mV is added to the possible −10 mV to +11.25 mV bridge offset voltage, there will be a problem. The sum of the output voltage becomes −13.9 mV to +15.15 mV with the magnetic portion. The output signal is only about 25% of the total variation. The value of Voff may be as high as 300%.

When the compass is turning around in a circle in a horizontal plane, the readings from the X and Y output will plot a circle, which is shown in Figure 1.a. Note here the circle centred about the O (0,0) point; there is no offset voltage interference with the readings output [4]. The readings are obtained from a Honeywell high accuracy magnetometer.

If the sensors' output mixed offset data, driving the HMC1052 in a circle would produce the bias shown in Figure 1.b. Note here that the X,Y plot is offset from the O (0,0) point to the O' point. This distortion can be determined systematically and applied to subsequent X, Y readings to eliminated the effects of the offset voltage.

4 BRIDGES' OFFSET VOLTAGE ELIMINATION METHOD

Traditionally Shunt Resistance Method, Amplifier Bias Nulling and Switching Feedback Method are used to compensate bridge offset voltage. To do this accurately, all magnetic signal stimulus should be removed. This can be done in either a helmholtz coil set or in a well shielded enclosure to remove the effects of earth's field on the sensor, that demand actual environment. When the sensitivity and linear of the sensor changes at different temperature, those methods can't dynamically eliminate the bias voltage of bridge. In such a condition, the amplifier in the circuit also bring the offset voltage and bias voltage. Therefore, according to the AMR sensor bridge bias characteristics, more effective methods were need to eliminate offset voltage.

The switch feedback method can eliminate the bridge offset voltage, as described in [5]. Although this method can eliminate the bias voltage signal in a dynamic, but the circuit will greatly increase the cost and the complexity of system design system.

Zhai Tao of Harbin Engineering University shows offset strap current method to eliminate offset voltage, which uses offset strap on its AMR sensor compensating for bridge offset voltages, as described in [6]. This strap consists of a spiral of metalization placed near the bridge to couple a magnetic field induced by current flowing through the strap. Not only does this strap make a nice current sensor, but the magnetic field produced can sum with an external sensed field to cancel the offset voltage. But it adjusts the flexibility to be bad, not only it may bring the error, moreover it's also not easy to control.

There is also a digital subtraction method has more applications in designing electronic compass [7,8]. Because it does not need to change hardware, ADC resolution is sacrificed to accommodate the dynamic range of sensor bridge offsets.

In order to obtain better compensation effects, the bias current method and the digital subtraction method were combined to eliminate the sensors' bridge offset voltage, which is known as the set-reset type digital subtraction. The method mainly uses the offset current strap and the set-reset strap which are in the AMR sensor, the bridge offset voltage is compensated mainly through the offset strap current, and then uses the set-reset switching on a one-shot basis, which is based on the completion of the remaining bias voltage.

The way to eliminate the bridge offset voltage is to make stable magnetic field measurements of the bridge output voltage in between each set and reset field application. Since the external field components of the bridge output voltage will flip polarity, the set and reset bridge output voltages can be subtracted and the result divided by two to calculate the bridge offset. Figure 2 shows the set-reset type digital subtraction scheme block diagram.

The set-reset type digital subtraction method uses a bias current to make a bridge out of a lower Voff, which is called "offset residual voltage." Then, uses the property that toggling the sensor element between "set" and "reset" conditions via the set-reset strap element; allows the field induced sensor output to reverse polarity, but the bridge offset voltage remains a consistent bias on the output. Figure 3 shows the toggling routine graphed with typical output levels.

The sensor offset residual voltage (Voff) can be eliminated by using a simple digital subtraction

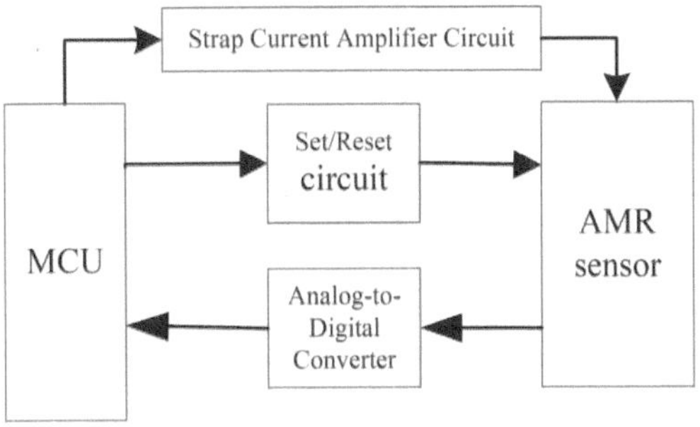

Figure 2. The set-reset type digital subtraction design scheme.

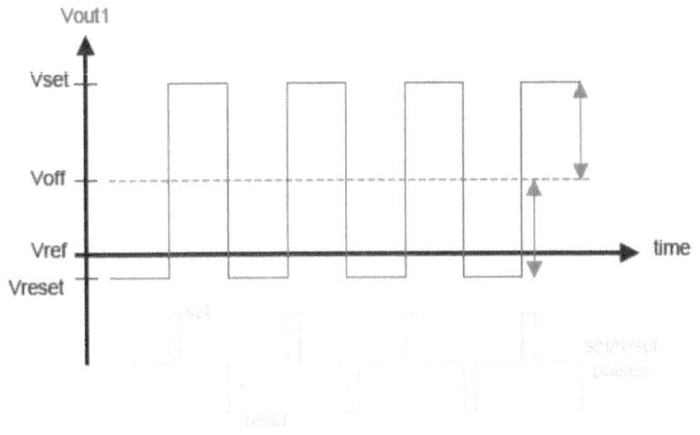

Figure 3. The setreset toggling sensor output.

technique. First, apply a set pulse, measure the output and store it as "Vset"(Figure 3.). Then apply a reset pulse and store that reading as "Vreset."

The two-output subtraction can eliminate the bias voltage of the Voff sensor bridges, and can be used to calculate the magnetic field strength H. The sensitivity of the system is K:

$$V_{set} = v_{off} + K \cdot H \tag{3}$$

$$V_{reset} = v_{off} - K \cdot H \tag{4}$$

Using Equation (3) to subtract Equation (4) can eliminate Voff:

$$\begin{cases} v_{off} = (V_{set} + V_{reset})/2 \\ H = (V_{set} - V_{reset})/(2K) \end{cases} \tag{5}$$

This shows, that with compensation based on the bias current, the set/reset after the measurement value can use Equation (5) for effective correction of the output of the AMR sensor. The method not only eliminates the bridge offset voltage, but it also eliminates the amplifier offset and drift of the sensor and amplifier.

Note that Equation (5) has no Voff term. This method not only effectively calibrates the output of the AMR sensor bridge, but also eliminates the amplifier offset as well. Another advantage is that the temperature drift of the amplifier output voltage and the hard iron disturbance is eliminated, as if it were not present [4].

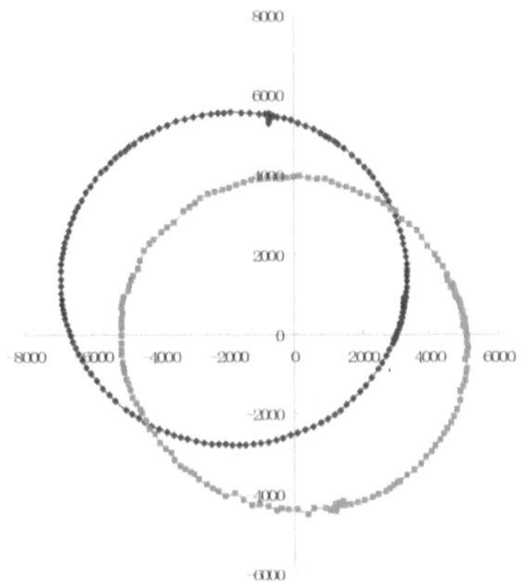

Figure 4. Magnetic sensor outputs (X, Y) rotated horizontally.

5 TEST RESULTS

In the application of an electronic compass system, once the X and Y magnetic readings are in the horizontal plane, Equation (6) can be used to determine the heading:

$$\alpha = \arctan(H_Y / H_X) \tag{6}$$

From Equation (6), it can be seen that if the offset voltage is not eliminated and the offset voltage is greater than the maximum voltage, the electronic compass heading will display a small range. When the maximum bias voltage is 11.25 mV, which is given in [2], the maximum error will reach nearly 300°.

On the horizontal plane, by making the compass turn around in a circle, 200 readings could be obtained. According to the measured data, the earth's field magnitude and heading error curve are shown in Figure 4 and Figure 5.

Figure 4 shows a sine and cosine output response of the X and Y direction in rotation. Note here that after eliminating the offset voltage, the X, Y plot is from the (−2000, 1990) point to the (0,0) point, which is shown as the blue diamond line to the purple circle line. In Figure 5, the blue diamond line is a heading angle error curve that shows the random error being from 0.5° to 70.9°, when the sensor output mixed the offset voltage. The purple box line shows the heading angle error after the set-reset type digital subtraction, it can control in 6.7°.

The set-reset type digital subtraction method can effectively reduce the offset voltage and the effect of the course correction angle. At the same time, after the elimination of the offset voltage, the heading error is still large. This is due to the fact that the electronic compass is not only affected by the sensor bridge offset voltage, but also by the soft iron effects in the working environment, and other factors.

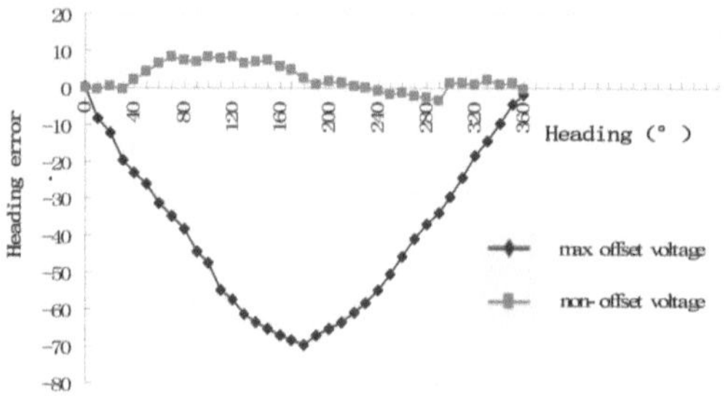

Figure 5. Error curves of heading.

6 CONCLUSION

In the geomagnetic detection device based on the AMR sensor, the sensor bridge offset voltage is one of the most important factors. A new method of eliminating the offset voltage has been discussed in this paper. For applications which are low-cost, the digital subtraction method may be a preference. If high measurement accuracy is required in the application occasion, then t combing with the offset strap of an AMR sensor and set-reset strap method will preserve the highest resolution. The set-reset type digital subtraction method can eliminate the offset voltage, and the method for the bridge circuit of bias voltage on bi-directional incentive elimination also plays a very good model role. In addition, this method also has high engineering application value.

ACKNOWLEDGEMENT

This work was supported by the Guangxi Key Laboratory of Automatic Detecting Technology and Instruments (YQ14203). Thanks to the project "Research on the construction and error compensated of the strapdown AHRS based on Multi-Sensor Microsystems", which was supported by the Natural Science Foundation of Guilin University of Aerospace Technology (YJI303).

REFERENCES

[1] DONG P C, DONG H C. 2003. A New Compensator for Strain Gage Bridges and Hall Cells. *Instrument Standardization & Metrology*, Vol. 3: 39–41.

[2] Handling sensors bridge offset. 2003. Honeywell Application Note, AN212.

[3] Sensor Products Data Sheet, "1 and 2 Axis Magneto resistive Microcircuits", Honeywell SSEC, 5/99, <www.magneticsensors.com>

[4] Kubik J, Vcelak J, Ripka P. On cross-axis effect of the anisotropic magneto-resistive sensors. Sensors and Actuators A: Physical, 2006, 129: 15–19.

[5] M J Caruso, Application of Magneto Resistive Sensors in Navigation Systems, Sensors and Actuators A: 1997, SAE SP-1220, (Feb. 1997) 15–21.

[6] Caruso M. J. Applications of magnetic sensors for low-cost compass system. Record – IEEE PLANS, Position, Location and Navigation Symposium. San Diego, CA, 2000: 177–184.

[7] Vcelak J, Ripka P, Kubik J, et al. AMR navigation systems and methods of their calibration. Sensors and Actuators A: Physical, 2005, 123–124: 122–128.

[8] Vcelak J, Ripka P, Platil A, et al. Errors of AMR compass and methods of their compensation. Sensors and Actuators A: Physical, 2006, 129: 53–57.

Future Mechatronics and Automation – Yang (Ed.)
 ISBN 978-1-138-02648-3

Evaluation and influencing factors of urban land intensive use – a case study of Xianning City

XueHua Cui, ChengShun Song & WenXia Zhai

School of Resources Environment Science and Engineering, Hubei University of Science & Technology, Xianning, P.R. China

ABSTRACT: This paper selects Xianning City, a built-up area of Hubei Province, as the study area and establishes the evaluation index system of urban land intensive use from a system point of view, combining four aspects of social, economic, ecological, and environmental benefit elements. The degree of the urban land intensive use of Xianning City from 1999–2008 was quantitatively calculated and analysed using the multi-factor comprehensive evaluation method. In this paper, we carried out the grey correlation analysis method to analyse the influencing factors of urban land intensive use. The results showed: (1) the degree of urban land intensive use took on a steady upward trend in Xianning City from 1999-2008. The urban land intensive use was in a basic intensive state from 1999–2007, and then the urban land intensive use was in an intensive state in 2008; (2) economic drive is the main influencing factor which promotes the level of the intensive urban land use in Xianning City. Based on the assessment results, some corresponding policies and countermeasures were put forward.

Keywords: Urban land; intensive use evaluation; influencing factors; Xianning City

1 INTRODUCTION

Urban land intensive use is a kind of social and economic phenomenon; it is the inevitable requirement of the urban land use with the economic and social development to a certain stage, as well as the inevitable result of the urban land supply scarcity[1]. With the enhancement of urbanization levels, the problems with urban land use are gradually increasing; the overexpansion of urban space and low efficiency of land use has seriously restricted the sustainable development of the social economy. Due to the limited urban stock land and hindered expansion of urban space, China has to stick to adopting the intensive land-use method both from the perspective of land resource protection or urban development and reasonable use of urban land[2].

In order to deepen the understanding of the causes, internal mechanism and basic process of urban land intensive use, in recent years, some scholars began to study its influencing factors[3–6]. However, in these studies, the evaluation index and influencing factors are relatively simple for they have ignored the role of ecological factors in the urban land intensive use system, which may lead to less objective comprehensive and authentic findings.

This article chooses the built-up area in Hubei Xianning as the research object, and it establishes the index system from four aspects including society, economic, ecological, and environmental factors. It uses the multi-factor comprehensive evaluation method to measure the urban land intensive use levels between 1999 and 2008, and it uses grey correlation analysis to study the influencing factors in urban land intensive use. The study aims to reveal the inner mechanism of urban land intensive use, to develop a scientific understanding of urban land intensive use, and to realize the urban social and economic sustainable development, which has important practical value and far-reaching strategic significance. At the same time, it can provide decision support for reasonable utilization of land resources.

2 EVALUATION OF URBAN LAND INTENSIVE USE

2.1 *Establishment index system*

In order to evaluate the urban land intensive use level accurately and scientifically, the index selection follows the principles of comprehensiveness, hierarchy, systematics, independence, and operability [7–8]. Combined with the study of the regional social development level and the status of the ecological environment, it chooses twenty indicators which can reflect the urban land intensive use scientifically, and builds an evaluation index system of Xianning City's urban land intensive use (shown in Table 1). The data is taken from the Xianning statistical yearbook and China cities statistical yearbooks from 2000 to 2009.

2.2 *Determination weights*

This paper adopted the entropy method to determine the weighted value of indicators of the urban land intensive use evaluation index system[9]. The specific steps are shown as follows:

(1) The index proportion transformation:

$$P_{ij} = X_{ij} \Big/ \sum_{i=1}^{n} X_{ij}$$

(2) Calculation of the entropy index:

$$H(x_j) = -k\sum_{i=1}^{n} P_{ij} \ln P_{ij}\ ,\ j = 1,2,3,\cdots,m$$

(3) Calculation of the difference coefficient of indicators:

$$h_j = 1 - H(x_j),\ j = 1,2,3,\cdots,m$$

(4) Calculation weight of the index:

$$w_j = h_j \Big/ \sum_{j=1}^{m} h_j\ ,\ j = 1,2,3,\cdots,m$$

In the formula, X_{ij} is the original value in i year and the j index, P_{ij} is the standardized value in i year and the j index, m is the number of the index, and the value ranges from 1 to 20, n is the number of years the corresponding values are 1 to 10, k is the adjustment coefficient, and $k = 1/\ln n$.

Using the entropy method can only determine the weight of the second level indices. According to the entropy of additivity, we can get the weight of indicators at the next higher level in turn by the sum of all kinds of secondary index weights. According index weights of Xianning City's urban land intensive use evaluation as shown in Table 1.

2.3 *Standardization of indexes*

In order to make the evaluation index comparable, we need to remove the dimensions from the data in order to eliminate the influence of the difference of the indicators' dimension or the indicators' measurement scale. The positive and negative index processing method is[10]:

Positive index: $y_{ij} = (x_{ij} - m_i)/(M_i - m_i)$

Negative index: $y_{ij} = (M_i - x_{ij})/(M_i - m_i)$

In the formula, x_{ij} is the actual value in the index, y_{ij} is the standardized values in the index. i is the index number, the data range is 1 to 20, j is the year (1999–2008), the corresponding value is 1 to 10. M_j and m_i are the upper and lower of the i index. Population density, Engel coefficient, the amount of wastewater per unit area, the amount of emissions per unit area, the amount of industrial solid waste per unit area belong to the negative index, the rest are positive index.

2.4 *Evaluation model*

To evaluate the urban land intensive utilization, multiply the dimensionless index value by the corresponding index weight and then conduct cumulative sum, shown as follows:

$$\lambda = \sum_{i=1}^{n} (W_i \times F_i)$$

In the formula, λ is the sample of the urban land intensive utilization degree, W_i is the weight of the i index, F_i is the standardized value of the j index, and n is the index number. If the value of λ is greater, the area of urban land intensive utilization degree is higher, or vice versa.

2.5 *Evaluation criterion*

Based on the characteristics of urban areas, economic development level and land use features, the urban intensive land use evaluation of Xianning could be carried out by 3 criteria. In addition, some rules and regulations issued by the government have been taken into consideration, including the "Intensive Land Use Evaluation Procedures of Land Resources Survey" (Trial) "Stinted and Intensive Use Evaluation Procedures of Construction Land" (TD / T 1018-2008), "Regulations of Urban Land Classification and Planning of Construction Land" (GBJ 137-90). Level I means that, when the intensive degree is greater than or equal to 0.60, the urban land use is relatively intensive; Level II means that, when the intensive degree is greater than 0.60 and less than 0.30, the urban land use is basic intensive; Level III means that, when the intensive degree is less than or equal to 0.30, the urban land use is not intensive.

2.6 *Evaluation result*

Using the multi-factor comprehensive evaluation method, we can calculate the urban land use intensity of Xianning in the years from 1999 to 2008, and contrast the evaluation standard to get the intensive level (shown in Table 2).

The evaluation results indicate that the urban land use intensity level of Xianning City during 1999 to 2008 was in a basic intensive state. In addition to 2008, which was in a relatively intensive level, and the fact that the extent of urban land use showed a steady upward trend, its value increased from 0.3127 in 1999 to 0.7122 in 2008, which had gained 0.3995 and increased to 127.76%. The trajectory of the last decade of the urban land use index showed that intensity grew slowly between 1999 and 2004, and intensity growth had accelerated between 2005 and 2008.

For a long time, the socio-economic development of Xianning was at a low level, the intensity of land

Table 1. Indexes and their weights of urban land intensive use in Xianning City.

Evaluation target	Rule layer	Weight	Index layer	Indicators of status	Weight
Urban Land Intensive use	Social factors	0.2317	Population density (person/km^2)	–	0.0062
			Per capita residential area (m^2/person)	+	0.1276
			The number of buses owned by every ten thousand people (unit)	+	0.0662
			The number of hospital beds owned by every ten thousand people (piece)	+	0.0232
			Engel coefficient (%)	–	0.0085
	Economic factors	0.4070	Per capita disposable income (Yuan/Person)	+	0.0464
			Per area GDP (ten thousand yuan/hm^2)	+	0.0728
			Per capita GDP (Yuan/Person)	+	0.0950
			Per capita retail sales of social consumer goods (Yuan/Person)	+	0.1648
			The ratio of financial revenue in GDP (%)	+	0.0280
	Ecological factors	0.1554	Garden green areas (hm^2)	+	0.0706
			Public green areas (hm^2)	+	0.0016
			Per capita green areas (m^2/person)	+	0.0491
			Green coverage area in built-up areas (hm^2)	+	0.0187
			Green coverage rate in built-up areas (%)	+	0.0154
	Environmental factors	0.2059	The amount of wastewater per unit area (ht/km^2)	–	0.0296
			The amount of emissions per unit area (hm^3/km^2)	–	0.0757
			The amount of industrial solid waste per unit area (ht/km^2)	–	0.0777
			The treatment rate of wastewater (%)	+	0.0205
			The comprehensive utilization of industrial solid waste (%)	+	0.0024

Table 2. Urban land intensive use degrees in Xianning City from1999 to 2008.

	1999	2000	2001	2002	2003	2004	2005	2006	2007	2008
Intensive degree	0.3127	0.3331	0.3516	0.4167	0.4375	0.4344	0.4616	0.5116	0.5687	0.7122
Intensive level	II	II	II	II	II	II	II	II	II	I

use was small, and the intensity of urban land use was lower. However, from 2005, the economy of Xianning has developed rapidly and its economic strength has increased markedly. To improve people's livelihoods, the government increased investment in urban infrastructure construction; the city has been given a new look, public transportation has increased conveniently, health care standards have improved step by step, and the social benefits of urban land intensive use have been increasing. At the same time, Xianning City's economic development has made significant achievements; the industrial structure has optimized and upgraded constantly. Basic conditions for urban land use have gradually improved; the land use level has constantly improved, the total economy has increased substantially, and the economic benefits of urban land use have entered a period of rapid growth. With Xianning's economy increasing, the social and economic benefits of urban land intensive use enter a sustained growth phase.

Xianning has witnessed the advancement in its economics and the social and economic benefits based on the urban land intensive use show a steady upward trend. In order to build a unique brand, it hosts the annual International Hot Spring Culture Tourism Festival. Especially since 2008, the acceleration of ecological city construction and the significant increase in the afforestation area, ecological management area and per capita green area have intensified the environmental management, which enhanced the image and increased the environmental benefits of Xianning.

The index of Xianning's land intensive use has increased year by year; it is identical with the actual situation of socio-economic and ecological development. On the one hand, it is due to the strong support of national macroeconomic policies which have led to the rapid development of the social economy in Xianning City. On the other hand, it has also been strongly influenced by ecological benefit promotion. It has shown that the city's land use has transferred from the extensional expansion to the potential of connotation; the allocation of Xianning City's land resources has been gradually optimized. Along with the rapid development of the national economy in Xianning City, the level of intensive land use has entered a stage of sustained growth.

3 DRIVING FACTOR ANALYSIS OF URBAN LAND INTENSIVE USE

Based on the idea of the driving mechanism of urban land intensive use theory system, the urban land intensive use is a complicated system as a whole. Due to

Table 3. Urban land intensive use by gray relational in Xianning City.

Impact factor	Gray relational	Rank	Comprehensive influence factors	Gray relational	Rank
Population density	0.8344	16	Social development	0.8116	4
Per capita residential area	0.6873	19			
The number of buses owned by every ten thousand people	0.7949	17			
The number of hospital beds owned by every ten thousand people	0.8580	15			
Engel coefficient	0.8835	12			
Per capita disposable income	0.9673	1	Economic drive	0.8568	3
Per area GDP	0.9349	3			
Per capita GDP	0.9366	2			
Per capita retail sales of social consumer goods	0.5577	20			
The rate of financial revenue in GDP	0.8874	10			
Garden green area	0.9348	4	Ecological constraint	0.8988	1
Public green area	0.8941	9			
Per capita green area	0.8978	7			
Green coverage area in built-up are	0.8852	11			
Green coverage rate in built-up area	0.8823	13			
The amount of wastewater per unit area	0.8945	8	Environmental constraint	0.8627	2
The amount of emissions per unit area	0.9112	5			
The amount of industrial solid waste per unit area	0.7451	18			
The treatment rate of wastewater	0.8648	14			
The comprehensive utilization of industrial solid waste	0.8978	6			

the system accords with the concept of grey system categorization, the driving mechanism of urban land use research can be considered to be the study of the grey system. In order to study the relation between various driving factors and urban land intensive use, the grey correlation analysis methods are introduced to analysis research.

According to the calculation steps of the grey correlation degree[11], it is calculated that the correlative degree of land intensive use influence factors in Xianning City as shown in Table 3. The relational degree is a direct reflection that the factors of urban land intensive use affect the level of intensive land use, and the higher the correlation is, the greater the influence degree is, and vice versa. From Table 3, it is seen that those magnitudes of relational degree are bigger than 0.90, including per capita disposable income, per capita GDP, per area GDP, garden green area, and the amount of the emissions per unit area, which indicates that these five factors are highly correlated with urban land intensive use. Some relational degrees are between 0.90 and 0.80 and they are, the comprehensive utilization of industrial solid waste, per capita green area, the amount of wastewater per unit area, public green area, the rate of financial revenue in GDP, green coverage area in built-up area, the Engel coefficient, green coverage rate in built-up area, the treatment rate of wastewater, the number of hospital beds owned by every ten thousand people, and the population density, which indicates a greater impact on the intensive use of urban land. The number of buses owned by every ten thousand people and the amount of industrial solid waste per unit area are between 0.80 and 0.70, and these factors have a weak influence on the intensive use of urban land. Furthermore, both the relational degrees of per capita residential area and per capita retail sales of social consumer goods are less than 0.70, especially the latter which is only 0.557, so their impact on the intensive use of urban land is minimal.

From the comprehensive factors of urban land intensive use, the effect of ecological constraint is greater than the environmental constraint, the effect of the environmental constraint is greater than the economic drive, and the effect of the economic drive is greater than social development. The five influence factors of the ecological constraint act as the biggest leading roles in the urban land intensive use, and their influence is most obvious. The garden green area is very obvious and the influence of the green coverage rate in built-up areas is relatively weak, which reflects the low green coverage rate in built-up areas. Xianning City should further strengthen the construction of ecological environments so that the green coverage rate will be improved. The environmental constraint plays a an important leading role leading role in the urban land intensive use; the effect of the amount of emissions per unit area is obvious, and in second place, the effect of the comprehensive utilization of industrial solid waste is also outstanding. It reflects that Xianning should pay attention to environmental protection in the process of urban land intensive use, and reduce emissions and enhance the comprehensive utilization of industrial solid waste. The economy drive is relatively weak; the correlation value between per capita disposable income and urban land intensive use is greatest, followed by the per capita GDP and per area GDP. To improve the level of Xianning City's land intensive use, we must speed up the city's economic development and enhance its economic strength. Social development factors have minimal impact on urban land intensive use.

4 CONCLUSION AND SUGGESTION

4.1 *Conclusion*

Urban land intensive use is the result of mutual constraints. This article set up the index system from four aspects: social, economic, ecological, and environmental factors, measured urban land intensive use level by applying the method of multi-factor comprehensive evaluation from 1999 to 2008, and used grey correlation to analyse the influence factors of urban land intensive use.

The urban land use in Xianning was at the relatively intensive level in 2008, and at other times it remains at the basic intensive state. In this decade, it showed a steady upward trend and the growth rate during the 2005–2008 was higher than that during the 1999–2005.

The study of urban land intensive use driving factors shows that the change of urban intensive use in Xianning is mainly affected by ecological constraint, followed by environmental constraint; the influence of the economy is small, and the influence of social development is minimal.

4.2 *Suggestion*

To improve the level of urban land intensive use and to realize the goal of social and economic sustainable development in Xianning, we must be in accordance with the construction requirement of "resource saving and environment friendly", in order to create a garden city, a civilized city, a healthy city, and an excellent tourism city, and promote the sustainable use of land resources. First, we must plan scientifically, optimize the structure of urban land use, and improve the efficiency of land use. Second, we must strengthen the market mechanism, allocate land effectively, and promote land saving and intensive use. Third, we must strengthen the industrial structure adjustment and promote economic sustainable development. Fourth, we must strengthen the protection of the environment and improve the ecological environmental benefits of land use.

ACKNOWLEDGMENTS

This research was supported by the Ministry of Education of Humanities and Social Science research project (13YJC630136), Humanities and Social Science research project in Hubei Province Department of Education (14G366), Statistical Scientific research projects in Hubei Statistics Bureau (HB132-51), National College students' innovative entrepreneurial training program (201210927077).

REFERENCES

[1] Hua Longlei, Lei Guoping, Zhang hui. Analysis on Evaluation and Driving Factors on Intensive Landuse of the Coal-base City—A Case Study of Qitaihe City of Heilongjiang Province [J]. Research of Soil and Water Conservation, 2012, 19(1):212–216.

[2] Li Chi, Lv Bing. Influence factors of urban land intensive utilization and land use pattern [J]. Natural Resource Economics of China, 2007(8):7–9.

[3] Liu Sifeng, Dang, Yaoguo, Fang Zhigeng etc. Grey system theory and its application [M]. Beijing: Science press, 2005.

[4] Qu Liping. Zhang Liqin, Hu Weiyan. Factors Influencing Change in Urban Land Intensive Use: A Case Studyof Wuhan City [J]. Resource Science, 2010, 32(5):970–975.

[5] Shen Haiyuan, Chen Ying, Zhang Caiyun, etc. Evaluation of land use efficiency—A case study of Xi'an [J]. Chinese Journal of Soil Science, 2009, 40(2):209–212.

[6] Song Chengxun, Chen zhi, Zhai Wenxia, Intensive Utilization Evaluation of Residential Land Based on model Parcelsin in Xining City [J]. Journal of Xianning College, 2011, 31(6):114–116.

[7] Tong Xiangning, Yang Gangqiao, Li Meiyan. The evaluation index and method of urban land use efficiency—A case study in Wuhan City[J]. Journal of Huazhong Agricultural University (Social Science Edition), 2006(4):53–57.

[8] Wang Jiating, Ji Kaiwen. Dynamic mechanism study of urban land intensive use [J]. Urban Problem, 2008(8):9–13.

[9] Wang Yuqing, Song Ge. The benefits evaluation method and application of urban land use [J]. Scientia Geographica Sinica, 2006, 26(6):743–748.

[10] Wu Yuling, Qu Futian. Mechanism of Intensive Urban Land Use in China: Theoretical and Practical Study [J]. Resource Science, 2007, 29(6):106–113.

[11] Zhang Junfeng, Xu Mengjie. Evaluation on Interactive land use benefits and coupling relationship in urban expansion: A case of Nanjing City [J]. Journal of Nanjing Agricultural University (Social Sciences Edition), 2010, 10(3):63–69.

Future Mechatronics and Automation – Yang (Ed.)
© 2015 Taylor & Francis Group, London, ISBN 978-1-138-02648-3

Short-term wind power forecasting based on Elman neural networks

ShunHua Zhang & XiaoPin Yang
School of Mechanical and Electrical Engineering, Nanchang Institute Technology, Nanchang, China

ABSTRACT: Wind power forecasting is the most important guarantee for the stable operation and the dispatch system of wind farms. A reasonable set-up for the Elman neural network input of wind power prediction would be the number of neurons in the hidden layer, the transfer function of the hidden layer as the tansign function, and the output layer as the linear function. The actual measured data for a wind farm in Poyang Lake in Jiangxi Province is given as an example. The example shows that the Elman neural network prediction results meet the engineering requirements.

1 INTRODUCTION

The common characteristics of the traditional method for wind power prediction is very difficult to establish the analytical expression between the future of wind power and the various factors affecting wind power. This is one of the most important reasons of wind power prediction precision. An artificial neural network method does not need to write out the analytical expression between the future wind power and various influencing factors; one just research what factors should be used as input variables.

This avoids the difficulties of process simulation. In forecasting, one only need to determine which factors would be used, and does not need to know the analytical expression between these factors with wind power. Therefore, the method of an network artificial neural needs to become better at prediction accuracy and to be an effective way to shorten the calculation time of forecast, and thus, wind power prediction becomes one of the most promising applications of the artificial neural network method. By using a neural network method to forecast the Poyang Lake region in Jiangxi Province, the research on wind power shows that the artificial neural network for short-term wind power prediction is successful and can meet the accurate engineering requirements.

2 THE MAIN RESEARCH METHODS

2.1 *Elman neural networks*

The Elman network can be thought of as one which has a partial memory unit and the forward neural network of a local feedback connection. The Elman network is similar to the multilayer forward network multilayer structure. It is the main structure of feed- forward links, including the input layer, the hidden layer, and the output layer; the connection power can be corrected by learning. The feedback connection consists of a set of "structure" units, used to memories the output value of the moment before its connection weights are fixed. The structure of the Elman neural networks is shown in Figure 1.

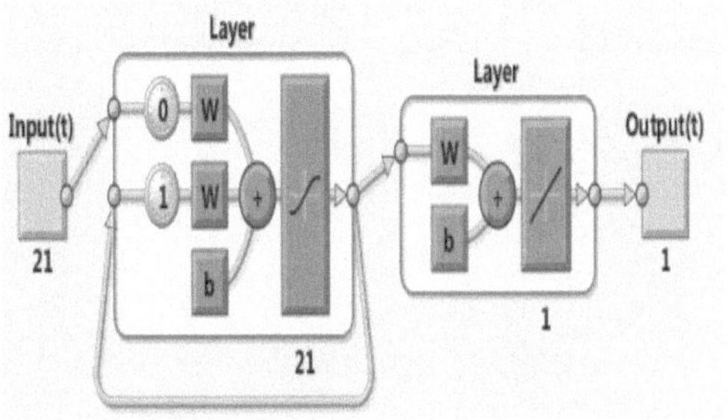

Figure 1. The structure of Elman neural networks.

In this kind of network there is, in addition to the ordinary hidden layer a special hidden layer, called the "link layer" (layer or contact unit); the layer from the hidden layer receives the feedback signal, and each one has a hidden layer of nodes associated with the matching layer node connection. The link layer is used to link together with the input layer and the hidden layer, the equivalent of state feedback. The transfer function of the hidden layer is a nonlinear function, such as the tansign function. The output layer uses the linear function.

2.2 *The input variable selection and networks setting*

The power of wind turbine capture can be expressed as:

$$p = \frac{1}{2}\rho A v^3 C_p \le p = \frac{1}{2}\rho A v^3 * 0.59 \quad (1)$$

In Equation 1: C_p is wind energy utilization coefficient, which means that in unit time, the use ratio of wind energy of absorbed by wind turbines. According to Bates limit, maximum Cp is 0.593. This means that only 59.3% of the wind energy is used for wind machine, even if there is no loss.

The twenty-one input vectors include:

the average, the minimum, the maximum wind speed at one minute, ten minutes, and twenty minutes before the predicted point.

The average temperature at one minute, ten minutes and twenty minutes before the predicted point.

The average, the minimum, the maximum wind power of one minute, ten minutes and twenty minutes before the predicted point.

The sample data are normalized and processed. In order to avoid a saturation phenomenon of neurons, the normalized function in MATLAB is used in the input layer. Function *premnmx(x)* in Equation 2 equivalent wind power value between [−1, 1], the Functions *postmnmx(x)* in Equation 3 is used to transform the actual wind power from [−1,1] in the output layer. In addition, the temperature of the normalized processing can also use the following two equations:

$$y = \frac{x - \frac{1}{2}(x_{\max} + x_{\min})}{\frac{1}{2}(x_{\max} - x_{\min})} \tag{2}$$

$$x = \frac{1}{2}(x_{\max} - x_{\min})y + \frac{1}{2}(x_{\max} + x_{\min}) \tag{3}$$

Where *xmax, xmin* are the maximum and the minimum sample data.

2.3 *The simulation analysis*

The simulation environment is MATLAB (2012) that is based on the data of the wind farm in the Poyang Lake area of Jiangxi Province on December 10, 2012. In this research, the time interval is one minute to predict twenty-five minutes in the future of wind power. The simulation results are shown in Figure 2. The forecasting error is shown in Figure 3. The mean absolute percentage error is defined as Equation 4:

$$\Delta L_{FM} = \frac{1}{N}\sum_{i=1}^{N}\frac{|L_{Fi} - L_{Pi}|}{L_{Pi}} \times 100\% \tag{4}$$

where $\Delta L_{FM}, L_{Fi}, L_{Pi}, N$ are the mean absolute percentage error, the forecasting wind power, the actual wind power, the number of forecasting points, respectively.

Through the forecasting simulation, the mean absolute percentage error ΔL_{FM} is less than 10%. The example shows that the Elman neural network prediction results meet the engineering requirements.

Table 1. The forecasting error of Elman neural networks.

Time (min)	The actual Power (kW)	The forecasting power (kW)	Error data (%)
1	745.55	665.16	−10.78
2	694.37	633.73	−8.73
3	679.92	543.39	−20.08
4	634.69	503.83	−20.61
5	658.73	522.58	−20.66
6	693.18	557.59	−19.56
7	596.94	623.25	4.407
8	677.83	553.43	−18.35
9	684.88	593.39	−13.35
10	581.41	555.76	−4.41
11	692.37	550.44	−20.49
12	752.52	689.75	−8.34
13	729.88	675.47	−7.45
14	714.92	676.55	−5.36
15	726.62	701.56	−3.44
16	755.67	757.56	0.24
17	734.60	694.31	−5.48
18	665.59	699.25	5.05
19	626.95	656.95	4.78
20	770.36	656.15	−14.82
21	762.26	722.16	−5.26
22	801.59	749.91	−6.44
23	802.54	815.06	1.55
24	730.25	687.63	−5.83
25	799.29	730.92	−8.55

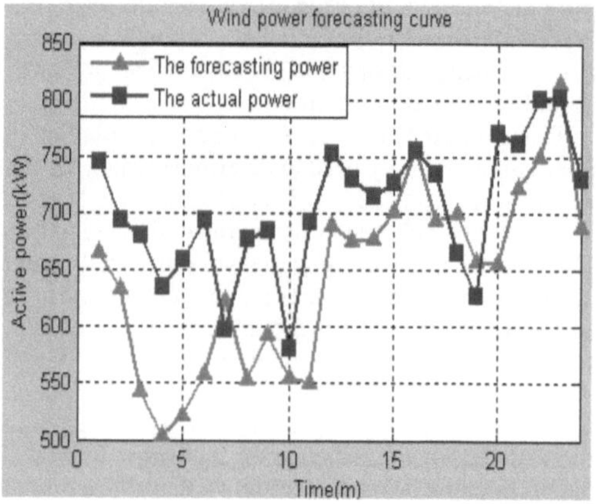

Figure 2. The wind power simulation forecasting results.

3 CONCLUSIONS

Very reasonable results have been obtained from the sample data in Table 1. They show that using the Elman neural network model to forecast the regional power network short-term wind power is effectively feasible. When the reasonable set-up for the input of the Elman neural network, the number of neurons in the hidden layer, the transfer function in hidden layer, the linear function in output layer, the error of wind power prediction would be within the allowed range. The actual measured data for a wind farm in Poyang Lake in

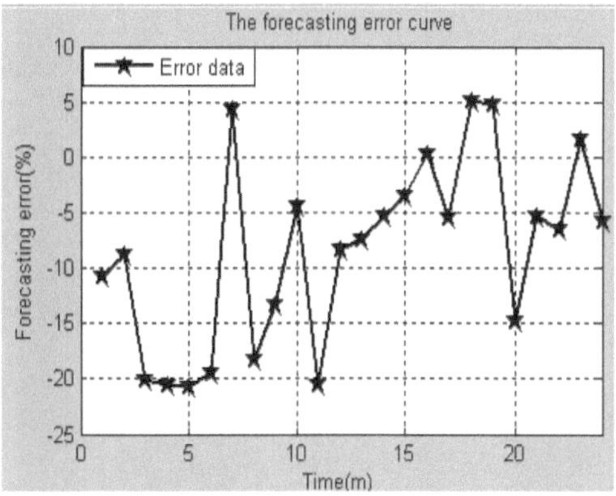

Figure 3. The wind power forecasting error.

Jiangxi Province is given as an example. The example shows that the Elman neural network prediction results meet the engineering requirements.

In short-term wind power forecasting, we must first consider the effect of meteorological factors, especially wind speed, but in different areas, the different meteorological conditions, the different wind turbines, and the different geographical positions will affect the meteorological factors function in short-term wind power forecasting. Secondly, the impact on air density, moisture, and atmospheric pressure should be considered in wind power forecasting. Therefore, in the neural network structure design, the model should include the actual wind power and the area, the meteorological factors, and the Elman neural network model setting. In fact, there is no fixed method or model that is suitable for all occasions and load forecasting, and which is superior to all other methods or models. A special method or model may be suitable for a particular occasion, because some special factors have a significant effect on that particular system. Therefore, for a variety of practical problems within the application, it should be dealt with seriously,such as input variable, the number of layer, the number of hidden layer, the number of cell in hidden layer, the transfer function in hidden layer and out layer; with concrete analysis of the concrete conditions, satisfactory forecasting effects can be obtained.

ACKNOWLEDGEMENTS

This work was financially supported by the Jiangxi Natural Science Foundation (20114BAB206036), the Science and Technology Project of Jiangxi Province Education Department (GJJ13769).

REFERENCES

Gaofeng Fan, Chun Liu, Weisheng Wang, et al. Wind power prediction based on multi-sector BP neural network model. *Chinese Society of Electrical Engineering Annual Meeting Papers*. Tianjin, China, 2009.

Jian Wang, Gangui Yan, Wei Song, et al. Review wind power prediction. *Journal of Northeast Dianli University*, 2011, 31(3):20–24.

Shuang Han. Short-term wind power prediction method research. Beijing: *North China Electric Power University*, 2008.

Ying Wang, Yunjun Wei. Review on forecasting model of wind speed and wind power. *Shanxi Electric Power*, 2011, 39 (11):18–21, 30.

Zack J W. Brower M C, Bailey B H. Validating of the fore wind model in wind forecasting application. *EUWEC Special Topic Conference Wind Power for the 21st Century*, Kassel Germany, 2010.

Zhiling Yang, Yongqian Liu. Short-term wind power prediction with particle swarm optimization. *Power System Technology*, 2011, 35 (5): 159–164.

Zhi Li, Xueshan Han, Li Han, et al. An ultra-short-term wind power forecasting method in regional grids. *Automation of Electric Power Systems*, 2010, 34(7): 90–94.

Future Mechatronics and Automation – Yang (Ed.)
© 2015 Taylor & Francis Group, London, ISBN 978-1-138-02648-3

Design of a multiple function intelligent car based on modular control

Chao Tan & Ling-Yun Wang
College of Electrical Engineering and Renewable Energy, China Three Gorges University, Yichang, China

Hong-Ming Zhao
Yichang Power Supply Company in State Grid Hubei Electric Power Company, Yichang, China

Chao Su
College of Electrical Engineering and Renewable Energy, China Three Gorges University, Yichang, China

ABSTRACT: This paper introduces a multifunctional intelligent car which was designed by an 8-bit micro-controller, Mega128. This multifunctional intelligent car has the function of tracing, metal detection, obstacle avoidance, phototactic, and distance measurement. This multifunction intelligent car system adopts modular control and it can track the route precisely by embedding the fuzzy control algorithm. This car is able to bypass the obstacle in advance by using the ultrasonic detecting obstacles. The phototactic system with adjustable sensitivity is designed to drive the car into the garage. The metal sensor in this system is used to detect the metal block, and the hall device is also used to measure the car driving distance. Finally, the number of metal blocks, running time, and distance will be displayed in the LCD screen. Meanwhile, the friendly human-machine interface is realized. The experiments show that the intelligent car system is able to complete variable function and achieve high stability and accuracy on the independent control module.

Keywords: Intelligent car, multiple function, modular control, tracking, obstacle avoidance

1 INTRODUCTION

The intelligent car is mostly designed as a single function tracing car, an obstacle avoidance car, or a remote control car, at present. This paper will design a multiple function intelligent car system with the function of tracking, metal detection, obstacle avoidance, light tracing, and distance measurement. Meanwhile, optimal control programs are used for each module that expects to get more precise control. This intelligent car system has the characteristics of high efficiency, high accuracy, and low cost.

The control methods of the intelligent car system are diverse such as conventional PID control method [1], fuzzy control method [2], image recognition method [3] and so on. All of these methods can achieve an ideal result for the car with single task control. In the designed intelligent car system proposed in this paper, the multitask feature of this car will be fully considered and the multifunctionality of the car will be achieved through the modular control method. Meanwhile, the fuzzy control algorithm is embedded on the main module so that the multifunction requirements of this car can be satisfied and the independent, stable operation of the control module can also be realized.

2 SYSTEM STRUCTURE DESIGN

Generally, the system design includes the hardware and software design. This system is provided by the power supply module, the drive module, the tracking module, a metal detector module, an ultrasonic obstacle avoidance module, a light tracing module, and the display module. The system hardware structure is shown in Figure 1.

2.1 *Driver module*

In this module, the driver chip L298N is used to control two DC motors and to change the speed of these two motors. The car is able to turn a corner or its running speed can be changed.

2.2 *Tracking module*

In this module, photoelectric sensors are used to detect the black line in the racing track. If the light emitting diode inside the photoelectric sensor is above the black line, the micro-controller will drive the car and turn to deviate the photoelectric sensors from the black line. Once they leave the black line, the micro-controller

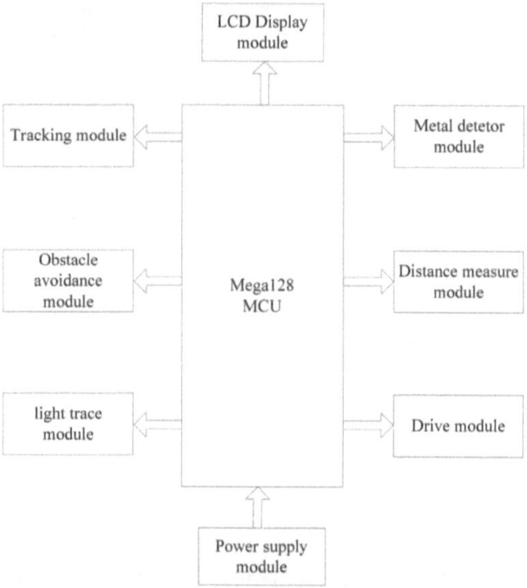

Figure 1. System structure diagram.

will drive the car straighter. In this loop, the car will be able to drive along the given black line.

Meanwhile, the fuzzy control algorithm is embedded in the tracking module in order to enable the car to adapt the route with different radian. Firstly, a fuzzy set X = {very partial left, partial left, left, centre, right, partial right, very partial right}, is established. Then, the fuzzy rules are defined as that when the first photoelectric sensors on the left side are above the black line, the car direction is considered to be right deviation; when the second photoelectric sensors on the left side are above the black line, the car direction is considered to be partial right; when the third photoelectric sensors on the left side are above the black line, the car direction is considered to be very right partial. If none of the photoelectric sensors are above the black line, the car direction is considered to be in the centre. In the same way, the fuzzy rule can be defined to the right side of the photoelectric sensor. Finally, the car can be able to adapt to the changeable routes through fuzzy control.

2.3 *Metal detection module*

As a result of fixing the metal sensors on the bottom of the car, when the car goes through the metal blocks, a micro-controller will receive a signal given by the metal sensors to control the buzzer to sound the alarm, and send the number of interruptions to the LCD, so that the number of metal blocks detected is displayed.

2.4 *Ultrasonic obstacle avoidance module*

In this module, the interruption capture function of Mega128 micro-controller is fully used. The pulse signal produced by the ultrasonic generator, which used to measure the distance from the car to the obstacle. The working process as follow: the falling edge trigger fashion is select as interruption fashion in micro-controller, when the interruption system triggered by receiver signal, which obtained from ultrasonic receiver sensor in the car, the timer is start working, at the same time the interruption fashion is set to the rising edge trigger. When the second interrupt event occurred, the time T1 of ultrasonic transmitted time in space can be obtained from timer, and interruption system will be closed until next measurement. Supposing the crystal of micro-controller is M, The frequency divider factor is N, the speed of ultrasonic in space is v, the time of ultrasonic transmission in space between the car with the obstacle is: $t = T1 \times M/(N \times 10^6)$, and the distance from the car to the obstacle is :$v \times t/2$.

2.5 *Silicon photocell phototaxis module*

In order to make this car work better, silicon photovoltaic cells are used as sensors in this module. When the light source is irradiated onto the surface of the silicon photovoltaic cells, the resistance value of the silicon photovoltaic cell will change; the greater intensity of the light, the smaller the resistance value. Installing three silicon photovoltaic cells on the typical position of the front of the car. Then, connecting these silicon photocells with the same reference voltage source, feeding the terminal voltage to the sampling channel of micro-controllers and transforming it to digital sample value. Judging from which direction the light intensity is greater by comparing the three sample values to decide if the car should go straight or turn a corner and then tend to the light source.

2.6 *Ranging module*

Installing two small magnets on the motor shaft and fix the Hall sensor on the vehicle body just above the rotary shaft. When the motor rotates, the magnet through the Hall element at the output end of the Hall sensor will generate electrical level hopping and record the electrical level hopping number. The travelling distance can be measured based on the travelling distance formula $L_2 = (N \times 2\pi R)/2$. Moreover, the measured distance can be displayed on the LCD screen.

3 SOFTWARE DESIGN

On the basis of the designed hardware, the function of the car can be controlled by software embedded in the micro-controller. A modular control method is adopted in this system in order to make each independent function as a separate subroutine. In the main program, these subroutines are invoked according to the logical sequence. So that the program will be more clear and the efficiency of the implementation of the program will be improved. Also the program modification and debug can be realized conveniently.

As the description in the tracking module in the main program, there are three essential subroutines.

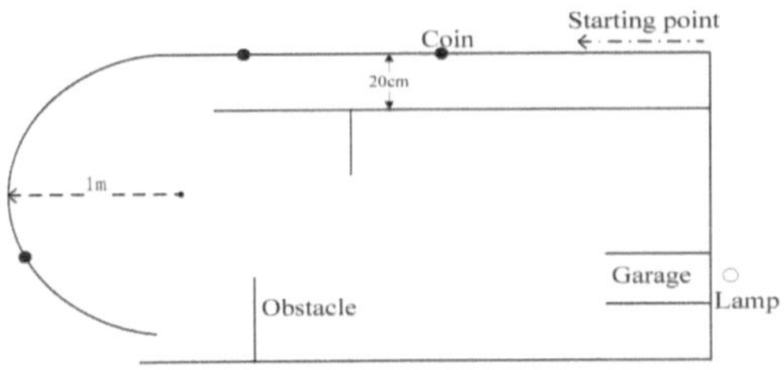

Figure 2. Tracing track for test.

Table 1. Test results of tracking module.

	Carton	Stone	Plank
number of tests	10	15	10
number of successes	9	11	10

The fuzzy control algorithm is used in Subroutine 1 to control the car to be driven along the tracking route. Subroutine 2 calculates the distance between the car and the obstacles and the car is controlled to avoid obstacles. In Subroutine 3, both left and right wheel of the car are controlled to rotate as light intensity of the silicon photovoltaic cell in order to make the car running into the garage successfully. The program adopts dynamic scanning and real-time monitoring so that the car is in the state of dynamic regulation.

4 PHYSICAL TESTING AND RESULTS ANALYSIS

In order to test the performance of the car, a racing track is designed according to the competition requirements shown in Figure 2. Select an open, weak natural light space as the test site. The car will drive along the black traction lines from the starting point, and after leaving the traction line, the car will avoid the obstacles and trace the light to drive into the garage. Several coins are placed under the traction line for detection, and as the car detects the coins while it passes along the road, the number of coins will be recorded and displayed on the LCD screen.

Due to the sensitivity adjustment module is added in the hardware of tracking system and the fuzzy control algorithm is adopted in the software design. Therefore, the tracking test results are better and the car can drive along the designed route. The test results are shown in Table 1.

In the obstacle avoidance module, since the ultrasonic module has requirements for the shape and surface area of the obstacle, when the obstacle does not meet the requirements, the ultrasonic module will not receive a reflected signal so that the obstacle avoidance will be failed. Furthermore, the people around the route can be also considered as a reflection source and will affect the test result. The test results are shown in Table 2.

Table 2. Test results of obstacle avoidance module.

	Straight	Curved corners	Serpentine bend
number of tests	10	10	15
number of successes	10	10	12

Table 3. Test results of phototaxis module.

	100 W Lamp	150 W Lamp	200 W Lamp
number of tests	15	15	15
number of successes	11	12	14

Table 3 shows the phototaxis module test results. The phototaxis module using silicon photovoltaic cell sampling, which has continuously variable promise comparison, is more accurate than the widely used photo-resistance that only compares high and low sensitivity, so the accuracy of the system is improved. As with the phototaxis module test, the same is true of the photo-resistance systems; the impact of natural light cannot be completely avoided.

5 CONCLUSION

This system uses an 8-bit micro-controller where a modular design concept is adopted to realize the design of the multifunctional intelligent car. In this system a variety of sensors are integrated, and multichannel signals are inputted into the micro-controller for analysing and judgements. Meanwhile, a friendly human-machine exchange interface is also designed within this system, which makes it more convenient to read out the number of detected metal blocks, the travelled distance and the running time. The use of the modular control method makes each part have less effect on the other modulars. Therefore, more modules can be added to this system to extend the other functions for secondary development. For example, the voice broadcasting function can be realized when the voice chip is embedded on this system. In addition, the car detection function can be also realized when a camera and wireless module are added.

REFERENCES

Echegaray S. & Wenbin L. 2008. The modular design and implementation of an intelligent cruise control system. IEEE International Conference on System of Systems Engineering, SoSE, Monterey, CA, United States, Vol. 1 (2008), p. 1–6.

Hanafi D. 2010. Comput. Eng. Appl. Vol.2 (2010), p. 60–63.

Ito K., Ichihara N., Inoue H., Fujita R. & Yasushi M. 2009. Car navigation system with image recognition. Digest

of Technical Papers – IEEE International Conference on Consumer Electronics, Las Vegas, Nv, United States, pp. 1–2, 2009.
Kyo S., Okazaki S., & Arai T. 2007. IEEE Trans. Comput. Vol. 56 (2007), p. 622–634.
Lupu C. & Lupu V. 2007. Multimodal biometrics for access control in an intelligent car. ISCIII'07: 3rd International Symposium on Computational Intelligence and Intelligent Informatics. Proceedings, Agadir, Morocco, Vol. 3 (2007), p. 261–267.
Srovnal V., Machacek Z., Hercik R., Slaby R. & Srevnal V. 2010. Intelligent car control and recognition embedded system. Proceedings of the International Multi conference on Computer Science and Information Technology, IMC-SIT, Wisla, Poland, Vol. 5 (2010), pp. 831–836.

Section 3: Intelligent robotics

Future Mechatronics and Automation – Yang (Ed.)
© 2015 Taylor & Francis Group, London, ISBN 978-1-138-02648-3

Research on virtual human motion generation using KernelPCA method

Xianqian Hu, Jiahong Liang, Quanping Liu & Yuewen Fu
National University of Defense Technology, Changsha, China

ABSTRACT: Aiming to the poor variability of the traditional methods for generating virtual human movements, we propose a method for generating virtual human movements based on KernelPCA method. Firstly the optimal kernel function is determined through optimization algorithm, which is used to generate a high-dimensional linear separable feature space. Then we reduce the dimension of the feature space by PCA algorithm. Finally we conduct linear mapping between the coefficients of the feature vector and the motion parameters. Through this method, users can obtain virtual human motion actions with different motion attributes and motion types based on a small amount of motion data.

1 INTRODUCTION

In recent years, with the continuous development of motion capture technology, the research on virtual human technology based on virtual human motion capture technology has become a hot topic in computer graphics, virtual reality and intelligent man-machine interfaces and other areas. These methods drive virtual human by recording real human motion data, by which the motions generated have a wealth of details, strong sense of reality, high efficiency and other merits. However, it is just a copy of motion and only able to generate a designated body movement captured beforehand. Moreover, the data captured just can be applied in specific environments. Therefore the technology is not suitable to be applied in changing scenes or with user interaction. To improve the utilization coefficient of motion capture data, researchers have proposed many example-based motion synthesis algorithms, which edit motion capture data properly, in order to improve their reusability and expand applicable range (Meng 2003, Guan 2005, Thalmann 2005 & Shilei 2010). Nevertheless, whether the motion edit techniques based on restrictions, signal processing, or hybrid interpolation, the following shortcomings always exist: they cannot directly produce movement with given characteristics, and cannot provide extrapolation.

However, principal component analysis (PCA) method has been widely used to edit and synthetize motion currently, whose superiority is the capability to decompose high-dimensional data by linear mapping, and then process the motion attributions by interpolation and extrapolation calculation in different feature spaces. For example, Thalmann (2007) and Clardon (2004) utilize PCA method to decompose motion data into the data with different physical attributions (velocity) and different motion types (walking, running), and then manipulate PCA coefficients to synthetize motion data with different attributions and types. Linear PCA cannot always detect out all structures of the given data, while KernelPCA uses appropriate nonlinear characteristics to extract more information which is more propitious to extract the complex nonlinear relationships of the raw data (Zhuoming 2010 & Junbao 2010). KernelPCA can also extract the features which are more favorable to classification. Considering that body movement is a very complex high-dimensional nonlinear system, this method can achieve higher recognition rate to virtual human's movements.

Consequently, this paper presents a method based on KernelPCA to generate virtual human motion. Firstly the optimal kernel function is determined through the optimization algorithm, which is used to generate a high-dimensional linear separable feature space. Then we reduce the dimension of the feature space by KernelPCA algorithm. Finally we conduct linear mapping between the coefficients of the feature vectors and the motion parameters, to generate virtual human's motion user desired in real time.

2 PCA AND KERNELPCA METHOD

2.1 *PCA method*

Principal component analysis (PCA) is an important statistical method to study how to transform the problem with multiple indicators into the problem with less composite indicators, which also can transform problems from a high-dimensional space into a low-dimensional space, thus simplifying the problems. In addition the composite indicators are uncorrelated with each other and contain most information of the original indicators.

The basic idea of PCA is concluded as follows: primarily transforming original component-relevant random variables into new irrelevant variables by the orthogonal transformation, i.e. transforming the covariance matrix of original variables into diagonal matrix, which changes the original variable system into a new orthogonal system that points to the orthogonal direction in which the sample points disperse into the widest range, and then reducing the dimension of the multidimensional variable system. In accordance with the views of the feature extraction, PCA is equivalent to an extraction method based on the minimum mean square error. This paper utilize the basic PCA idea dilated in the literature (Glardon 2004) to accomplish specific solution of PCA space.

2.2 *KernelPCA method*

Kernel-based principal component analysis (KemelPCA) is a new feature extraction method, which expands PCA non-linearly with the kernel technique. It is capable to capture the non-linear feature of the data effectively, but the kernel function will greatly affect the performance of this algorithm.

KernelPCA method doesn't calculate the eigenvectors of the sample's covariance matrix directly, instead, solve the eigenvalues and the eigenvectors of the kernel matrix, thus avoiding to calculate the eigenvectors in the entire feature space. Compared with other nonlinear feature extraction methods, it does not need to solve the nonlinear optimization problems.

KernelPCA conducts nonlinear transformation on a sample set $X = \{\mathbf{x}_1, \ldots, \mathbf{x}_m\}$, obtaining $\Phi(X)$, which maps X into a high-dimensional space F. For the new sample space, the covariance matrix is shown as formula (1).

$$C = \frac{1}{m}\sum_{i=1}^{m}\Phi(\mathbf{x}_i)\Phi(\mathbf{x}_i)^T \tag{1}$$

We can obtain eigenvalue λ and eigenvector v of C by PCA method (Glardon 2004). When $\lambda \neq 0$, v locates in the space expanded by $\Phi(X_i)$, so $\exists a_i$ $(i = 1, 2, \ldots, m)$ satisfies formula (2).

$$\mathbf{v} = \sum_{i=1}^{m} a_i \Phi(\mathbf{x}_i) \tag{2}$$

Conducting inner product operation at both sides of the equation $\lambda\mathbf{v} = \mathbf{Cv}$ with $\Phi(\mathbf{x}_i)$ yields formula (3).

$$\lambda < \Phi(\mathbf{x}_i), \quad \mathbf{v} > = < \Phi(\mathbf{x}_i), \quad \mathbf{Cv} > \tag{3}$$

Substituting formula (1) and (2) into formula (3) and simplifying it yields formula (4).

$$m\lambda\mathbf{a} = \mathbf{Ka} \tag{4}$$

where, $\mathbf{K} = (\Phi^T(\mathbf{x}_i)\Phi(\mathbf{x}_j))_{m\times m} = [k(x_i, x_j)]_{m\times m}$ is the kernel matrix, $\mathbf{a} = (a_1, a_2, \ldots, a_m)^{\mathbf{T}}$. $m\lambda$ and $\mathbf{a}$ are the eigenvalue and the eigenvector of kernel matrix K respectively. Supposing that the eigenvectors corresponding to the eigenvalues which are greater than 0 are $\mathbf{a}^{\mathbf{p}}, \mathbf{a}^{\mathbf{p+1}}, \ldots \mathbf{a}^{\mathbf{m}}$, in order to hold $\langle v^k, v^k \rangle = 1$, choosing $\mathbf{a}^{\mathbf{k}}$ which satisfys $\langle a^k, a^k \rangle = 1$, then the projection of the sample $\Phi(\mathbf{x})$ on $\mathbf{v}^{\mathbf{k}}$ is shown as formula (5).

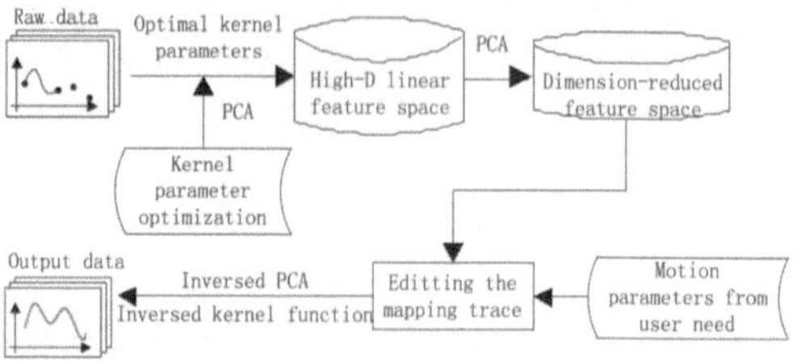

Figure 1. The procedure of KernelPCA-based motion generation and control algorithm.

$$g_k(\mathbf{x}) = \left\langle \mathbf{v}^{\mathbf{k}}, \Phi(\mathbf{x}) \right\rangle = \sum_{i=1}^{m} \mathbf{a}_{\mathbf{i}}^{\mathbf{k}} k(\mathbf{x}_{\mathbf{i}}, \mathbf{x}) \tag{5}$$

where $k = p, p+1, \ldots, m$. $g_k(\mathbf{x})$ is the nonlinear principal component corresponding to $\Phi(\mathbf{x})$. The vector $(g_1(\mathbf{x}), g_2(\mathbf{x}), \ldots, g_\ell(\mathbf{x}))$ deprived from all the projection values can be regard as the new feature of the sample X.

3 KERNELPCA-BASED MOTION GENERATION AND CONTROL

This paper proposes a virtual human motion generation and control algorithm based on KernelPCA. We firstly use kernel function to map the low-dimensional linear inseparable data into the high-dimensional linear separable data, and then utilize PCA method to reduce its dimension. The primary task of KernelPCA method is to choose the optimal kernel function and the optimal parameters according to the specific issues in order to achieve the best effect and improve computing performance of the algorithm. At first, we determine the optimal kernel function and the optimal parameters, and then analyze and process data in the feature space by PCA algorithm to obtain a dimension-reduced PCA space, at last interpolate and extrapolate the dimension-reduced PCA space to generate the motions user desired.

The procedure of KernelPCA–based virtual human motion generation and control algorithm is shown in Figure 1.

3.1 *Obtain and capture data*

If only the moving gesture of virtual human need to be considered, it can be represented by the angle vectors of other joints (not including the location and orientation of the root node), thus the action changes can be represented by the angle vector θ. When capturing data, we select three people with different heights as the objects. Every sample's speed ranges from 0 m/s to 2 m/s, and sampling is conducted at every 0.1 m/s. Four motion cycles (a walking cycle is designated from left foot lifting up to right foot falling down) in a moving sequence corresponding to one speed are picked out, while the motion cycle at 0 m/s is represented by four stand cycles. Therefore, the database consists of the motion data in 252 cycles. The data are normalized, so that each motion cycle contains a fixed number of frames.

3.2 *Determine the optimal kernel*

Different kernel functions bring the feature spaces mapped with different data distribution structures. Kernel function will directly affect the distribution of the sample data in the feature space thus affecting kernel learning algorithm's performance. This paper proposes a kind of kernel learning method to achieve the optimal kernel function, which can make the resulting feature space of better linear separability, thus capturing the non-linear data more effectively.

3.2.1 *The kernel function optimized*

This paper introduces the data-dependent kernel function proposed by Amari et al. (1999) as the objective kernel function, shown as formula (6):

$$k(a,b) = q(a)q(b)k_0(a,b) \tag{6}$$

where, $k_0(a, b)$ is the basic kernel function, such as Gaussian kernel or polynomial kernel function, a, b are the data samples, $q(.)$ is the factor function, which can be expressed as $q(a) = a_0 + \sum_{i=1}^{m} a_i k_1(a, e_i)$, where, $k_1(a, e_i) = e_i^{\|a-e\|^2/t_1}$, a_i is a combination coefficient, $e_1, e_2, \ldots, e_m$ are m vectors randomly selected from the training samples or selected according to the training samples' distribution. Indicating the above formula (6) in a matrix form yields formula (7):

$$K = QK_0Q \tag{7}$$

where, Q is a diagonal matrix, the entries on the diagonal, respectively, are $q(\mathbf{x}_1), q(\mathbf{x}_2), \ldots q(\mathbf{x}_N)$. Setting $\mathbf{q} = [q(\mathbf{x}_1), q(\mathbf{x}_2), \ldots q(\mathbf{x}_N)]^T$, $\mathbf{a} = [a_0, a_1, \ldots a_m]^T$ yields formula (8):

$$q = \begin{bmatrix} 1 & k_1(x_1, e_1) & \cdots & k_1(x_1, e_m) \\ 1 & k_1(x_2, e_1) & \cdots & k_1(x_2, e_m) \\ \vdots & \vdots & \ddots & \vdots \\ 1 & k_1(x_N, e_1) & \cdots & k_1(x_N, e_m) \end{bmatrix} \begin{bmatrix} a_0 \\ a_1 \\ \vdots \\ a_m \end{bmatrix} = K_1 a \tag{8}$$

According to formula (6), the elements k_{ij} in the kernel matrix can be obtained as formula (9):

$$k_{ij} = (k_0)_{ij} \mathbf{a}^T (k_1)_i^T (k_1)_j \mathbf{a} \tag{9}$$

where, $(k_1)_i$ is the ith row of the matrix $\mathbf{K}_1$.

3.2.2 *Establish the optimal objective function*

According to the above formula, to seek optimized kernel function is equal to find out the optimal combination coefficient of the constrained optimization equation among selected combination coefficients. This paper establishes the optimal objective function with the maximum interval criterion according to the literature (Saul et al. 2004), shown as formula (10):

$$\begin{aligned} \Gamma &= \frac{1}{2}\sum_{i=1}^{N}\sum_{j=1}^{N}\left\|\Phi(x_i) - \Phi(x_j)\right\|^2 = \\ &\frac{1}{2}\sum_{i=1}^{N}\sum_{j=1}^{N}(k_{ii} + k_{jj} - 2k_{ij}) = N\sum_{i=1}^{N}k_{ii} - \sum_{i=1}^{N}\sum_{j=1}^{N}k_{ij} \end{aligned} \tag{10}$$

Substituting formula (7) and formula (8) into formula (10) yields formula (11).

$$\begin{aligned} \Gamma &= N1_N^T Q(I.*K_0)Q1_N - 1_N^T QK_0Q1_N \\ &= Nq^T(I.*K_0)q - q^T * K_0 q \\ &= a^T K_1^T[(NI - 1_{N\times N}).*K_0]K_1 a = a^T V a \end{aligned} \tag{11}$$

where I is the identity matrix, $\mathbf{1}_{\mathbf{N}\times\mathbf{N}}$ is the matrix whose elements are 1, V is a positive semi-definite matrix.

3.2.3 *Constraint conditions*

If maximizing Γ infinitely, the geometry of the training samples may be undermined. For the preservation of the local geometry of the input data, here we introduce a binary adjacency matrix S, used to describe the local relationship between the data. If the data $\mathbf{x}_j$ is one of the nearest k data away from $\mathbf{x}_i$, then $S_{ij} = 1$, otherwise $S_{ij} = 0$. According to the literature (Saul et al. 2004) the mapping space with a partial interval constraint should satisfy the equation (12):

$$\left\|\Phi(\mathbf{x}_i) - \Phi(\mathbf{x}_j)\right\|^2 = k_{ii} + k_{jj} - 2k_{ij} = \left\|\mathbf{x}_i - \mathbf{x}_j\right\|^2 \tag{12}$$

Substituting formula (9) into the above equation yields formula (13):

$$\begin{aligned} &\mathbf{a}^T[(k_0)_{ii}(\mathbf{k}_1)_i^T(\mathbf{k}_1)_i + (k_0)_{jj}(\mathbf{k}_1)_j^T(\mathbf{k}_1)_j - \\ &2(k_0)_{ij}(\mathbf{k}_1)_i^T(\mathbf{k}_1)_j]\mathbf{a} = \mathbf{a}^T C_{ij}\mathbf{a} = \left\|\mathbf{x}_i - \mathbf{x}_j\right\|^2 \end{aligned} \tag{13}$$

3.2.4 *Seek out the optimal combination coefficient*

Generally the number of constraint equations is more than the number of combination coefficients, which makes $\mathbf{a}$ overdetermined. Here we relax the constraint conditions, so the optimization problem is expressed by Equation (14):

$$\mathbf{a}^* = \arg\min_{\mathbf{a}} -\mathbf{a}^T\mathbf{V}\mathbf{a} \quad \text{s.t.} \quad \mathbf{a}^T C_{ij}\mathbf{a} \le \left\|\mathbf{x}_i - \mathbf{x}_j\right\|^2 \tag{14}$$

Then we can obtain the optimal kernel function by solving formula (14) with genetic algorithms (Goldberg 1989).

Finally, we make experimental analysis of a set of raw data, the feature space mapped by the basic

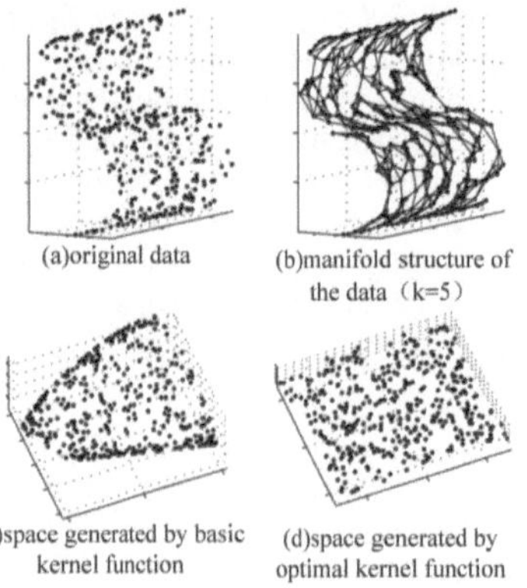

Figure 2. The feature space mapped by the basic kernel and the optimization kernel function.

kernel and the optimization kernel function shown in Figure 2.

The spatial distribution of the data are obvious in graph (c) and (d). Among these graphs, only the space in graph (d) is in a manifold structure, and more similar to a two-dimensional linear space.

3.3 *Reduce space's dimension by PCA*

To improve the computational efficiency and achieve the purpose of generating motions in real time, this paper will make the PCA analysis of the high-dimensional feature space of every sample data, generate the corresponding sub PCA space, establish the corresponding relationship between the speed value of each sample and its principle component, and conduct interpolation and extrapolation operation, thus reducing the space dimension and the calculation amount.

The PCA space is achieved by analyzing and processing the raw data with PCA algorithm (Glardon 2004), shown in formula (15):

$$\Phi(\theta) \cong \bar{\Phi}(\theta) + \sum_{i=1}^{m} \alpha_i e_i = \bar{\Phi}(\theta) + \alpha E \tag{15}$$

where, the vector $\bar{\Phi}(\theta)$ is the average vector of all n-dimension vectors $\Phi(\theta)$, $\alpha = (\alpha_1, \alpha_2, \ldots, \alpha_m)$ is the principle component's coefficient matrix, $E = (e_1, e_2, \ldots, e_m)$ is the principle component matrix of $\Phi(X)$.

A new PCA space, called sub-PCA space, can be generated by using PCA method on $\alpha = (\alpha_1, \alpha_2, \ldots, \alpha_m,)$. Therefore, the coefficient matrix α with respect to $\Phi(\theta)$ can be expressed as Equation (16):

$$\alpha \cong \bar{\alpha} + \sum_{i=1}^{n} \beta_i f_i = \bar{\alpha} + \beta F_\theta \tag{16}$$

where, $\beta = (\beta_1, \beta_2, \ldots \beta_n)$ is the coefficient matrix of the new principle component, $F_\theta = (f_1, f_2, \ldots f_n)$ is the

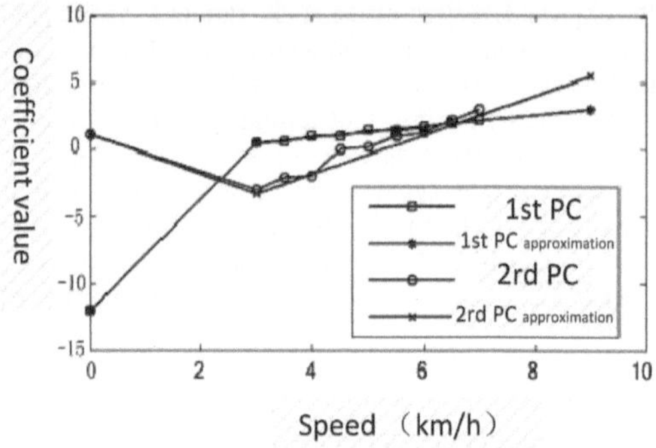

Figure 3. The effect of applying approximate function (PC refers to the principle component).

principle component matrix, n is less than the data space dimension of $\Phi(\theta)$, $\bar{\alpha}$ is the mean of the principle component coefficient matrix of $\Phi(\theta)$.

The linear approximation function will be constructed in any-dimension data space in order to accomplish interpolation and extrapolation operation. We establish a linear function $B_i(V) = B_i = m_i V + b_i$ with respect to the coefficient vector β and the corresponding velocity value V (not including zero speed). This function is laid on the foundation that formula (17) reaches the minimum.

$$\sum_{j=1}^{nbs} (\beta_{ij} - d_{ij})^2 = \sum_{j=1}^{nbs} (\beta_{ij} - m_i s_i - b_i)^2 \tag{17}$$

where, β_{ij} is the actual coefficient value, d_{ij} is the approximate coefficient value, *nbs* refers to the number of the speed values, j is the index corresponding to different speed. Figure 3 shows the experimental results of applying approximate function to an object, from which we can see a pretty approximation between various principle components. Therefore, at a given speed, using the formula (16) and (17) can realize not only the interpolation operation, but also the extrapolation operation.

3.4 *Motion redirection and time bending*

When capturing data from different objects, we must ensure the diversity of postures, at the same time, we should pay more attention on the height difference of them. Murry (1967) has demonstrated that "leg related angles" have very similar trajectories in the sagittal plane, when adults moving in the same velocity. Speed V is defined to be the ratio of the actual velocity and the height of the hip (H), shown as formula (18):

$$V = v / H \tag{18}$$

After obtaining speed V, the motion is normalized through time bending method, which yields the data samples with fixed frame number in every motion cycle. Time bending method introduced is determined by a walking cycle function connecting speed with

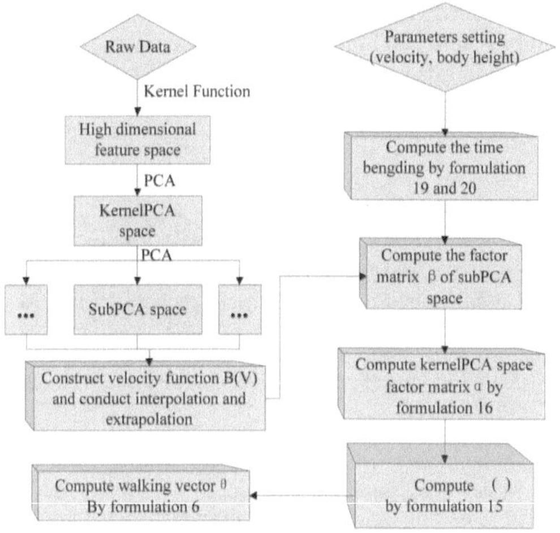

Figure 4. Virtual human motion generation process with KernelPCA.

frequency (Inman 1981). The above method is called Inman rules, which is applied in our experiments. We assemble the data so that they have the approximate property with the function ax^b, to conform to the basic requirements of Inman rules. For all the data captured during the walking process, the statistical frequency function is expressed as formula (19):

$$f = 0.85V^{0.4} \tag{19}$$

Thus, the walking engine is capable to adjust the speed at any time, and estimate the stage when walking (Boulic 1990). In general, for a given time variable Δt, the change of the stage $\Delta\phi$ is determined by formula (20):

$$\Delta\varphi = \Delta t \,/\, walking\_cycle = \Delta t \cdot f \tag{20}$$

where, f is calculated by formula (19). Therefore, our time bending method is a function about the speed value, which is "normalized" based on the human leg's height, thus realizing motion redirection.

Through the above analysis, virtual human motion generation process with KernelPCA is shown in Figure 4.

4 RESULTS AND ANALYSIS

In order to verify the feasibility of virtual human motion generation method based on KernelPCA, we further develop a dynamic simulation environment on matlab. In this environment, we utilize the existing motion capture data of three objects in different heights under different speeds to achieve motion generation by our method. As shown in Figure 5, when expressing information more than 85%, this algorithm generally needs 2 to 5 principle components.

Additionally, compared with the PCA-based virtual human motion generation method, it is found

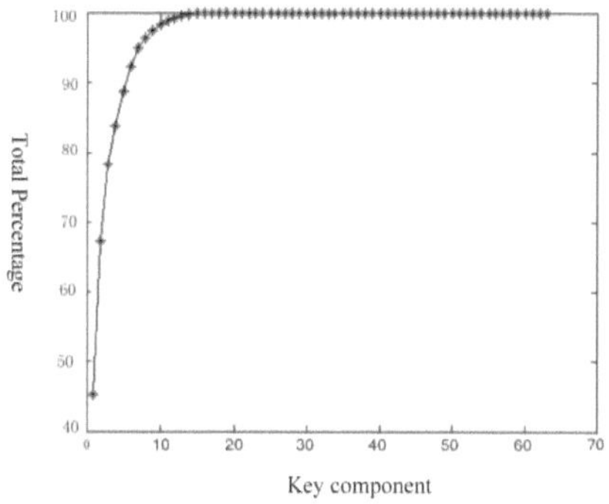

Figure 5. The proportion of the principal components holds in data information.

Table 1. The computation time ratio and the respective recognition rate of KernelPCA and PCA.

Movement types	Walking	Running	Recognition rate (k = 5)	
Number of training samples	100	100	KPCA	90.5%
KernelPCA/PCA	13.21	6.90	PCA	81.3%

that the calculation speed of the feature extraction by KernelPCA is slower than that by PCA, but the recognition rate of KernelPCA is higher. That is to say, under the same feature extraction vector, the motion generated by KernelPCA method is more authentic, and filled with more moving details. Table 1 shows the computation time ratio and the respective recognition rate of KernelPCA and PCA under 100 training samples with different movement types. (Training sample library is the basic movement data downloaded from Graphics Lab of Carnegie Mellon University).

When the matlab program is running, we modify the speed and height parameter in real-time. Concluding from the results obtained, this algorithm is able to control virtual human with different types to complete relatively smooth movement, with preferable action fidelity. Figure 6 shows the straight-walking of virtual human generated under different speeds and different hip heights.

According to the above method, we capture data under different movement types, and then use Poser, DI-GUY to generate three-dimensional human body model, next, add and modify the freedom of virtual human in Creator, finally, generate the basic movements in VegaPrime. Virtual human's walking and running movements generated by this method are shown in Figure 7, which shows KernelPCA-based method is capable to produce more authentic virtual human movements.

5 CONCLUSIONS

This paper conducts a research on virtual human walking movement generation using KernelPCA approach,

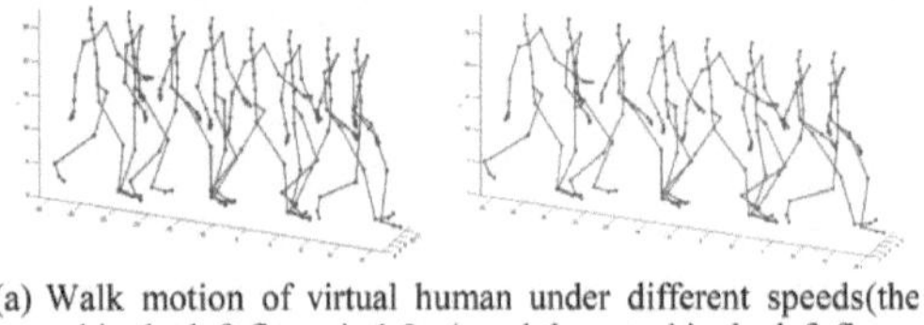

(a) Walk motion of virtual human under different speeds(the speed in the left figure is 1.0m/s and the speed in the left figure is 1.5m/s)

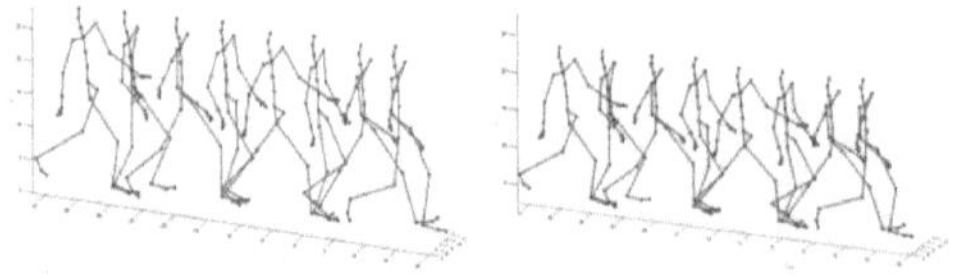

(b) Walk motion of virtual human with different hip heights

Figure 6. Walking movements of virtual human under different speeds and different hip heights.

(a) Walking movement

(b) Running movement

Figure 7. KernelPCA-based virtual human walking and running movement generation.

and generates motion animations with different attributes through mapping the eigenvector coefficients to the motion attributes. Meanwhile it makes a preliminary comparison between KernelPCA and PCA, showing the feasibility and the efficiency to recognize movement feature of KernelPCA algorithm. At last, it is indicated from the experiment results that the proposed virtual human motion generation algorithm is able to generate lifelike animation, which provides an low-level method to the research on generating high-level complex behaviors.

REFERENCES

Amari, S, Wu, S. 1999. Improving Support Vector Machine Classifiers by Modifying Kernel Functions. *Neural Networks* 12(6): 783–789.

Boulic R. et al. 1990. A global human walking model with real-time kinematic personification. *The Visual Computer* 6(6):344–358.

Goldberg, D.E. 1989. *Genetic Algorithms in Search, Optimization and Machine Learning.* Boston: Addison-Wesley Longman Publishing Co.

Guan Luo et al. 2005. Virtual human technology research Overview. *Computer engineering* 31(17): 7–9.

Glardon, P. et al. PCA-based Walking Engine using Motion Capture Data. *In Proceedings of the Computer Graphics International*,2004: 292–298

Inman, V. et al. 1981. *Human Walking.* Baltimore: Wiliams & Witkins.

Junbao, Li & Gao, Huijun. 2010. The Kernel optimization algorithm based on data-dependent kernel function. *Pattern recognization and Artificial Intelligence* 23(3):300–306.

Murray, M.P. 1967. Gait as a total pattern of movement. *Am J Phys Med* 46(1):290–333.

Meng, Xu et al. 2003. Study on virtual huamn motion contol technology. *Academic journal of system simulation* 15(3): 338–342.

Shilei, Li. 2010. Reseach on virtual human motion generation and control hybrid griven by data and model. National University of Defense Technology.

Saul, L.K. et al. 2004. Think globally, fit locally: unsupervised learning of low dimensional manifolds. *Journal of Machine Learning Research* 4(2): 119–155.

Thalmann, D., Raupp Musse, S. 2007. *Crowd Simulation.* London: Springer.

Thalmann, N.M. & Thalmann, D. 2005. Virtual Humans: Thirty Years of Research, What Next. *The Visual Computer* 21(12): 997–1015.

Zhuoming, Du et al. 2010. Research on KPCA method process and application. *Computer engineering and application* 46(7): 8–10.

Future Mechatronics and Automation – Yang (Ed.)

The research and realization of digital library landscape based on OpenGl

Wei-Guang Liu & Yu-Xian Hui
Zhong Yuan Institute of Technology, Zhengzhou, Henan, China

ABSTRACT: By comparison of virtual modeling technology, this paper gives graphic rendering technology combining with third-party software, and finally designs the method of visual and virtual scene based on OpenGl. Zhongyuan institute of technology library landscape is designed in 3d digital library building landscape drawing and various real-time, human-computer interaction operations are realized by using OpenGl. This simulation method of programming has high efficiency, is easy to operate, and can make users to observe effect from multi-angle, multi-dimensional. Meanwhile, it also can be used in game development, terrain modeling, urban planning, product designing and many other areas. The experiments show that the method given in the paper has the characteristic of more manageable operation than before and high performance.

1 INTRODUCTION

With the development of multimedia technology, visualization technology and computer graphics technology, virtual reality technology has become more and more important. By the technology, mechanical engineer can be freed from 2D planar, step into 3D world and get 3D model of mechanical parts. Military commanders can face with battlefield terrain generated by 3D technology, make 3D planes, warships, tanks with sense of reality and analyze combat scheme. Virtual reality technology has three important features, immersion, interaction, imagination.

Nowadays, there are mainly three kinds of methods in the virtual scene modeling technology, based on the graphics rendering, image modeling and hybrid modeling technology.

Graphics rendering technology is making full use of computer graphics techniques to model and render for virtual scene. First, thinking of the real world abstractly, and establishing mathematical model and the corresponding geometry elements' attribute. After that, according to the given site and direction of observation, the computer is used to realize polygon processing, coloring, blanking, illumination and projection of a series of drawing process, which finally created the need visual scene. It has a drawback, produce a large amount of data, which give the rendering speed, data storage, transmission a big challenge.

The essence of image modeling technology is the application of texture mapping technology in computer graphics. This technology can be directly used to create 3D scene, rendering quickly, having strong sense of reality, avoiding complicated modeling work and possessing small amount of data. However, its drawback is poor human-computer interaction, which can only go on real-time browsing instead of real-time control.

Hybrid modeling method can not only avoid amount of calculation of complex scene geometry mode, but also satisfy the real-time requirements. In fact, it brings a great of interactive problems, such as two kinds of scene seamless connection and interaction when the scene changes.

In addition, virtual scene can be also established by 3D scanners whose precision is very high and that can be used in professional applications. Nevertheless, its price is very "professional" because a set of 3D scanner takes tens of thousands, yet, not ordinary users can afford.

The appropriate expression of 3D model not only affects the storage of data, but also be related to effect and efficiency on virtual scene. Therefore the details and the complexity of the equilibrium model should be paid attention to during process of modeling. In this paper, virtual scene will be built by utilizing the technology graphics rendering and third party software (Google SketchUp and 3Ds Max). Because these third modeling tools can perform data optimization and coplanar treatment on the established model, reduce the number of geometric drawing, operate simply, make model's appearance more exquisite and support a variety of patterns of data into and out of other software.

2 3D DRAWING PROCESS BASED ON OPENGL

As shown in the Fig. 1, Geometric vertex data include model set of vertices, line, polygons mage data sets

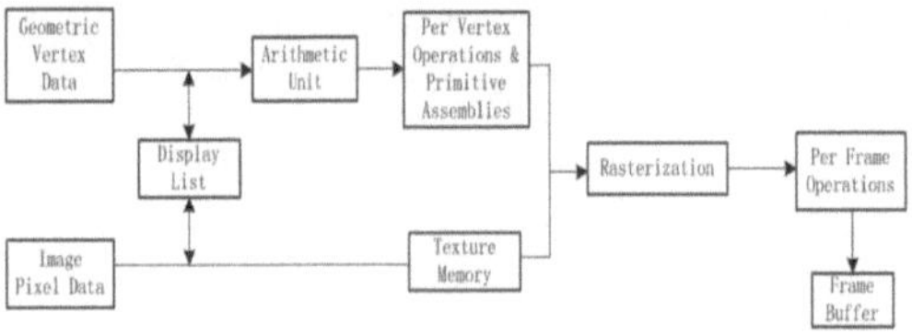

Figure 1. The working process of OpenGL.

contain pixel and bitmap set. Although the processing mode is different between the image pixel data and geometric vertex data, they are all through the grating, element slice by slice processing until the last raster data are written to the frame buffer.

Including geometric vertex data and pixel data, all data in OpenGl can be stored in the display list and get immediate treatment. Display list technology plays an important role in the OpenGl library.

According to OpenGL library, all the geometry units should be described as vertexes so that vertex operator and calculation operations can calculate and operate each vertex. Then, these vertexes form raster pattern pieces; as to the pixel data, the pixel operations are stored in a texture assembly in memory. Like geometric vertex operations, they are finally shaping raster graphics tablets element formation. In the end of whole operation process, those tablets elements are operated one by one, so that the final pixel values in frame buffer can display on the screen.

3 THE RESEARCH ON 3D LIBRARY LANDSCAPE

The design models of 3D library landscape include library model, trees, lawn, vehicle and pavement.

First of all, to shoot the field Library landscape, and after a series of photos, use these pictures and the function "matching image modeling" included in Google SketchUp 8 to establish models and mapping texture. That software is easy to build model and modify, has the ability to make stylized and realistic renderings, can achieve the visual design and optimize the model. After all the models are completed, they are imported into 3DSMAX to generate triangular mesh data model. OpenGL is a graphics interface instead of generated model, and is provided with grid data to paint. Generally speaking, the model grids are generated by 3D modeling software (3DS Max, Maya...), then exported in some format. In the exported document, there are kinds of data that are provided to OpenGL. In this paper, the "obj" format document will be exported from 3DSMAX and called by VC++ and OpenGL to carry on drawing.

What's more, UVW map can be realized in 3DSMAX. Taking the library of model as the example, the near map can be combined to map texture once instead of repeatedly. So, the software is more convenient to chartlet and provide more complex model to OpenGL that can draw it. There are 3D library

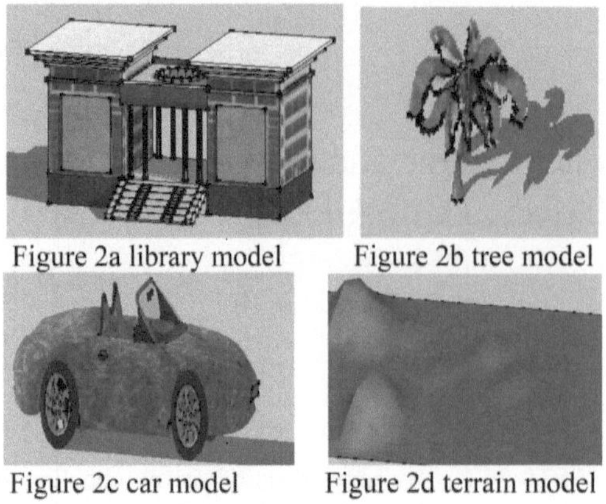

Figure 2. a, b, c, d single module example.

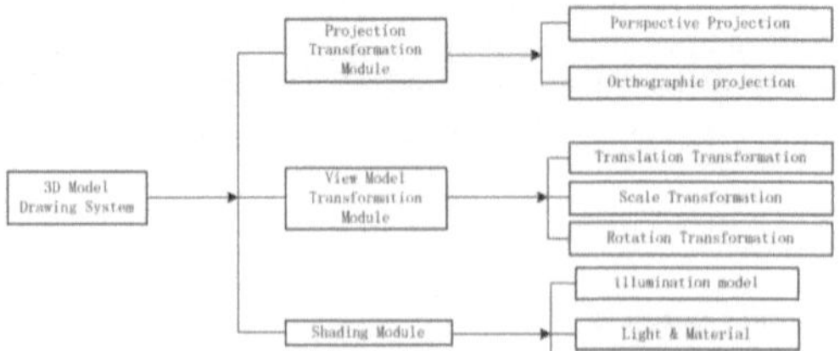

Figure 3. The 3D function module based on OpenGL.

landscape models which are as follows. The library is designed as Zhongyuan institute of technology library.

After 3D model are designed, they will be called and painted by VC++ and OpenGL. The function module of planar graph is shown in Figure 3.

4 THE REALIZATION OF 3D LIBRARY LANDSCAPE

4.1 *The realization of projection transformation module*

OpenGL supports two types of the projection transform, namely the perspective projection and orthogonal projection. The projection is realized by using matrices which are operated by the function glMatrixMode(GL_PROJECTION) before that, the current matrix should be set to the identity matrix: glLoadIdentity().

The results that prospective projection builds are similar to the photos (the farther the distance is, the smaller object in distant position is; the closer the distance is, the larger object in distant position is). If you want to make the current view window set for the projective projection, you can use the function glFrustumf (float left, float right, float bottom, float top, float Near, float Far) and gluPerspective (float fovy, float aspect, float Near, float Far). There are some codes about the prospective projection transform as follows.

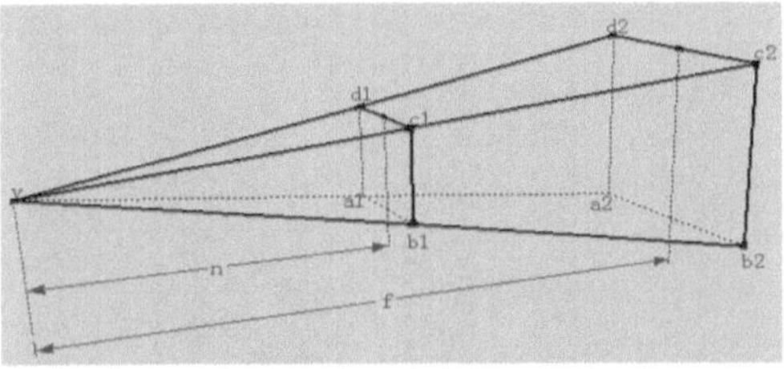

Figure 4. 3D prospective graph.

asp=float (screen_w) /screen_h; //w, h respectively express the screen width and height
glMatrixMode (GL_PROJECTION);
glLoadIdentity ();
gluPerspective (45, ASP, 1001000);

As we see in the Fig. 4, there is a 3D frustum of the prospective projection. Suppose (left) P1 represents the near plane of and (right) plane is named P2; there is a parallel between two rectangular plane, the distance from the viewpoint to the near plane is n, and the far plane is f; in the near plane, a1 is located in the lower left corner, its 3D coordinate is (l, b, n) and coordinate of vertex c1 is (r, t, n). So the matrix named P can expresses this prospective projection.

$$P = \begin{bmatrix} \frac{2n}{r-l} & 0 & \frac{r+l}{r-l} & 0 \\ 0 & \frac{2n}{t-b} & \frac{t+b}{t-b} & 0 \\ 0 & 0 & \frac{-(f+n)}{f-n} & \frac{-2fn}{f-n} \\ 0 & 0 & -1 & 0 \end{bmatrix} \quad (1)$$
$$(r \neq l, t \neq b, f \neq n)$$

Orthographic projection is equivalent to the observed results at infinity and it is an ideal state. While for the computer, orthographic projection is likely to get better operating speed. The current view window can be transformed to Orthographic projection space by the function glOrtho (float left, float right, float bottom, float top, float Near, float Far).

4.2 *The realization of view model transformation*

In the world coordinate system, the final rendering of the object position is up to view model transformation. The transformation defines the location of the observation point, focus and upward direction, and then frustum location also is fixed.

The view can be changed by the function gluLookAt(), whose prototype is as follows.

Void gluLookAt (GLdouble eyex, GLdouble eyey, GLdouble eyez, GLdouble centerx, GLdouble centery, GLdouble upx, GLdouble upy, GLdouble upz)

When landscapes will be operated to meet users' need, there are three functions that can complete these operation: glTranslate(), glRotate(), glScale(). In addition, the function glMultiMatrix() can also manage it. proceedings, and not as an independent document. Please do not revise any of the current designations.

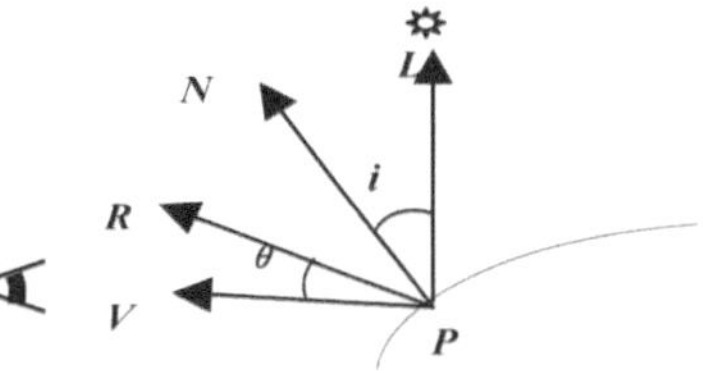

Figure 5. The direction of vector involved in the calculation of phong model.

4.3 *The realization of shading module*

The final shading effect in 3D landscape is up to the light source, object material attribute, illumination model and texture mapping.

4.3.1 *The definition of illumination model*

According to rules of Optical Physics, the light model is a combined formula computing brightness of the light and color from observer's eyes that each point on the surface of objects project to. There are some illumination model, such as Cook-Torrace model, Whitted model and Phong model. But Phong model is easy to compute and is very useful in computer graphics. What's more, it has been supported in business graphics hardware.

Phong model can be defined as:

$$\begin{cases} I = I_a K_a + \sum \left[I_d K_d \cos\theta + I_s K_s \cos^n \alpha \right] \\ \theta, \alpha \in \left[0, \frac{\pi}{2}\right] \\ K_d + K_s = 1 \\ K_a \cdot K_d, K_s \in [0,1] \end{cases} \quad (2)$$

Vertex shading need to make the summation $(\sum)$ of all the special light source and each light source will bring these light reflection intensity including the brightness of ambient light intensity (I_a), diffuse reflection light intensity (I_d) from the object surface and specular reflection of light intensity (I_s). At the same time, according to the above light intensity, there are corresponding coefficients: K_a, K_d, K_s. There is an incident angle (θ), an angle of the specular direction and gaze direction, and a convergence exponent (n) of specular reflection (relate to smooth surface of objects). As is shown in Fig. 5, the Direction of Vectors Involved in the Calculation of Phong Model are as follows.

L is a vector from a point (P) of the object to light source; there is a normal vector (N), R represents the direction of specular reflection; V is the direction of viewpoint. Therefore,

$$\cos\theta = L \bullet N, \ \cos\alpha = R \bullet V \quad (3)$$

By the function glLightModel(), the global environment, the observation point, light model and so on can be defined.

4.3.2 *The attibution of light and material*

OpenGL can simulate real light effect by using ambient light, diffuse light, specular light, emissive light.

The former three kinds of light are contained in light properties while the last belongs to material properties. These attributions can be set by the following codes.

```
GLfloat light_a[], light_d[], light_s[], light_position[];
glLightfv(GL_LIGHT0,GL_AMBIENT,light_a);
// Set the ambient light
glLightfv(GL_LIGHT0,GL_DIFFUSE,light_d);
// Set the diffuse light
glLightfv(GL_LIGHT0,GL_SPECULAR,light_s);
// Set the specular light
glMaterialfv(GL_FRONT,GL_SHININESS,shiness);
// Set the material attribution
```

4.3.3 *Texture mapping*

In addition to the light and material processing of objects, adding texture features to the surface of the object makes it look more real. This is called "texture technology".

A general approach of texture mapping is defined:

- Defining texture mapping
- Control texture mapping
- Confirm the method of texture mapping
- Defining texture coordinates

In general, the discrete method is more commonly used in texture definition method. Its function is glTexImage2D().

There are a lot of problems when project Texture image onto the object, such as how the texture correspond to pixel on the screen and how texture mapping realizes texture scaling and texture repeat etc. The defining function is glTexParameter(), function of texture mapping method is glTexEnv() and defining texture coordinates function is glTexCoord().

4.4 *GLUI—Interactive interface library*

The user can control the simulative landscape through interaction operation. Interactive operation is relatively easy while the rotation operation is an exception. Because of three-dimensional virtual landscape and two-dimensional screen space where mouse can move, if only take advantage of the difference between before and after the mouse movement to rotate the landscape, obviously, object can be rotated in two-dimensional space. So, 3D scene can't be rotated to arbitrary and expected angles.

The effective method to solve this problem is to use the mouse tracking ball. The user's mouse with spherical motion in 3D scene, it is like that using a hand in space to rotate the ball arbitrarily. So, 3D scene can be rotated to any desired viewing angle. Most importantly, GLUI provides a rotating support for mouse ball and a node translation, selection, moving, scaling, 3D coordinate acquisition etc. Those functions lay the foundation for the development of 3D interactive application. As we see in Fig. 6, the different control of GLUT window is shown.

GLUI is a C++ library of GUI, providing GUI controls, such as buttons, check boxes, radio buttons, labels, panel, text fields etc. It can create a user' interface, go on human-computer interaction and is independent instead of relying on GLUT. In addition, all the system related problems can be dealt with, such as window, keyboard and mouse management.

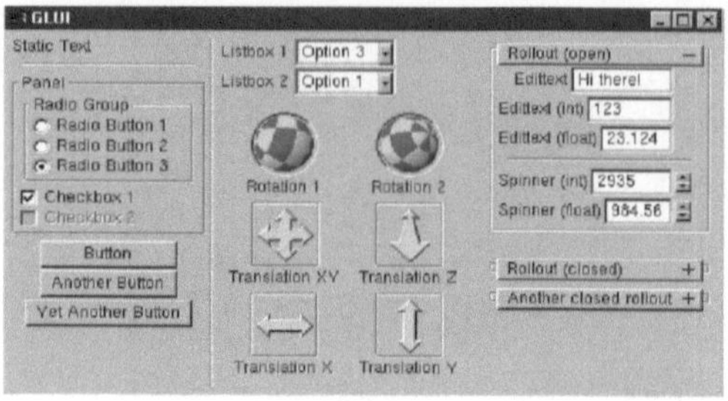

Figure 6. GLUI function window.

Figure 7a before rotation, the front landscape.

Figure 7b after rotation, the side landscape.

Figure 7. a,b comparative landscape map between the front and the side.

The following landscape can be realized by win7, VS2010 and OpenGl.

According to View model transformation and GLUI, these operations including reset, scaling, rotation, translation can be achieved. Take the rotation for example, comparative landscape map between the front and the side is shown in Fig. 7.

There are other functions in this procedure, for example, choosing projection method, automatic rotation, stopping rotation and hit the light. In the end, as shown in Fig. 8, 3D digital library landscape has been realized.

5 CONCLUSIONS

This paper introduces in detail the implementation of 3D library building landscape and real-time interactive operation by using OpenGl. The virtual landscape is vivid, has better visual effect. The experiment shows

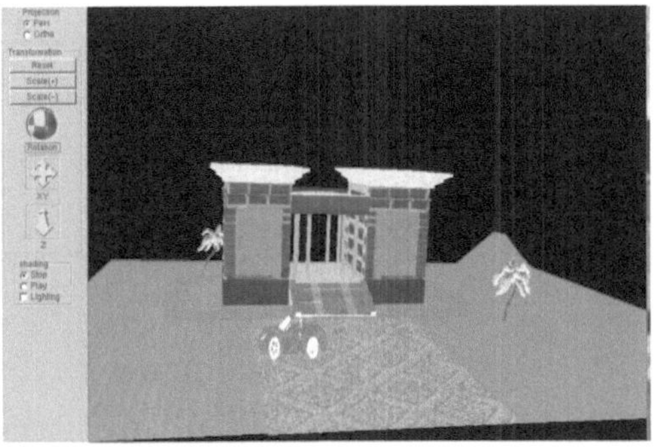

Figure 8. 3D digital library landscape.

that the method given in the paper has the characteristic of more manageable operation than before and high performance, which meets related users needs by multi angle and multi azimuth. Meanwhile, this simulation method can also provide certain reference value for many areas of terrain modeling, city planning, product design, medicine etc.

REFERENCES

Burdea G. 1993. Virtual Reality System and Application. Electro 1993 Interztional Coference, Edison, NJ.

He Xu-jia. 2013. Real-time Interaction Processing of Medical Image Based on OpenGl. Computer Applications and Software 30(4): 48–49.

Hu Gu-fei. 2011. Analysis and Implementation of Texture Mapping Based on Android Platform. Information Technology.

Jin Yu & Chen Jian-sun. 2005. Method for Research and Performance Evaluation of Distributed Virtual Simulation Scenario Generation. Computer Simulation 22(8): 178–179.

Li Zhan-li & Liu Yun-ting. 2011. Visual Simulation of Tree Motion under the Action of a Strong Wind. Journal of Computer-Aided Design & Computer Graphics 23(8):1369–1370.

Liu Zhou-zhou. 2013. Key Technology Research on Universal Visual Simulation Platform Based on OpenGl. Electronic Design Engineering 21(13): 12.

Luo Jun-song & Deng Fei. 2013. Research on 3D Visualization Technology Based on Qt. Computer Measurement & Control 21(2): 799.

Niu Ruo-xi. 2013. Visual Simulation of Satellite in Orbit. Computer Simulation 30(8):52–53.

Zhang Qian. 2013. Simulation Research on sphere generation algorithm based on OpenGL. Computer Simulation 28(4): 226–228.

Future Mechatronics and Automation – Yang (Ed.)
© 2015 Taylor & Francis Group, London, ISBN 978-1-138-02648-3

A class of memory guaranteed cost control of T-S fuzzy system

Yanhua Wang
Chaoyang Teachers College, Chaoyang, Liaoning, China

Xiqin He
University of Science and Technology Liaoning, Anshan, Liaoning, China

Zhihua Wu & Chunguang Wang
Chaoyang Teachers College, Chaoyang, Liaoning, China

ABSTRACT: Based on the uncertain state of T-S fuzzy system, it was designed of the memory fuzzy feedback controller to achieve guaranteed cost control of the system. Firstly, by constructing a Lyapunov function and based on the Laypunov stability theory, it was proved that the closed-loop system is asymptotically stable through the application of this memory feedback controller, at the same time, gave an upper-bound for the given guaranteed cost control. Secondly, by using the linear matrix inequality (LMI) technique in MATLAB, the feedback controller gain would found out. Finally, the simulation example is given to illustrate the validity of the proposed method.

1 INTRODUCTION

In 1972, Chang and Peng [1], first proposed guaranteed cost control, they designed a guaranteed cost state feedback controller to achieve that the closed-loop system was asymptotically stable, at the same time, and to guarantee the performance indexes of the system after the deterioration of less than the predetermined target. This concept was put forward, it has been the concern of many scholars in this field, and has done a lot of research and exploration of [2–6]. The literature [2] discussed the guaranteed cost control of neutral systems with time-varying delay parameters uncertain, in [3], it was studied the optimal guaranteed cost control of generalized nonlinear, in [4], In the condition of the additive disturbance controller, they used LMI method to considered the non fragile guaranteed cost control for singular systems with time delay uncertain. The neutral delay systems with memory state feedback guaranteed cost controller was studied in [5]. The problem of guaranteed cost control was researched for a class of neutral delay systems with nonlinear parameter perturbations in [6]. In [7], they considered the design of T-S fuzzy memory non fragile guaranteed cost controller. But, about a memory guaranteed cost controller in T-S fuzzy systems, there was a little research. So this paper would research the memory guaranteed cost controller of the state uncertainty T-S model.

2 PRELIMINARIES AND PROBLEM FORMULATION

2.1 *Relevant lemmas*

Lemma 1 [8]: for all $x, y \in R^n$, we have for and constant $\varepsilon > 0$ and any positive definite $S \in R^{n\times n}$ that

$$2x^T y \le \varepsilon x^T Sx + \varepsilon^{-1} y^T S^{-1} y$$

Lemma 2: for all $x, y \in R^n$ and any symmetric positive definite matrice $S \in R^{n\times n}$, that

$$2x^T Sy \le x^T Sx + y^T Sy$$

Proof: according to lemma 1 and contract matrix, this lemma can be proved.

Lemma 3 [10]: let U, V and F be real matrices of appropriate dimensions with $F^T F \le I$, then for any scalar $\varepsilon > 0$, we have

$$UFV + V^T F^T U^T \le \varepsilon UU^T + \varepsilon^{-1} V^T V$$

Lemma 4 [11]: for any real vector x, y and matrices L, X, Y, Z of appropriate dimensions, if X, Z are symmetric positive definite and $\begin{bmatrix} X & Y \\ Y^T & Z \end{bmatrix} \ge 0$, we have

$$-2x^T Ly \le \inf_{X,Y,Z} \begin{bmatrix} x \\ y \end{bmatrix}^T \begin{bmatrix} X & Y-L \\ Y^T - L^T & Z \end{bmatrix} \begin{bmatrix} x \\ y \end{bmatrix}.$$

2.2 *Problem formulation*

Consider an state uncertain T-S fuzzy model, the i-th rule is represented as follows:

$$IF: \theta_1 is \mu_{i1}, and, \cdots, and \theta_p is \mu_{ip}$$

$$THEN: \ \dot{x}(t) = (A_i + \Delta A_i)x(t) + B_i u(t) \quad (1)$$

where $x \in R^n$ and $u \in R^m$ are the state and control input, respectively; A_i and B_i are constant real matrices with appropriate dimensions; θ_j and μ_{ij} are respective the premise variables (which are the functions of state variable) and the fuzzy sets. It is assumed that the premise variable are independent of the input variable; r is the number of the *IF-THEN* rules; p is the number of the premise variables; ΔA_i represents the parameter perturbations in A_i, and satisfies $\Delta A_i = D_i F_i(t) E_i$, with $F_i^T F_i \le I$, where the matices D_i and E_i are known with appropriate dimension, $i = 1, 2, \ldots, r$.

The overall fuzzy model is achieved by fuzzy blending (aggregation) of each individual rule (model) as follows:

$$\dot{x}(t) = \sum_{i=1}^{r} h_i(\theta)[(A_i + \Delta A_i)x(t) + B_i u(t)] \quad (2)$$

The following memory state feedback fuzzy controller is taken, the i-th rule is represented as follows:

IF: θ_1 is μ_{i1}, and, ..., and θ_p is μ_{ip}

THEN: $u(t) = K_{1i}x(t) + K_{2i}x(t-\tau)$

where $K_{1i}, K_{2i} \in R^{m\times n}$ are the feedback gain matrix of the i-th rule, for $i = 1, 2, \ldots, r$.

The controller of the overall fuzzy model is described by

$$u(t) = \sum_{i=1}^{r} h_i(\theta)[K_{1i}x(t) + K_{2i}x(t-\tau)], i = 1, 2, \cdots, r \quad (3)$$

The closed-loop system with state feedback by (3) and (2) can be described by the following equation:

$$\begin{aligned}\dot{x}(t) = \sum_{i=1}^{r} h_i(\theta) \sum_{j=1}^{r} h_j(\theta)[(A_i + \Delta A_i)x(t) \\ + B_i(K_{1j}x(t) + K_{2j}x(t-\tau))]\end{aligned} \quad (4)$$

where $\overline{A}_i = A_i + \Delta A_i$, we have

$$\dot{x}(t) = \sum_{i=1}^{r} h_i(\theta) \sum_{j=1}^{r} h_j(\theta)[\overline{A}_i x(t) + B_i(K_{1j}x(t) + K_{2j}x(t-\tau))] \quad (5)$$

For the closed-loop system (4), the following quadratic performance of guaranteed cost function is defined by:

$$J = \int_0^{\infty} [x(t)^T R x(t) + u(t)^T S u(t)] dt \quad (6)$$

In the paper, we design a fuzzy memory state feedback controller (3), that can not only make the closed-loop (4) that is asymptotically stable for all the uncertainties in the system, and can satisfy quadratic performance of guaranteed cost function.

3 MAIN RESULTS

Theorem: system (4) is asymptotically stabilizable by (7) and (8) if there exist some positive constants $\varepsilon_i (i = 1, 2, \ldots, r)$, some symmetric positive definite matrix $\overline{P}, \overline{Q}, \overline{X}, \overline{Z}$, the feedback controller for the fuzzy system (2) is a guaranteed cost controller, and the cost function has upper bounded J^*.

$$J \le J^* = x(0)^T P x(0) + \int_{-\tau}^{0} x(t)^T Q x(t) ds + \int_{-\tau}^{0}\int_{\beta}^{0} \dot{x}(\alpha)^T Z \dot{x}(\alpha) d\alpha d\beta$$

$$\overline{\Lambda} = \begin{bmatrix} \overline{\Lambda}_{11} & \overline{\Lambda}_{12} \\ \overline{\Lambda}_{12}^T & \overline{\Lambda}_{22} \end{bmatrix} < 0 \quad (7)$$

$$\text{where } \overline{\Lambda}_{11} = \begin{bmatrix} \Lambda_5 & \Delta_2 & \tau\Delta_3 & \overline{P}E_i^T & \varepsilon_i D_i \\ \Delta_2^T & -\overline{Q} & \tau\overline{K}_{2j}^T B_i^T & 0 & 0 \\ \tau\Delta_3^T & \tau B_i \overline{K}_{2j} & -\tau\overline{Z} & 0 & \tau\varepsilon_i D_i \\ E_i\overline{P} & 0 & 0 & -\varepsilon_i I & 0 \\ \varepsilon_i D_i^T & 0 & \tau\varepsilon_i D_i^T & 0 & -\varepsilon_i I \end{bmatrix}$$

$$\overline{\Lambda}_{12} = \begin{bmatrix} \overline{K}_{1j}^T & \overline{P} & 0 \\ 0 & 0 & \overline{K}_{2j}^T \\ 0 & 0 & 0 \\ 0 & 0 & 0 \\ 0 & 0 & 0 \end{bmatrix}, \quad \Delta_2 = B_i\overline{K}_{2j} - \overline{Y}, \quad \Delta_3 = \overline{P}A_i^T + \overline{K}_{1j}^T B_i^T$$

$$\overline{\Lambda}_{22} = \begin{bmatrix} -0.5S^{-1} & 0 & 0 \\ 0 & -R^{-1} & 0 \\ 0 & 0 & -0.5S^{-1} \end{bmatrix}$$

$$\Lambda_5 = A_i\overline{P} + \overline{P}A_i^T + B_i\overline{K}_{1j} + \overline{K}_{1j}^T B_i^T + \tau\overline{X} + \overline{Y} + \overline{Y}^T + \overline{Q}$$

$$\begin{bmatrix} \overline{X} & \overline{Y} \\ \overline{Y}^T & \overline{Z} \end{bmatrix} \ge 0 \quad , \quad \overline{P} \ge \overline{Z} \quad (8)$$

Proof: we use the following Lyapunov function candidate for the system (4):

$$V(x(t)) = V_1(x(t)) + V_2(x(t)) + V_3(x(t)) \quad (9)$$

Where $V_1(x(t)) = x(t)^T P x(t)$

$$V_2(x(t)) = \int_{-\tau}^{0}\int_{t+\beta}^{t} \dot{x}(\alpha)^T Z \dot{x}(\alpha) d\alpha d\beta$$

$$V_3(x(t)) = \int_{t-\tau}^{t} x(s)^T Q x(s) ds$$

The time derivative of V along the trajectory of (5) is given by

$$\dot{V}(x(t)) = \dot{V}_1(x(t)) + \dot{V}_2(x(t)) + \dot{V}_3(x(t))$$

where $\dot{V}_1(x(t)) = 2x(t)^T P \dot{x}(t)$, from(5)

$$\begin{aligned}\dot{V}_1(x(t)) &= \sum_{i=1}^{r} h_i(\theta) \sum_{j=1}^{r} h_j(\theta)[2x(t)^T P \overline{A}_i x(t) \\ &+ 2x(t)^T P B_i (K_{1j}x(t) + K_{2j}x(t-\tau))] \\ &= \sum_{i=1}^{r} h_i(\theta) \sum_{j=1}^{r} h_j(\theta)[2x(t)^T P(\overline{A}_i + B_i K_{1j} + B_i K_{2j})x(t) \\ &- 2x(t)^T P B_i K_{2j} \int_{t-\tau}^{t} \dot{x}(\alpha) d\alpha]\end{aligned}$$

According to lemma 4,

$$\le \sum_{i=1}^{r} h_i(\theta) \sum_{j=1}^{r} h_j(\theta)\{2x(t)^T P(\bar{A}_i + B_i K_{1j} + B_i K_{2j})x(t)$$

$$+\int_{t-\tau}^{t} \begin{bmatrix} x(t) \\ \dot{x}(\alpha) \end{bmatrix}^T \begin{bmatrix} X & Y - PB_i K_{2j} \\ (Y - PB_i K_{2j})^T & Z \end{bmatrix} \begin{bmatrix} x(t) \\ \dot{x}(\alpha) \end{bmatrix} d\alpha\}$$

$$= \sum_{i=1}^{r} h_i(\theta) \sum_{j=1}^{r} h_j(\theta)[x(t)^T (P\bar{A}_i + \bar{A}_i^T P + PB_i K_{1j}$$

$$+K_{1j}^T B_i^T P + \tau X + Y + Y^T)x(t) - 2x(t)^T$$

$$*(Y - PB_i K_{2j})x(t-\tau) + \int_{t-\tau}^{t} \dot{x}(\alpha)^T Z \dot{x}(\alpha) d\alpha]$$

$$\dot{V}_2(x(t)) = \tau \dot{x}(t)^T Z \dot{x}(t) - \int_{t-\tau}^{t} \dot{x}(\alpha)^T Z \dot{x}(\alpha) d\alpha$$

$$\dot{V}_3(x(t)) = x(t)^T Q x(t) - x(t-\tau)^T Q x(t-\tau)$$

So, $\dot{V}(x(t)) = \dot{V}_1(x(t)) + \dot{V}_2(x(t)) + \dot{V}_3(x(t))$

That ,we have

$$\dot{V}(x(t)) + x(t)^T R x(t) + u(t)^T S u(t) \tag{10}$$

from (3) and (5),

$$\le \sum_{i=1}^{r} h_i(\theta) \sum_{j=1}^{r} h_j(\theta)\{ \begin{bmatrix} x(t) \\ x(t-\tau \end{bmatrix}^T$$

$$*\left(\begin{bmatrix} \Lambda_1 & PB_i K_{2j} - Y \\ K_{2j}^T B_i^T P - Y^T & -Q \end{bmatrix} + \tau \begin{bmatrix} (\bar{A} + B_i K_{1j})^T \\ K_{2j}^T B_i^T \end{bmatrix} \right.$$

$$\left. *Z \begin{bmatrix} (\bar{A} + B_i K_{1j}) \\ B_i K_{2j} \end{bmatrix}^T \right) \begin{bmatrix} x(t) \\ x(t-\tau \end{bmatrix} + (K_{1j} x(t)$$

$$+K_{2j} x(t-\tau))^T S(K_{1j} x(t) + K_{2j} x(t-\tau))\} \tag{11}$$

where $\Lambda_1 = P\bar{A}_i + \bar{A}_i^T P + PB_i K_{1j} + K_{1j}^T B_i^T P$

$$+\tau X + Y + Y^T + Q + R$$

According to lemma 4, we have

$$(K_{1j} x(t) + K_{2j} x(t-\tau))^T S(K_{1j} x(t) + K_{2j} x(t-\tau))$$

$$\le 2x(t)^T K_{1j}^T S K_{1j} x(t) + 2x(t-\tau)^T K_{2j}^T S K_{2j} x(t-\tau)$$

from (11),

$$\le \sum_{i=1}^{r} h_i(\theta) \sum_{j=1}^{r} h_j(\theta)\{ \begin{bmatrix} x(t) \\ x(t-\tau \end{bmatrix}^T$$

$$*\left(\begin{bmatrix} \Lambda_2 & PB_i K_{2j} - Y \\ K_{2j}^T B_i^T P - Y^T & -Q + 2K_{2j}^T S K_{2j} \end{bmatrix} \right.$$

$$\left. +\tau \begin{bmatrix} (\bar{A} + B_i K_{1j})^T \\ K_{2j}^T B_i^T \end{bmatrix} Z \begin{bmatrix} (\bar{A} + B_i K_{1j})^T \\ B_i K_{2j} \end{bmatrix}^T \right) \begin{bmatrix} x(t) \\ x(t-\tau \end{bmatrix}\}$$

$$= \sum_{i=1}^{r} h_i(\theta) \sum_{j=1}^{r} h_j(\theta) \begin{bmatrix} x(t) \\ x(t-\tau \end{bmatrix}^T \Lambda \begin{bmatrix} x(t) \\ x(t-\tau \end{bmatrix} \tag{12}$$

where $\Lambda_2 = P\bar{A}_i + \bar{A}_i^T P + PB_i K_{1j} + K_{1j}^T B_i^T P + \tau X + R$

$$+Y + Y^T + Q + 2K_{1j}^T S K_{1j}$$

applying the Schur complement [9], Λ is equivalence to (13),

$$\Lambda = \begin{bmatrix} \Lambda_2 & PB_i K_{2j} - Y & \tau(\bar{A}_i + B_i K_{1j})^T \\ K_{2j}^T B_i^T P - Y^T & -Q + 2K_{2j}^T S K_{2j} & \tau K_{2j}^T B_i^T \\ \tau(\bar{A}_i + B_i K_{1j}) & \tau B_i K_{2j} & -\tau Z^{-1} \end{bmatrix} \tag{13}$$

By $\Delta A_i = D_i F_i(t) E_i$, (13) is equivalent to the following,

$$= \Pi + \begin{bmatrix} PD_i \\ 0 \\ \tau D_i \end{bmatrix} F(t) \begin{bmatrix} E_i \\ 0 \\ 0 \end{bmatrix}^T + \left(\begin{bmatrix} PD_i \\ 0 \\ \tau D_i \end{bmatrix} F(t) \begin{bmatrix} E_i \\ 0 \\ 0 \end{bmatrix}^T \right)^T$$

According to lemma 3, we have

$$\le \Pi + \varepsilon_i \begin{bmatrix} PD_i \\ 0 \\ \tau D_i \end{bmatrix} \begin{bmatrix} PD_i \\ 0 \\ \tau D_i \end{bmatrix}^T + \varepsilon_i^{-1} \begin{bmatrix} E_i \\ 0 \\ 0 \end{bmatrix} \begin{bmatrix} E_i & 0 & 0 \end{bmatrix} \tag{14}$$

where

$$\Pi = \begin{bmatrix} \Lambda_3 & PB_i K_{2j} - Y & \tau(A_i + B_i K_{1j})^T \\ K_{2j}^T B_i^T P - Y^T & -Q + 2K_{2j}^T S K_{2j} & \tau K_{2j}^T B_i^T \\ \tau(A_i + B_i K_{1j}) & \tau B_i K_{2j} & -\tau Z^{-1} \end{bmatrix}$$

where $\Lambda_3 = PA_i + A_i^T P + PB_i K_{1j} + K_{1j}^T B_i^T P + \tau X$

$$+Y + Y^T + Q + R + 2K_{1j}^T S K_{1j}$$

applying the Schur complement [9] again, (14) is equivalent to (15),

$$= \begin{bmatrix} \Delta_{11} & \Delta_{12} \\ \Delta_{12}^T & \Delta_{22} \end{bmatrix} \tag{15}$$

where

$$\Delta_{11} = \begin{bmatrix} \Lambda_3 & PB_i K_{2j} - Y & \tau(A_i + B_i K_{1j})^T \\ K_{2j}^T B_i^T P - Y^T & -Q + 2K_{2j}^T S K_{2j} & \tau K_{2j}^T B_i^T \\ \tau(A_i + B_i K_{1j}) & \tau B_i K_{2j} & -\tau Z^{-1} \end{bmatrix}$$

$$\Delta_{12} = \begin{bmatrix} E_i^T & PD_i \\ 0 & 0 \\ 0 & \tau D_i \end{bmatrix}, \quad \Delta_{22} = \begin{bmatrix} -\varepsilon_i I & 0 \\ 0 & -\varepsilon_i^{-1} I \end{bmatrix}.$$

Pre- and post-multiplying (15) by $diag\{P^{-1}, P^{-1}, I, I, \varepsilon_i\}$, and using the Schur complement [9] again, (15) is equivalent to the following,

$$\bar{\Lambda} = \begin{bmatrix} \bar{\Lambda}_{11} & \bar{\Lambda}_{12} \\ \bar{\Lambda}_{12}^T & \bar{\Lambda}_{22} \end{bmatrix}, \tag{16}$$

$$\text{Where } \bar{\Lambda}_{11} = \begin{bmatrix} \Lambda_4 & \bar{\Delta}_{12} & \bar{\Delta}_{13} & \bar{P}E_i^T & \varepsilon_i D_i \\ \bar{\Delta}_{12}^T & -\bar{Q} & \tau \bar{K}_{2j}^T B_i^T & 0 & 0 \\ \bar{\Delta}_{13}^T & \tau B_i \bar{K}_{2j} & -\tau \bar{Z} & 0 & \tau\varepsilon_i D_i \\ E_i \bar{P} & 0 & 0 & -\varepsilon_i I & 0 \\ \varepsilon_i D_i^T & 0 & \tau\varepsilon_i D_i^T & 0 & -\varepsilon_i I \end{bmatrix}$$

$$\bar{\Delta}_{12} = B_i \bar{K}_{2j} - \bar{Y}, \quad \bar{\Delta}_{13} = \tau(\bar{P}A_i^T + \bar{K}_{1j}^T B_i^T)$$

$$\Lambda_4 = A_i \bar{P} + \bar{P}A_i^T + B_i \bar{K}_{1j} + \bar{K}_{1j}^T B_i^T + \tau \bar{X} + \bar{Y} + \bar{Y}^T + \bar{Q}$$

$$\bar{\Lambda}_{12} = \begin{bmatrix} \bar{K}_{1j}^T & \bar{P} & 0 \\ 0 & 0 & \bar{K}_{2j}^T \\ 0 & 0 & 0 \\ 0 & 0 & 0 \\ 0 & 0 & 0 \end{bmatrix}, \quad \bar{\Lambda}_{22} = \begin{bmatrix} -0.5S^{-1} & 0 & 0 \\ 0 & -R^{-1} & 0 \\ 0 & 0 & -0.5S^{-1} \end{bmatrix}$$

$$\bar{P} = P^{-1}, \bar{Q} = P^{-1}QP^{-1}, \bar{Y} = P^{-1}YP^{-1}, \bar{Z} = Z^{-1},$$

$$\bar{X} = P^{-1}XP^{-1}, \bar{K}_{1j} = K_{1j}P^{-1}, \bar{K}_{2j} = K_{2j}P^{-1}$$

by (7), $\bar{\Lambda} < 0$, so $\dot{V}(x(t)) + x(t)^T Rx(t) + u(t)^T Su(t) < 0$, we have $\dot{V}(x(t)) < 0$, the closed-loop (4) that is asymptotically stable, therefore, when $t \to \infty$, $x(t)^T Px(t) \to 0$, $\int_{t-\tau}^{t} x(s)^T Qx(s)ds \to 0$, $\int_{-\tau}^{0}\int_{t+\beta}^{t} \dot{x}(\alpha)^T Z\dot{x}(\alpha)d\alpha d\beta \to 0$, then $V(x(t)) \to 0$, and quadratic performance (6) satisfies

$$J = \int_0^\infty [x(t)^T Rx(t) + u(t)^T Su(t)]dt < -\int_0^\infty \dot{V}(x(t))dt$$

$$= V(x(0)) - V(x(\infty)) = V(x(0))$$

The guaranteed cost function has upper bounded J^*.

$$J \le J^* = V(x(0)) = x(0)^T Px(0) + \int_{-\tau}^{0} x(t)^T Qx(t)ds$$

$$+ \int_{-\tau}^{0}\int_{\beta}^{0} \dot{x}(\alpha)^T Z\dot{x}(\alpha)d\alpha d\beta$$

In the proof, when applying lemma 4, we ensure $\begin{bmatrix} X & Y \\ Y^T & Z \end{bmatrix} \ge 0$, from (8),

$$\begin{bmatrix} \bar{X} & \bar{Y} \\ \bar{Y}^T & \bar{Z} \end{bmatrix} = \begin{bmatrix} P^{-1}XP^{-1} & P^{-1}YP^{-1} \\ P^{-1}Y^T P^{-1} & P^{-1}Z^{-1}P^{-1} \end{bmatrix} \ge 0 \qquad (17)$$

$$\bar{P} = P^{-1} \ge \bar{Z} = Z^{-1} \qquad (18)$$

Pre- and post-multiplying (17) by $diag\{P, P\}$ results in the following,
$\begin{bmatrix} X & Y \\ Y^T & PZP \end{bmatrix} \ge 0$, from (18), we have $P \le Z$, so

$$0 \le \begin{bmatrix} X & Y \\ Y^T & PZP \end{bmatrix} \le \begin{bmatrix} X & Y \\ Y^T & ZZ^{-1}Z \end{bmatrix} = \begin{bmatrix} X & Y \\ Y^T & Z \end{bmatrix},$$

to ensure that the conditions of Lemma 4.

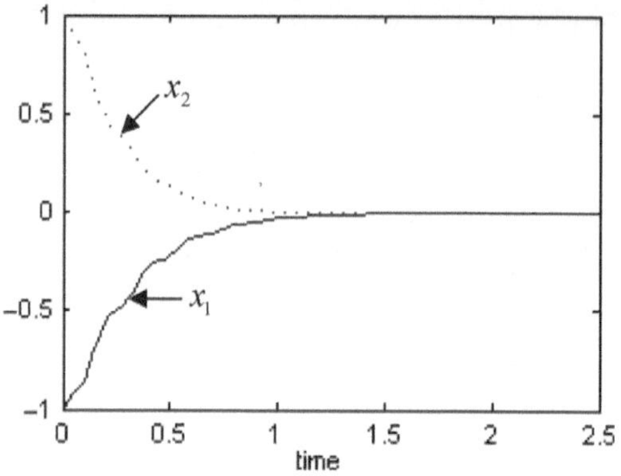

Figure 1. State output, where "—" is the state x_1 output, "..." is the state x_2 output.

4 NUMERICAL EXAMPLES

Considering the global fuzzy system (4) is described by the two rules, where,

$$A_1 = \begin{bmatrix} -1 & 0 \\ 2 & 1 \end{bmatrix}, A_2 = \begin{bmatrix} 2 & 1 \\ 0 & -1 \end{bmatrix}, B_1 = B_2 = \begin{bmatrix} 1 & 0 \\ 0 & 1 \end{bmatrix},$$

$$D_1 = D_2 = \begin{bmatrix} 2 & 0 \\ 0 & 2 \end{bmatrix}, E_1 = E_2 = \begin{bmatrix} 1 & 0 \\ 0 & 1 \end{bmatrix}, R = \begin{bmatrix} 1 & -0.1 \\ -0.1 & 1 \end{bmatrix},$$

$$S = \begin{bmatrix} 2 & 0.2 \\ 0.2 & 2 \end{bmatrix}, \quad \tau = 0.3.$$

From Theorem 1, one feasible solution to LMIs is computed to be,

$$\bar{P} = \begin{bmatrix} 0.0168 & -0.0017 \\ -0.0017 & 0.0200 \end{bmatrix}, \bar{Q} = \begin{bmatrix} 0.0146 & -0.0033 \\ -0.0033 & 0.0211 \end{bmatrix}$$

$$K_{11} = \begin{bmatrix} -2.3469 & -1.6281 \\ -0.3279 & -3.3785 \end{bmatrix}, K_{21} = \begin{bmatrix} -0.0427 & 0.0921 \\ 0.0610 & -0.6115 \end{bmatrix},$$

$$K_{12} = \begin{bmatrix} -4.6792 & -0.4307 \\ -1.0782 & -2.0303 \end{bmatrix}, K_{22} = \begin{bmatrix} -0.0644 & 0.0977 \\ 0.1376 & -0.1648 \end{bmatrix},$$

$$\varepsilon_1 = 0.012, \varepsilon_2 = 0.0117, J^* = 99.7672.$$

The initial conditions are given as: $x(0) = [-1 1]^T$, $F_1(t) = F_2(t) = \sin(10\pi t)\begin{bmatrix} 1 & 0 \\ 0 & 1 \end{bmatrix}$, we choose the following membership functions:

$$\mu_1(t) = 1/(1 + \exp(-2x_1)), \quad \mu_2(t) = 1 - \mu_1(t).$$

The state output of the system are shown in figure obtained by Matlab.

5 CONCLUSIONS

In the paper, based on the uncertain state of T-S fuzzy system, it was designed of the memory fuzzy feedback controller to achieve guaranteed cost control of

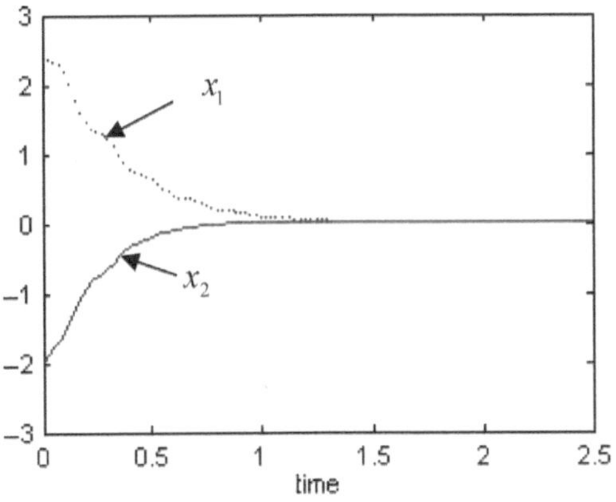

Figure 2. Contoller $u(t)$ input, where "..." is contoller input of the state x_1, "—" is contoller input of the state x_2.

the system. Firstly, by constructing a Lyapunov function and based on the Laypunov stability theory, it was proved that the closed-loop system is asymptotically stable through the application of this memory feedback controller, at the same time, gave an upper-bound for the given guaranteed cost control. Secondly, by using the linear matrix inequality (LMI) technique in MATLAB, the feedback controller gain would found out. Finally, illustrative examples have also been presented to demonstrate the whole design procedure in using our approaches. Simulation results show that the system is asymptotically stable.

REFERENCES

Afang Hang, Lihuan Chen, LMI-based Guaranteed Cost Controller Designs for T-S Fuzzy Memory Non-fragile systems, Journal of Southwest University (Natural Science Edition), 2012(11), pp. 125–131.

Bao-Yan Zhu, Qing-Ling Zhang, Optimal Guaranteed Cost control for T-S Fuzzy Descriptor systems with Uncertain Parameters [J]. Systems Engineering-theory & Practice, 2004, 24 (12): 48–57.

Cao, Y.Y. Sun Y.X. and Lam, J. Delay-dependent robust H_∞ control for uncertain systems with time-varying delays, IEE Proceeding online no 1998. 19–51.

Chand S, Peng T. Adaptive Guaranteed Cost Control of Systems with Uncertain Parameters [J]. IEEE Transactions on Automatic Control, 1972, 17(4):474–483.

Chang-Hua Lien, Ruobust Observer-based control of systems with state perturbations via LMI approach, IEEE Transactions on Automatic Control, 2004, 49(8):1365–1370.

Chen B, Lamj, Xu S. Memory State Feedback guaranteed Cost Control for Neutral Delay Systems [J]. International Journal of Innovative Computing, Information Control, 2006, 2(2): 293–303.

Jinlin Xu, Baida qu, Baoguo xu, Guaranteed cost for networked control systems with uncertain time-delay, Computer engineering and applications, 2014, 50(5), pp. 239–242.

Liberzon D. Switching in systems an control. Boston: Birkhauser, 2003.

Wenzi Li, Baowei Wu, Xu Quan, Non-fragile Guaran teed Cost Control for Uncertain Singular Systems With Time-delay [J], Journal of Yunnan Normal University (Natural Sciences Edition) 2008, 28(2):5–10.

Yafei Zhang, Jinling Liang, Guaranteed cost control for a class of delayed neutral systems with nonlinear perturbations, Proeeedings of the 31st Chinese Control Conference, July 25–27, 2012, Hefei, China.

Yu Ker-Wei, Lien Chang-Hua, Delay-Dependent Conditions for Guaranteed Cost Observer-Based Control of Uncertain Neutral Systems with Time-Varying Delays [J]. IMA Journal of mathematical control and information, 2007, 24(2); 383–394.

Future Mechatronics and Automation – Yang (Ed.)
© 2015 Taylor & Francis Group, London, ISBN 978-1-138-02648-3

Application of improved BP neural network in fiber grating pressure measuring system

Q.G. Zhu & M. Yuan
College of Information Science and Engineering, Heibei Key Laboratary of Especial Fiber and Fiber Sensor, YanshanUniversity, Qinhuangdao, Heibei Province, China

C.F. Wang & Y.Y. Gao
Liren college of Yanshan University, Qinhuangdao, Hebei Province, China

ABSTRACT: To overcome the shortcomings of BP network such as uncertainty of the structure, slow training speed and poor global searching capability, the grey correlation analysis method and the particle swarm optimization with genetic operators (GA-PSO) has been combined to optimize the BP neural network. Firstly, the grey correlation analysis method was used to determine the node number of the hidden layer in the neural network. And then, GA-PSO was used to optimize initial weights and threshold values of the network. Finally, in order to demonstrate the performance of the network, the simulation test has been performed in the neural network model mentioned above to eliminate the pressure sensor's temperature error according to the measurement data of the optical grating pressure sensor. From the comparison analysis between the improved BP network model and the conventional BP model, it has been shown that the improved BP network model has integrated the advantages of BP algorithm, genetic algorithm and PSO algorithm, which can bring it strong generalization ability and achieve the optical fiber grating pressure sensor's temperature compensation.

1 INTRODUCTION

With the development of artificial intelligence theory, computer neural network has gained more and more popularity. BP neural network, because of its strong nonlinear mapping, high self-learning and high self-adaptive ability, is widely used in sensor calibration and compensation[1–3]. However there are weaknesses lie behind it, for structure of BP network is difficult to determine, and unreasonable structure may lead to long training time, which would reduce the generalization ability of network[4,5]. Moreover, modification of the weights and threshold values, which belongs to linear search optimization, goes towards the inverse side of the error function gradient. The training process is influenced by the initial weights and threshold greatly, and is easy to fall into local minimum and premature convergence[6].

The advantages of intelligent algorithm, such as direct object manipulation, stochastic optimization, parallel computing and global search characteristics, made it a practical method to improve the convergence speed and accuracy, and the generalization ability of the network [7–9]. There are two ways in intelligent optimization algorithm and neural network: one is using intelligent algorithm to optimize network structure, then the network parameters. The other is using intelligent algorithm to optimize weights and threshold values on the basis of determination of the network structure. According to analysis, using intelligent algorithm to optimize network structure has several drawbacks, such as complex process, long training time, and network structure got from each training differs from each another. Therefore, the second way is adopted in this paper. Firstly, the grey correlation analysis theory[10] was used to determine the node number of the hidden layer in the neural network. And then, GA-PSO was used to optimize initial weights and threshold values of the network, BP algorithm is used to accurately train the network structure. Some improvement are also made for the problem that premature phenomenon is easy to appear in the training process of PSO.

2 DETERMINATIONS OF NEURAL NETWORK STRUCTURE

Correlation analysis in grey system theory analyzes the correlation degree of each factor in the system mainly by comparing the geometry relation of data sequence. Node number of the hidden layer in the neural network is affected by many factors, and quantitative relationship between each factor is uncertain, so the system shows great grayness. Therefore, grey correlation analysis method is used to determine the node number of the hidden layer to optimize the network structure.

Suppose there are n learning samples in network,

$$y_i = (y_i(1), y_i(2), \cdots, y_i(n)) \ (i = 1,2,\cdots,m) \quad (1)$$

where, y_i is the output of sample n learned by node i in hidden layer.

$$Y = (Y_1, Y_2 \ldots, Y_n) \quad (2)$$

Y is the output of n learning samples learned by output layer neurons. Y is reference sequence, Y is comparison sequence.

The steps of using grey correlation analysis method to determine node number of the hidden layer in the neural network are listed as follow:

(1) Train the network to set accuracy.
(2) Calculate the correlation coefficient:

$$\varepsilon_i(k) = \frac{\Delta_{\min} + \Delta_{\max}}{\Delta_i(k) + \rho\Delta_{\max}} \quad (3)$$

Where, $\Delta_i(k) = |Y_k - y_i(k)|$, $\Delta_{\min}$ and $\Delta_{\max}$ are the minimum and maximum difference of absolute value of all comparative sequence in each moment. ρ is resolution ratio which value is 1 normally.

(3) Calculate the correlation and delete the nodes with lower correlation.

Correlation is calculated by mean value method:

$$\gamma_i = \frac{1}{n}\sum_{k=1}^{n}\varepsilon_i(k) \quad (4)$$

When a certain hidden node's effect on the network output is small enough, the impact on network can be neglected. Suppose ε is 0.5, delete the nodes that meet the condition $\gamma_i < \varepsilon$, thus network topology structure is optimized.

3 PARTICLE SWARM OPTIMIZATION BASED ON GENETIC ALGORITHM

3.1 *Standard Particle Swarm Optimization*

When PSO is used to solve optimization problems, the answers are corresponded to a single particle in space. Each particle has its own position and velocity, as well as a fitness value determined by optimal function. In the process of optimization, particles update their position and velocity by tracking individual extreme pbest and global extreme value gbest. In the standard particle swarm algorithm, velocity and position update equations are:

$$V_{id}^{t+1} = w \times v_{id}^t + c_1 r_1 \times (pbest_{id}^t - x_{id}^t) + c_2 r_2 \times (gbest_{gd}^t - x_{id}^t) \quad (5)$$

$$X_{id}^{t+1} = X_{id}^t + V_{id}^t \quad (6)$$

where, V_{id}^t and V_{id}^{t+1} are the d dimension velocity of particle i in generation t and t+1; w is inertia weight; c_1 and c_2 are the acceleration factor; r_1 and r_2 are random values in(0,1); X_{id}^t and X_{id}^{t+1} are the d dimension position of particle i in generation t and t + 1; $pbest_{id}^t$ is the d dimension of particle i 's t generation individual extreme; $gbest_{gd}^t$ is the d dimension position of global extreme in generation t.

3.2 *Particle swarm optimization with genetic operators*

For the problem that premature "aggregation" of particles may lead to local optimal, particle swarm optimization based on genetic operators is proposed, which increases particle diversity through genetic operation, to get global optimal solution.

Control function of genetic operation is defined as P:

$$P = \frac{iter}{Maxitter} \quad (7)$$

Maxiter and *iter* ($0 < iter < Maxiter$) are the maximum number of iterations and current generation number. Randomly generate a uniformly distributed random matrix n between 0~1,

$$n = rand(Maxiter, 1) \quad (8)$$

Compare $n(iter, 1)$ with P under the current iterations, if $n(iter, 1) < P$, genetic operations such as selection, crossover and mutation are processed before conventional particle swarm operation; if not, only conventional particle swarm operation is processed. As iteration increases, P increases, as well as the probability of $n(iter, 1) < P$. In later iteration, the particles were genetic operated with probability close to 1.

In order to keep the diversity of particle swarm and improve the probability of progenies, particles with excellent and poor performance are selected as parent particles. Each particle is sorted by fitness from good to bad, and makes up position matrix $A_{n\times d}$ and velocity matrix $B_{n\times d}$ the first N (N < n/2) particles of $A_{n\times d}$, $B_{n\times d}$ are chosen to compose particle matrix of excellent performance $A_{N1\times d}$, $B_{N1\times d}$, the last N particles of $A_{n\times d}$, $B_{n\times d}$ are chosen to compose particle matrix of poor performance $A_{N2\times d}$, $B_{N2\times d}$. Then, on the basis of poor performance particles, progeny particles are produced by the random crossover between two groups of particles ($A_{N1\times d}$ and $A_{N2\times d}$, $B_{N1\times d}$ and $B_{n\times d}$). In order to increase the diversity of particles, mutation is processed in a small probability after crossover. Calculate the fitness value of progeny particles and compare it with that of parent particles, higher fitness half can turn into next generation.

4 IMPROVED PARTICLE SWARM ALGORITHMS USED TO OPTIMIZE NEURAL NETWORK

Initial weight and threshold of traditional neural network are randomly set up, adjustments are made

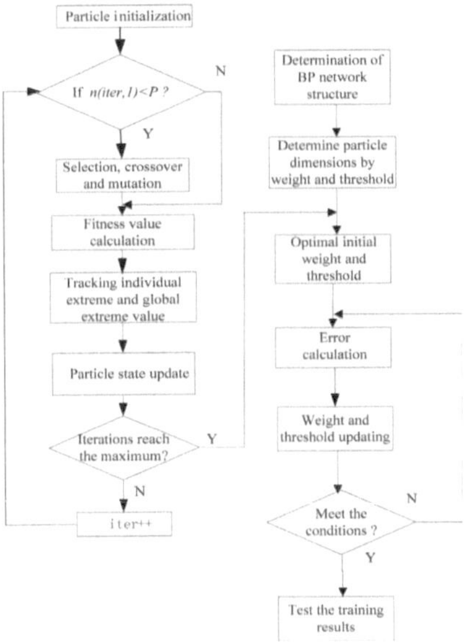

Figure 1. Diagram of algorithm.

according to certain rules during the training, thus better weight and threshold distribution can be got. However, BP neural threshold distribution may influence the generalization ability of the network, so particle swarm algorithm is used to optimize initial weight and threshold.

In order to use improved PSO to optimize initial weight and threshold, it is needful to define network state as particle position vector $\overrightarrow{X}$. First, initialize position vector $\overrightarrow{X}$ and velocity vector $\overrightarrow{V}$, and then search the optimal position through improved PSO, to make the value of following fitness function smallest.

$$E = \frac{1}{n}\sum_{i=1}^{n}\sum_{j=1}^{c}(q_{j,i} - y_{j,i})^2$$

where, n is the number of training samples, c is the number of output layer neurons. $q_{j,i}$ is the ideal output of node j in sample i; $y_{j,i}$ is the actual output of node j in sample i. Diagram of using improved PSO to optimize neural network is shown in Figure 1.

5 RESULTS AND ANALYSIS

In order to verify the validity of the proposed neural network model, we apply it to optical fiber grating pressure sensor system. Since the measurement of optical fiber grating pressure sensor system is easily get affected by temperature, so temperature compensation of the sensor is required. In this paper, sensor's current measure pressure and current temperature are taken as input samples of neural network. Weight and threshold of BP neural network are improved by GA-PSO, the output of the neural network after training can approach sensor nominal pressure, thus the optical fiber grating pressure sensor's temperature compensation is realized, and the measurement accuracy and reliability are improved.

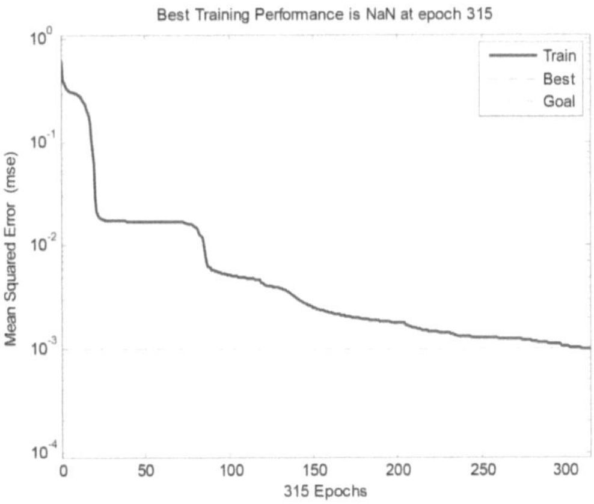

Figure 2. Training curve of BP algorithm.

Record temperature changes from 20 degrees to 70 degrees, step length is 2 degrees, nominal pressure ranges changes from 1 MP to 15 MP, step length is the pressure value measured by optical fiber grating pressure sensor under 0.5 MP. The data was normalized that 80% of the data are taken as training samples, the rest of data are test samples.

Network is a three-layer structure, the node numbers of input layer and the output layer are determined by the actual situation, takes 2 and 1 respectively. The node number of hidden layer is determined by grey correlation analysis method, and is eventually determined to be 5. The transfer function of hidden layer is tangsig, transfer function of output layer is purelin, and training function is trainlm.

The three-layer neural network is respectively trained by BP algorithm, GA-BP algorithm, PSO-BP algorithm and GA-PSO-BP algorithm, and then tests the network after training.

The algorithm parameters are set as follow: target precision is 0.001; the maximum training step is 1000. BP algorithm's learning rate is 0.01; population size of GA algorithm is 20, the maximum iteration number is 20. Selection is processed by roulette wheel method, crossover probability is 0.9, mutation probability is 0.01; The particle number of PSO is 20, $c_{start} = 3.5$, $c_{end} = 0.5$. The parameter numbers to be optimized are $2 \times 5 + 5 \times 1 + 5 + 1 = 21$, so there are 21 dimensions in each particle. Parameter settings in GA-BP, PSO-BP and GA-PSO-BP are the same as the above. The training curves of four kinds of algorithms are shown in Figure 2 to Figure 5.

We can see from the pictures that GA-PSO-BP has a better performance compared with other algorithms. PSO-BP's convergence speed is fast, but this algorithm is easy to fall into local minimum and stagnation. Compared with PSO-BP, GA-BP has a lower training speed, but it can jump out of local minimum more easily. GA-PSO-BP combines the advantages of GA algorithm

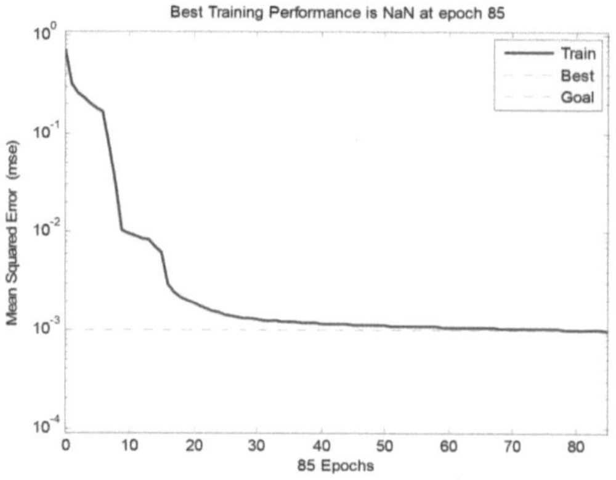

Figure 3. Training curve of PSO-BP algorithm.

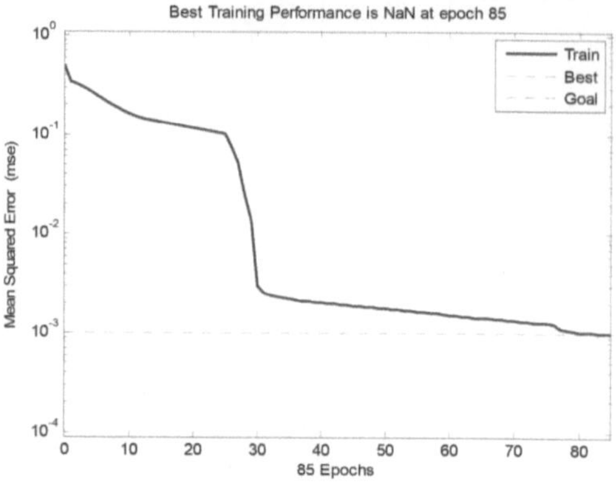

Figure 5. Training curve of GA-PSO-BP algorithm.

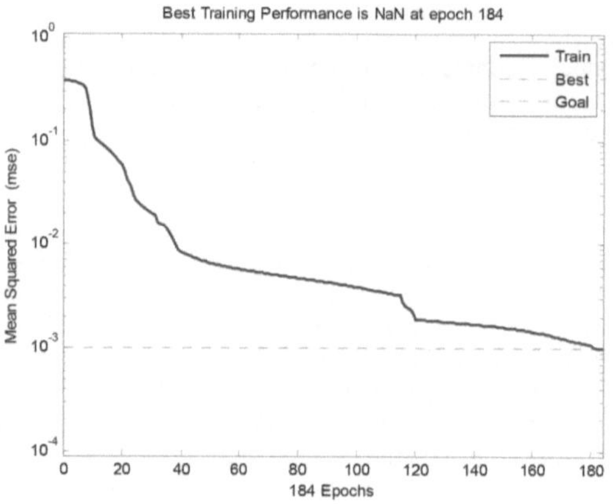

Figure 4. Training curve of GA-BP algorithm.

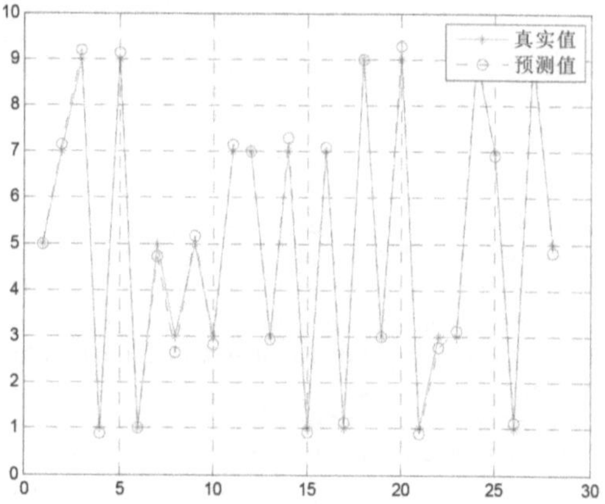

Figure 6. Results comparison.

and PSO algorithm, the training speed is fast and it is not easy to fall into local minimum, which makes it more stable compared with other algorithms.

The network is trained for 10 times, respectively by BP algorithm and GA-PSO-BP algorithm. According to the statistical results of convergence steps and training accuracy, the minimum and maximum convergence steps of network that trained by BP algorithm are 53 and 1000, average convergence steps are 462. Through simulation on training samples, the minimum and maximum errors are 0.2509 and 0.2809, average error is 0.2742. Through simulation on test samples, the minimum and maximum errors are 0.1436 and 0.1561, and average error is 0.1484. In comparison, the minimum and maximum convergence steps of network which trained by GA-PSO-BP algorithm are 53 and 140, average convergence steps are 94. Through simulation on training samples, the minimum and maximum errors are 0.2460 and 0.2682, average error is 0.2533. Through simulation on test samples, the minimum and maximum errors are 0.1288 and 0.1353, and average error is 0.1325.

It can be concluded from comparison that GA-PSO-BP algorithm is more stable, has better training results. Accuracy of training samples and test samples has been improved a lot.

Input the test samples into network trained by GA-PSO-BP, the results are shown in Figure 6. Where, "*" is nominal pressure, "o" is pressure predicted by neural network. From the picture we can see that improved neural network can restrain the temperature effect on fiber grating pressure measuring system.

6 CONCLUSIONS

In this paper, BP neural network is used to achieve optical fiber grating pressure sensor's temperature compensation. For the drawbacks of BP neural network such as slow convergence, easy to fall into local minimum, a new thought is proposed, that is using grey correlation analysis method to determine network structure and using GA-PSO-BP to train network. The results show that application of BP network in the measurement system of optical fiber grating pressure sensor is feasible, and compared to simple BP network, performance of GA-PSO-BP is more stable and reliable.

ACKNOWLEDGEMENTS

This work was financially supported by the national natural Science Foundation of China (61201112 and 61172044), the Natural Science Foundation for Young Scientists of Hebei Province, China (F2013203250, F2012203169).

REFERENCES

[1] Yan, S.A. et al. 2013. The research of fiber Bragg grating in the application of the winding temperature measurement. *Optical Technique* 39(3): 200–203.

[2] Islam, T. & Saha, H. 2007. Study of long-term drift of a porous silicon humidity sensor and its compensation using ANN technique. *Sensors and Actuators* 133: 472–479.

[3] Futane, N.P. & Chowdhury, S.R. 2010. ANN based CMOS ASIC design for improved temperature-drift compensation of piezoresistive micro-machined high resolution pressure sensor. *Microelectronics Reliability* 50: 282–291.

[4] Peng, J.W. et al. 2013. Temperature compensation for humidity sensor based on improved GA-BP neural network. *Chinese Journal of Scientific Instrument* 34(1): 153–160.

[5] Qiao, J.F. & Han, H.G. 2010. Dynamic optimization structure design for neural networks: review and perspective. *Control Theory & Applications* 27(3): 350–357.

[6] Zhong, Y. & Wang, B.W. 2002. BP Network Sequence Prediction Model Based on Genetic Algorithm. *Systems Engineering and Electronics* 24(4): 8–11.

[7] Li, W. et al. 2012. An Effective Backpropagation Algorithm for Optimizing BP Neural Network Based on Rough Set and Modified Genetic Algorithm. *Journal of Northwestern Polytechnical University* 30(4): 601–606.

[8] Zhang, S.D. et al. 2011. Triangulation and PSO-BP Neural Network Used in Star Pattern Recognition. *Opto-Electronic Engineering* 38(6): 30–37.

[9] Xue, S.H. et al. 2010. Modeling with Independent Component Analysis and GA-BP Network in Near Infrared Fast Checking. *Acta Metrologica Sinica* 31(3): 285–288.

[10] Tang, W.H. 2005. The Study of the Optimal Structure of BP Nueral Network. *Systems Engineering* 25(8): 95–100.

Future Mechatronics and Automation – Yang (Ed.)
© 2015 Taylor & Francis Group, London, ISBN 978-1-138-02648-3

Mobile robot vision location based on improved BP-SIFT algorithm

Q.G. Zhu, J. Wang, X.X. Xie & W.D. Chen
College of Information Science and Engineering, Heibei Key Laboratary of Especial Fiber and Fiber Sensor, Yanshan University, Qinhuangdao, Heibei Province, China

ABSTRACT: Several approaches to object recognition make extensive use of local image information extracted in interest points, known as local image descriptors. The most advanced methods perform a statistical analysis of the gradient information about the interest point, which often relies on the computation of the image derivatives with pixel differencing methods. In this paper, we show the advantages of using smooth derivative filters instead of pixel differences in the performance of a well known local image descriptor. The method is based on the use of odd Gabor functions, whose parameters are selectively tuned to as a function of the local image properties under analysis. We perform an extensive experimental evaluation to show that our method increases the distinctiveness of local image descriptors for image region matching and object recognition. Furthermore, we propose an image registration algorithm named BP-SIFT, where we formulate key points matching of SIFT descriptors as a global optimization and provide a suboptimum solution using belief propagation (BP). Experimental results show the significant improvement over conventional SIFT based on matching with reasonable computation complexity.

1 INTRODUCTION

Visual navigation of the mobile robot mainly through the vision to perceive a scene, it extracts the image feature and describes the environmental information. Generally, image features are the description of image properties and initial features obtained by mobile robot vision[1]. Image features can be divided into global and local features. The global features are extracted from the whole frame image, they are sensitive to the view angle and illumination, and meanwhile, they need a large amount of calculation[2]. The local feature focuses on the description of the local features in the image. It can effectively resist geometric transformation and its redundancy is low[3,4].

Some extensions of the SIFT descriptor have been proposed recently to improve matching properties or to reduce computational complexity. It is a local feature descriptor based on the invariant technique and the linear scale space, which keeps invariant to image scale, rotation, and even affine transformation, so it is widely used in robot location, navigation and mapping[5–7]. The SIFT concatenates the first order x and y image derivatives of each sub region, PCA performs a principal component analysis of the data selection and reduce the vector dimension[8]. A main objective of PCA-SIFT is to keep the SIFT matching properties, reducing the size of descriptor. On the other hand, the Gradient location-orientation histogram (GLOH) is an extension of SIFT that computes the histogram using a log-polar spatial grid and reduces the descriptor size using PCA[8]. The SIFT algorithm combines with the Contourlet transformation[9]. Although in some degree it improves the accuracy. Linear classifier method of clustering local feature based on SIFT is complex[10,11]. In order to improve the matching results, the GLOH uses a more robust spatial grid to compute the gradient histogram[12]. The methods mentioned above have shown better performance than SIFT in some experimental conditions.

2 SMOOTH DERIVATIVE FILTER

Based on the visual navigation of the mobile robot, the position of robot changes, as is known to all, the obtained images of the same scene can have the transformation of scale, rotation and illumination. First, we review the SIFT local descriptor computation, and then present a modification of the SIFT descriptor. We finally use Gabor functions as smooth filters to improve the distinctiveness of the image derivatives. The SIFT feature is a feature that the extreme points of three consecutive images of Gauss difference. Gauss differential extremum point in the image is the gray value of each pixel which is larger or smaller than it around, so the extracted points from the extreme points in the SIFT feature have good stability[13–15].

2.1 *SIFT feature extraction*

In the traditional algorithm of the SIFT descriptor, image region is represented with the concatenation of gradient orientation histograms which is relative to several rectangular sub regions. The function used is the scale-space Difference of Gaussians (DoG), and the image regions are selected by the local extreme at DoG. In order to get the local descriptor, the derivatives I_x and I_y of the image I with pixel differences are computed by the regions which are scale-normalized.

$$I_x(x,y) = I(x+1,y) - I(x-1,y) \tag{1}$$

$$I_y(x,y) = I(x,y+1) - I(x,y-1) \tag{2}$$

Image gradient magnitude and orientation for every pixel are computed:

$$M(x,y) = \sqrt{I_x(x,y)^2 + I_y(x,y)^2} \tag{3}$$

$$\Theta(x,y) = \tan^{-1}\frac{I_y(x,y)}{I_x(x,y)} \tag{4}$$

The image is divided into 16 sub regions. Each sub region contains 8 direction information, the maximum value is set as the direction of feature points, which keep the rotation invariance. To provide orientation invariance, we can computer the orientation of the image region, and the orientation is given by the highest peak of the gradient orientation histogram of the image region.

2.2 *SIFT algorithm based on smooth derivative filter*

The traditional SIFT descriptor applies the pixel difference method into feature description. The method has the characteristic of high pass filtering on image spectrum, amplifying the high frequency range, which is mainly composed by noise. However, the high frequency mingles an enormous amount of noise, and the calculation of $I_x(x,y)$ and $I_y(x,y)$ makes the SIFT feature description methods sensitive to noise, and meanwhile, it does not have the noise immunity. In order to reduce the sensitivity to noise, it is common to combine a low-pass filter with a high-pass derivative. In this paper, we propose to use Gabor filter to improve SIFT descriptor.

Gabor filter functions are defined as the multiplication of a exponential function and a Gaussian function, the formula is:

$$g_{x,y,\theta} = \frac{1}{2\pi\sigma_1\sigma_2} * \exp\left(-\frac{(x\cos\theta + y\sin\theta)^2}{2x\sigma_1^2} - \frac{(y\cos\theta - x\sin\theta)^2}{2x\sigma_2^2}\right) * \exp\left(i\frac{2\pi}{\lambda}(x\cos\theta + y\sin\theta)\right) \tag{5}$$

where σ_1 and σ_2 are the Gaussian envelope standard deviations, θ is the filtering orientation, λ is its wavelength, and oriented along directions θ and $\theta + \pi/2$ respectively.

The first order image derivatives I_x and I_y are calculated by the odd part of Gabor filter. Respectively, the orientations will be $\theta = 0$ and $\theta = \pi/2$ for the horizontal and the vertical direction. We can replace $\delta = \sigma_1 = \sigma_2, \lambda = 6\delta$ to make the Gabor filter approximate to the Gaussian derivatives and the formula follows:

$$g_{x,y,\theta} = \frac{1}{2\pi\delta^2}\exp\left(-\frac{x^2+y^2}{2\delta^2}\right)\sin\left(\frac{2\pi}{6\delta}(x\cos\theta + y\sin\theta)\right) \tag{6}$$

where $\theta = 0$ computers I_x, and $\theta = \pi/2$ computers I_y.

The calculation of δ is a process of selecting filter gradient magnitude. It is not difficult to find that derivative in horizontal and vertical direction in the gradient magnitude is easier to calculate. The standard of scale selection is tending to the smaller scale values. According to the analysis, we can calculate the scale values which the Gabor function can obtain the largest energy in the horizontal and the vertical direction respectively, and from these we can select the smaller as the final scale values of Gabor function, the calculation formulas are as follows:

$$\hat{\delta}_x = \arg\max_{\delta}\left|I * g_{x_i,y_i,\theta=0}(\delta)\right|, \tag{7}$$

$$\hat{\delta}_y = \arg\max_{\delta}\left|I * g_{x_i,y_i,\theta=\frac{\pi}{2}}(\delta)\right|, \tag{8}$$

$$\hat{\delta}(x_i,y_i) = \min(\hat{\delta}_x, \hat{\delta}_y), \tag{9}$$

$$I_x(x_i,y_i) = (I * g_{x_i,y_i,0}(\hat{\delta}))(x_i,y_i), \tag{10}$$

$$I_y(x_i,y_i) = (I * g_{x_i,y_i,\frac{\pi}{2}}(\hat{\delta}))(x_i,y_i). \tag{11}$$

where (x_i, y_i) is a point in the scale-normal region, and $\hat{\sigma}$ is the adequate filter width at position (x_i, y_i).

3 IMPROVED BP-SIFT ALGORITHM

Based on the autonomous navigation of mobile robot vision, the SIFT descriptor can accurately match the feature points in the image through Gabor filter, however, the traditional method does not take into account the geometric relationship between the feature points. When the two key points have the same local image information but the geometric distance is far, it easily leads to mistake matching. Geometric information is one of the most important image features, and it has been widely used in image registration. As we known, belief propagation(BP) can be used to formulate key point matching of SIFT descriptors as a global optimization problem and to provide a suboptimum solution, therefore we use the SIFT algorithm based on the Gabor filter to extract the image feature descriptors, meanwhile, combine the belief propagation algorithm to improve the matching accuracy.

3.1 *SIFT feature points matching*

The main purpose of SIFT algorithm is to determine the position of feature point that is invariant to size, rotation and illumination in the image. The reference image is denoted as I_1, and the target image is denoted as I_2. The $D_1(i)$ is recorded as descriptor vector of ith in I_1, and the $D_2(j)$ is recorded as descriptor vector of jth in I_2. Respectively, $x_{D1}(i)$ and $x_{D2}(j)$ are the corresponding locations of the key points in I_1 and I_2.

The aim of matching is to find the corresponding feature point in the image I_1 and I_2. Optimal matching is to calculate the minimum value of the Euclidean distance between feature points. This is to say, for a descriptor $D_1(i)$, the optimal matching $j(i)$ should satisfy the formula:

$$\hat{j}(i) = \arg\min_j \|D_1(i) - D_2(j)\|_2 \tag{12}$$

Traversing all the feature points to make the equation above minimum, however, there may be not exist feature points that can be matched with. To identify such a case, Lowe supposed to use the second minimum distance from the target descriptor to measure the probability of a match. The first stage is to obtain every feature vector's distance, and make the feature distance in accordance with the ascending sort; Then take the ratio the nearest feature distance to the sub-nearest, if the ratio is less than a threshold, the nearest points are selected as the candidate matching points. The formula is as follows:

$$\left\|D_1(i) - D_2(\hat{j}(i))\right\|_2 < T\|D_1(i) - D_2(j)\|_2, \quad \forall j \neq \hat{j}(i) \tag{13}$$

where T is independent of image I_1 and I_2.

The set of descriptors in I_1 is denoted by I_{D1}, the vector of I_1 mapping to I_2 is denoted by $m = [m_1, \cdots, m_{|I_{D1}|}]$. The penalty function $\phi(m)$ is defined as: the sum of difference between the distance from one key point to another in image I_1 and the distance between the corresponding points in I_2. The formula is as follows:

$$\varphi(m) = \sum_{i\in I_{D_1}} \sum_{j\in I_{D_1}} \varphi(i, j; m(i), m(j)) \tag{14}$$

$$\varphi(i, j; i', j') = \left\| \sqrt{\|x_{D_1}(i) - x_{D_1}(j)\|_2} - \sqrt{\|x_{D_2}(i') - x_{D_2}(j')\|_2} \right\|_2 \tag{15}$$

Therefore, the global optimization problem as follows:

$$\hat{m} = \arg \underset{m}{max} \left(\psi(m) + \varepsilon\phi(m)\right) \tag{16}$$

$$\psi(m) = \sum_{i\in I_{D_1}} \vartheta(i, m(i)) \tag{17}$$

$$\psi(i, i') = \|D_1(i) - D_2(i')\| \tag{18}$$

Of which: ε is the Lagrange multiplier that can be adjusted by the prior information.

3.2 *Belief propagation (BP)*

It is not difficult to find that the equation above appears to have exponential complexity, and it is possible to discard some useful information that does not satisfy the equation. The belief propagation algorithm has been widely used in signal processing application. The main idea is taking the discrete optimal problem approximate to the global optimum using iterative transmission information[16,17].

In each iteration, each local feature estimates the optimal solution, and two local adjacent features will be exchange belief. Each local feature will be merged into the next iteration of the belief calculation process. When the all local feature points converge to the most probable belief or the iteration times reaches the fixed predefined limit, the iteration is end.

3.3 *SIFT matching algorithm based on belief propagation*

Even if the objective function presents exponential growth, optimization problem still does not change, therefore (16) was improved:

$$\hat{m} = \arg \underset{m}{max}(\exp(-\psi(m))\exp(-\varepsilon\phi(m))) \tag{19}$$

Further consolidation is

$$\hat{m} = \arg\max_m \prod_{i\in I_{D_1}} b_{Des_i}(m_i) \prod_{i\in I_{D_1}, j\in I_{D_1}} b_{Dist_{i,j}}(m_i, m_j) \tag{20}$$

$$b_{Des_i}(m_i) = \exp(-\psi(i, m_i)/C_{Des}) \tag{21}$$

$$b_{Dist_{i,j}}(m_i, m_j) = \exp(-\vartheta(i, j; m_i, m_j)/C_{Dist}) \tag{22}$$

$C_{Des}/C_{Dist} = \kappa, b_{Des_i}(m_i)$ can be approximated to the belief that the ith key point in I_1 matches with the key point m_j in I_2. $b_{Dist}(m_i, m_j)$ presents the information that the jth key point in I_1 matches with the m_jth key points in I_2. C_{Des} and C_{Dist} adjust the prior matching possibility and geometric information, and the coefficient is inversely proportional to the matching belief.

If the descriptors are far away between each other, the correlation between them is almost zero, so the correlation between the corresponding descriptors is also small. Therefore, we have the optimization function and only consider i and j are the near points in *Dist*. Based on the above analysis, we use the max product algorithm to improve the BP algorithm, the formula is:

$$\hat{m} = \arg \underset{m}{max}\left(\prod_{i\in I_{D1}} b_{Des_i}(m_i) \prod_{i\in I_{D1}, j\in V\leftarrow(i)} b_{Dist_{i,j}}(m_i, m_j) \right) \tag{23}$$

We define $\mu_{i,j}^{(n)}j(m_j)$ is the ith key point's belief of the nth iteration in the condition of the jth key point correctly matching with m_j. The information transfers in the iteration process. The algorithm is as follows:

Figure 1. The platform of pioneer 3TM.

Figure 2. Monocular camera Canon VC-C50i.

(1) $\mu_{i,j}^{(0)}(m_j)$ is initialized to a constant. Set $n = 1$;
(2) For $i \in I_{D1}, j \in I_{D2}$, the iteration updates messages:

$$\tau_1 \max_{m_i} b_{Dist_{i,j}}(m_i, m_j) b_{Des_i}(m_i) \prod_{j' \in N(i) \setminus j} \mu_{i,j'}^{(n-1)}(m_i) \rightarrow \mu_{i,j}^{(n)}(m_j)$$

(3) for all $i \in I_{D1}$, calculate the belief:

$$\tau_2 b_{Des_i}(m_i) \prod_{j' \in N(i)} \mu_{i,j'}^{(n)}(m_i) \rightarrow b_i^{(n)}(m_i)$$

$$\hat{m}_i = \arg\max_{m_i} b_i^{(n)}(m_i)$$

(4) $n \leftarrow n + 1$, and jump to step (2). When the iteration reaches to a maximum value, end the iteration.

Among them: τ_1 and τ_2 are normalization constants. $b_i(m_i)$ indicates the possibility of i matching to m_i. If $b_i(\hat{m}_i)$ less than a threshold p_{th}, discard the key point.

4 SIMULATION ANALYSES

SIFT feature extracting and matching experiments are validated in an indoor environment using the pioneer 3TM mobile robot platform, and are tested in MATLAB software. Monocular camera Canon VC-C50i is installed on the top of the mobile robot, and its maximum resolution is 704 × 576 pixels, rotation range is ±100° in horizontal direction, rotation range is ±30° in vertical direction. Mobile robot and vision camera respectively as shown in figure 1 and figure 2:

On the MATLAB simulation, we contrast the SIFT algorithm and the improved SIFT algorithm to scale, rotation and illumination matching results.

(a) SIFT (b) Improved SIFT

Figure 3. The matching images moving backwards 0.5 m.

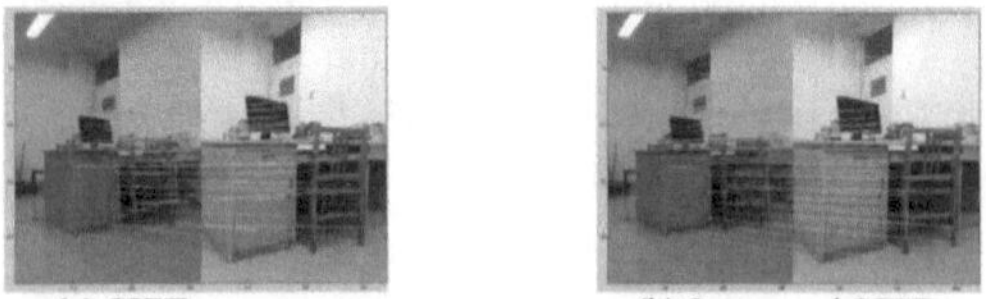

(a) SIFT (b) Improved SIFT

Figure 4. The matching images moving backwards 1 m.

Table 1. The scale transformation data.

	Backward 0.5 m		Backward 1 m	
	SIFT	Improved SIFT	SIFT	Improved SIFT
Mistake/all	4/62	1/74	5/53	2/69
Correct rate (%)	93.54	98.65	90.56	97.10

(1) Scale transformation
We take experiments in the case of mobile robot moving backward for 0.5 meters and 1 meter. When the images exist similarity regions, there would emerge false matching. The SIFT algorithm matching results are as shown in Figure 3 (a) and 4 (a). The improved SIFT algorithm would acquire image geometry information, keep the space constraint, and excuse the mistake matching that having the similar local structure in the impossible region. The results are shown in Figure 3(b) and 4(b). The improved algorithm is better than the original algorithm in match and enhanced 5.1% and 6.5% respectively, as shown in table 1.
(2) The rotation transformation
We take experiments in the case of mobile robot respectively rotating 30° and 45°. SIFT algorithm in small rotation angle, is almost similar with the improved algorithm, as shown in Figure 5, but in the case of rotation angle large, the SIFT algorithm mismatch significantly increases, as shown in Figure 6 (a). The SIFT combined with the belief propagation algorithm considers the relation of geometric information in the image, so there are constraint each other between key points and the corresponding points in matching image. The matching accuracy is improved clearly, as shown in Figure 6 (b). The improved algorithm

(a) SIFT (b) Improved SIFT

Figure 5. the matching images rotating 30°.

(a) SIFT (b) Improved SIFT

Figure 6. the matching images rotating 45°.

Table 2. The matching data of rotation transformation.

	Rotation30°		Rotation 45°	
	SIFT	Improved SIFT	SIFT	Improved SIFT
Mistake/all	6/86	2/92	10/35	3/42
Correct rate(%)	93.02	97.83	71.43	92.86

(a) SIFT (b) Improved SIFT

Figure 7. The matching images in different illumination.

is better than the original algorithm and enhanced 4.8% and 21.4% respectively, as shown in table 2.

(3) Illumination transformation

When the illumination is different, SIFT algorithm matching rate is low, as shown in Figure 7 (a). The improved algorithm has higher accuracy and robustness. After Gabor filtering, it removes the noise that mobile robot vision system obtaining images doped in. After smoothing the image, we combine with the belief propagation. Belief propagation does not affect the correct matching when the light is changed. Considering the geometric information of images, the correct rate of image matching is evidently better than SIFT algorithm. Experiments are taken in different luminance, the improved algorithm is enhanced 8.3% in matching rate than the original algorithm, as shown in table 3.

Table 3. The data of illumination transformation.

	Images in different luminance	
	SIFT	Improved SIFT
Mistake/all	4/48	0/51
Correct rate(%)	91.67	100

5 CONCLUSION

We propose a modification of SIFT using smooth derivative filter, and improve the descriptor. It convolves the Gabor function with the obtained image, and acquires the image feature description, besides, effectively removes the noise in the image. Combining with the belief propagation algorithm, we take the geometric information of the images into consideration. Through the image matching in the real scene, experiments show that the improved algorithm can effectively remove the error matching points, and improve the matching accuracy.

ACKNOWLEDGEMENTS

This work was financially supported by the national natural Science Foundation of China(61201112 and 61172044), the Natural Science Foundation for Young Scientists of Hebei Province, China (F2013203250, F2012203169).

REFERENCES

[1] Reza, J.A. et al. 2012. A New Illumination Invariant Feature Based on SIFT Descriptor in Color Space. *Procedia Engineering* 41: 305–311.

[2] Liang, J.N. et al. 2012. Image matching based on orientation–magnitude histograms and global consistency. *Pattern Recognition45* (10): 3825–3833.

[3] Tuan Hue Thi. et al. 2012. Structured learning of local features for human action classification and localization.*Image and Vision Computing* 30(1): 1–14.

[4] Zhang, L. & Wang, H.L. 2009. Matching of Interesting Points Based on Improved SIFT Algorithm. *Journal of Electronics & Information Technology* 31(11): 2620–2625.

[5] LOWE, D.G. 2004. Distractive image features from scale-invariant keypoints. *International Journal of Computer Vision* 60(2): 91–110.

[6] P. Weinzaepfel. et al. 2011. Reconstructing an image from its local descriptors. *Proceedings of the IEEE Computer Society Conference on Computer Vision and Pattern Recognition*. Colorado Springs.

[7] Leonardo, C. et al. 2012. A Bayesian approach for object classification based on clusters of SIFT local features. *Expert Systems with Applications* 39(2): 1679–1686.

[8] Zhang, Y. & Wei, K.B. 2010. Research on Wide Baseline Stereo Matching Based on PCA-SIFT. *International Conference on Advanced Computer Theory and Engineering(ICACTE)*. Chengdu.

[9] Chen, S.R. et al. 2013. Contourlet-SIFT Feature Matching Algorithm. *Journal of Electronics & Information Technology* 35(5): 1215–1221.

[10] Florent, P. & Christopher, D. 2007. Fisher kernels on visual vocabularies for image categorization. *IEEE Conference on Computer Vision and Pattern Recognition.* Minneapolis.

[11] Josip, K. et al. 2011. Modeling spatial layout with fisher vectors for image categorization. *IEEE International Conference on Computer Vision*. Barcelona.

[12] Huang, Q.H. et al. 2009. A new adaptive interpolation algorithm for 3D ultrasound imaging with speckle reduction and edge preservation. *Computerized Medical Imaging Graphics* 33(2): 100–110.

[13] Liao, K.Y. et al. 2013. An improvement to the SIFT descriptor for image representation and matching. *Pattern Recognition Letters* 34(11): 1211–1220.

[14] Zhong, S. et al. 2013. A real-time embedded architecture for SIFT. *Journal of Systems Architecture* 59(1): 16–29.

[15] Zhang, Q. et al. 2012. Particle Filter Object Tracking Based on Harris-SIFT Feature Matching. *Procedia Engineering* 29: 924–929.

[16] Olivier, B. et al. 2009. Fast nonparametric belief propagation for real-time stereo articulated body tracking. *Computer Vision and Image Understanding* 113(1): 29–47.

[17] Haidawati, N. et al. 2012. Singular value decomposition based fusion for super-resolution image reconstruction. *Signal Processing: Image Communication* 27(2): 180–191.

Future Mechatronics and Automation – Yang (Ed.)
© 2015 Taylor & Francis Group, London, ISBN 978-1-138-02648-3

Direct adaptive fuzzy sliding mode control for a class of uncertain MIMO nonlinear systems

Shenglin Wen & Ye Yan
National University of Defense Technology, Changsha, China

ABSTRACT: This paper presents a direct adaptive fuzzy sliding mode control scheme for a class of uncertain multi-input multi-output nonlinear systems with unknown control gain matrix and external disturbance. We show that the only restriction on the unknown control gain matrix is that it is nonsingular. No extra constraint is required and the condition that the matrix is to be symmetric positive definite is also not needed. Within the direct control scheme, fuzzy adaptive systems are used to approximate the ideal control law. Moreover, an adaptive switch control term is appended to eliminate the fuzzy approximation error. By using the Lyapunov theory, all adaptive learning laws in the proposed control system are derived and the stability of the closed-loop system can be guaranteed. Finally, the effectiveness of the proposed control scheme is demonstrated through the simulation of an uncertain nonlinear system.

1 INTRODUCTION

The control of general uncertain nonlinear system has been a widely investigated problem because of its wide applications in practical systems. Many practical systems may be so complex or, even, model-free that to construct a mathematical model or identify its parameters is difficult or even impossible. Adaptive fuzzy control allows incorporating linguistic fuzzy information from human operators and can provide universal nonlinear approximator. Therefore, fuzzy control is a very good candidate for control of uncertain nonlinear dynamic systems (Li and Tong, 2003; Phan and Gale, 2008; Chen et al., 2008).

For uncertain nonlinear systems, adaptive fuzzy control can be classified into two categories, direct and indirect adaptive fuzzy control (Labiod and Guerra, 2010). In the direct adaptive scheme, the fuzzy system is used to approximate an unknown ideal controller. On the other hand, the indirect scheme uses fuzzy systems to estimate the plant and then construct a control law based on these estimates. Thus, it can be said that direct adaptive fuzzy control is structurally simpler than indirect adaptive fuzzy control. However, direct adaptive fuzzy control scheme usually require more constraints on the control gain matrix (Rong and Zhao, 2013).

In addition to the well-known necessary condition that the control gain matrix be nonsingular for a controllable plant, some extra constraints on the matrix are usually needed for stability and convergence analysis for direct adaptive fuzzy control(Sofiane and Khaber,2012; Labiod and Guerra, 2007; Chen, 2012; Chen et al., 2009; Essounbouli and Hamzaoui, 2006). Besides, compared with uncertain single-input single-output (SISO) nonlinear systems, the problem of adaptive fuzzy control of uncertain multi-input multi-output (MIMO) nonlinear systems is more difficult because of the coupling that exists between the control inputs and the outputs (Shi, 2013). In this paper, we will show that the only necessary requirement is the controllability condition that the control gain matrix is nonsingular. No extra constraint is required and the condition that the matrix is to be symmetric positive definite is also not needed. Moreover, we also introduce use of a sliding mode control system to further enhance our proposed direct adaptive fuzzy controller.

Pure direct adaptive fuzzy control is simple, but they also have disadvantages. It is often that adaptive fuzzy control is combined with another control technique to compensate for approximation error. In general, there exist approximation errors when approximating nonlinear functions by fuzzy systems. These approximation errors may effect and deteriorate the stability and performance of adaptive fuzzy control systems. To overcome this problem, previous researchers have proposed combining AFC with another controller. An indirect adaptive fuzzy controller combined with a fuzzy sliding mode controller is proposed in Nekoukar and Erfanian (2011). The fuzzy sliding mode controller is designed to compensate for the approximation errors. Much of the current research (Yu et al., 2000; Tong and Li, 2003) proposes adaptive fuzzy control with a sliding mode control term. The sliding mode control term is designed using some known bounds of approximation errors. The term is then added to the control output to compensate for the effect of approximation errors. However, the bounds of approximation errors are normally hard to obtain in practice. Thus, they take a step further by proposing some adaptive mechanisms to estimate these bounds online.

In this paper, the tracking control problem has been addressed for a class of MIMO uncertain nonlinear systems. The basic idea is to use fuzzy systems to adaptively construct an unknown ideal controller. A switch control term is used to eliminate the fuzzy approximation error. The parameter adaptive laws and the overall closed-loop system stability are studied by using a Lyapunov approach. With respect to the research works of Labiod and Guerra (2010), Labiod and Guerra (2007), Nekoukar and Erfanian (2011), the main contributions of this paper are:

(1) In Labiod and Guerra (2010), Labiod and Guerra (2007), extra restriction should be taken to guarantee that the control gain matrix is needed to be symmetric positive definite. However, in this paper, the only necessary requirement for the control gain matrix is the controllability condition that it is nonsingular. Thus, the constraint on the control gain matrix is relaxed.
(2) In Nekoukar and Erfanian (2011), the number of parameter updated online for fuzzy systems is n^2+n, in which the estimation of the unknown nonlinear function needs n fuzzy systems and the estimation of the control gain matrix needs n^2 fuzzy systems. In the proposed control scheme, only n fuzzy systems are needed. Therefore, the proposed direct adaptive fuzzy control scheme is structurally simpler than indirect ones.

2 PROBLEM STATEMENT

Consider the following MIMO uncertain nonlinear system

$$\begin{aligned} y_1^{(r_1)} &= f_1(x)+\sum_{j=1}^{m} g_{1j}(x)u_j+d_1' \\ &\vdots \\ y_m^{(r_1)} &= f_m(x)+\sum_{j=1}^{m} g_{mj}(x)u_j+d_m' \end{aligned} \tag{1}$$

where $\boldsymbol{x}=[y_1,\cdots,y_1^{(r_1-1)},y_2,\cdots,y_2^{(r_2-1)},\cdots,y_p,\cdots,y_m^{(r_m-1)}]^T$ is the state vector which is available for measurement $\boldsymbol{y}=[y_1,y_2,\cdots,y_m]^T$ is the output vector, $\boldsymbol{u}=[u_1,u_2,\cdots,u_m]^T$ is the control input vector, $f_i(\mathbf{x})(i=1,2,\cdots,m)$ are smooth unknown nonlinear functions, $g_{ij}(\boldsymbol{x})(i,j=1,2,\cdots,m)$ are unknown continuous nonlinear functions, and $\boldsymbol{d}'=[d_1',d_2',\cdots,d_m']^T$ are unknown external disturbances.

Define $\boldsymbol{r}=[r_1,r_2,\cdots,r_m]^T$, $\boldsymbol{y}^{(\mathbf{r})}=[y_1^{(r_1)},y_2^{(r_2)},\cdots,y_m^{(r_m)}]^T$ and

$$\boldsymbol{f}(\boldsymbol{x})=[f_1(\boldsymbol{x}),f_2(\boldsymbol{x}),\cdots,f_m(\boldsymbol{x})]^T$$

$$\boldsymbol{g}(\boldsymbol{x})=\begin{bmatrix} g_{11}(\boldsymbol{x}) & \cdots & g_{1m}(\boldsymbol{x}) \\ \vdots & \ddots & \vdots \\ g_{m1}(\boldsymbol{x}) & \cdots & g_{mm}(\boldsymbol{x}) \end{bmatrix}$$

The dynamic system equation (1) can be rewritten in the following form:

$$\boldsymbol{y}^{(r)}=\boldsymbol{f}(\boldsymbol{x})+\boldsymbol{g}(\boldsymbol{x})\boldsymbol{u}+\boldsymbol{d}' \tag{2}$$

Note that $\boldsymbol{f}(\boldsymbol{x})\in R^m$, $\boldsymbol{g}(\boldsymbol{x})\in R^{m\times m}$ and $\boldsymbol{d}'\in R^m$ are unknown in this study. As the system (2) is required to be controllable, we make the following reasonable assumption.

Assumption 1: The unknown control gain matrix $\boldsymbol{g}(\boldsymbol{x})$ is assumed to be nonsingular, i.e., $\boldsymbol{g}^{-1}(\boldsymbol{x})$ exists, and bounded for all $\boldsymbol{x}\in \boldsymbol{U_x}$, where $\boldsymbol{U_x}\subset R^m$ is some compact set.

Equation (2) can be rewritten as

$$\boldsymbol{0}=-\boldsymbol{g}^{-1}(\boldsymbol{x})\boldsymbol{y}^{(r)}+\boldsymbol{g}^{-1}(\boldsymbol{x})\boldsymbol{f}(\boldsymbol{x})+\boldsymbol{u}+\boldsymbol{g}^{-1}(\boldsymbol{x})\boldsymbol{d}' \tag{3}$$

By adding $\boldsymbol{y}^{(r)}$ to both side of equation (3), we obtain

$$\boldsymbol{y}^{(r)}=\boldsymbol{y}^{(r)}-\boldsymbol{g}^{-1}(\boldsymbol{x})\boldsymbol{y}^{(r)}+\boldsymbol{g}^{-1}(\boldsymbol{x})\boldsymbol{f}(\boldsymbol{x})+\boldsymbol{u}+\boldsymbol{g}^{-1}(\boldsymbol{x})\boldsymbol{d}' \tag{4}$$

Then we rewrite equation (4) as follows

$$\boldsymbol{y}^{(r)}=\boldsymbol{F}(\boldsymbol{x})+\boldsymbol{u}+\boldsymbol{d} \tag{5}$$

where

$$\boldsymbol{F}(\boldsymbol{x})=\left(\boldsymbol{I}_m-\boldsymbol{g}^{-1}(\boldsymbol{x})\right)\boldsymbol{y}^{(r)}+\boldsymbol{g}^{-1}(\boldsymbol{x})\boldsymbol{f}(\boldsymbol{x})$$

$$\boldsymbol{d}=\boldsymbol{g}^{-1}(\boldsymbol{x})\boldsymbol{d}'$$

Define reference trajectory $\boldsymbol{x}_d=[x_{d1},x_{d2},\cdots,x_{dm}]^T$, the control objective is to determine a direct adaptive fuzzy sliding mode controller such that the system output $\boldsymbol{y}$ can track the reference trajectory $\boldsymbol{x}_d$.

3 DIRECT ADAPTIVE FUZZY SMC DESIGN

3.1 *The ideal controller*

Define the tracking error as

$$\begin{aligned} e_1 &= x_1-y_{1d} \\ &\vdots \\ e_m &= x_m-y_{md} \end{aligned} \tag{6}$$

For each subsystem, we can define integral sliding surface as follows:

$$\begin{aligned} s_i &= y_1^{(r_1-1)}-\int_0^t\left(y_{d1}^{(r_1)}-k_{11}e_1^{(r_1-1)}-\cdots k_{1,r_1}e_1\right)dt \\ &\vdots \\ s_m &= y_m^{(r_m-1)}-\int_0^t\left(y_{dm}^{(r_m)}-k_{m1}e_m^{(r_m-1)}-\cdots k_{m,r_m}e_m\right)dt \end{aligned} \tag{7}$$

The time derivatives of the sliding surface of each subsystem are

$$\begin{aligned} \dot{s}_1 &= e_1^{(r_1)}+k_{11}e_1^{(r_1-1)}+\cdots+k_{1,r_1}e_1 \\ &\vdots \\ \dot{s}_m &= e_m^{(r_m)}+k_{m1}e_m^{(r_m-1)}+\cdots+k_{m,r_1}e_m \end{aligned} \tag{8}$$

If the nonlinear function $\boldsymbol{F}(\boldsymbol{x})$ and the lumped uncertainty $\boldsymbol{d}$ can be exactly known, then an ideal controller can be designed as follows

$$\boldsymbol{u}^{*}=-\boldsymbol{F}(\boldsymbol{x})-\boldsymbol{d}+\boldsymbol{v} \tag{9}$$

where $\boldsymbol{v}=[v_1,v_2,\cdots,v_m]^T$, $v_1,v_2,\cdots,v_m$ are given as follows

$$\begin{aligned} v_1 &= y_{1d}^{(r_1)}-k_{11}e_1^{(r_1-1)}-\cdots-k_{1,r_1}e_1 \\ &\vdots \\ v_m &= y_{md}^{(r_m)}-k_{m1}e_1^{(r_m-1)}-\cdots-k_{m,r_1}e_m \end{aligned} \tag{10}$$

Substituting the ideal controller equation (9) into equation (5), gives the error dynamic equation

$$\begin{aligned} e_1^{(r_1)}+k_{11}e_1^{(r_1-1)}+\cdots+k_{1,r_1}e_1 &= 0 \\ &\vdots \\ e_m^{(r_m)}+k_{m1}e_m^{(r_m-1)}+\cdots+k_{m,r_1}e_m &= 0 \end{aligned} \tag{11}$$

In equation (11), if $k_{ij}(i=1,\cdots m;j=1,\cdots r_i)$ is chosen such that all polynomials are Hurwitz, it implies that $\lim_{t\to\infty} e_i(t)=0$.

According to the above analysis, the control law (9) is easily obtained if $\boldsymbol{F}(\boldsymbol{x})$ and $\boldsymbol{d}$ are known. However, in this paper, $\boldsymbol{F}(\boldsymbol{x})$ is assumed to be unknown, the lumped uncertainty $\boldsymbol{d}$ is generally unknown in practical applications, so that $\boldsymbol{u}^*$ in equation (9) is unavailable. In this case, a direct adaptive fuzzy sliding mode control system is proposed to achieve trajectory tracking control. In this control system, a fuzzy system is used to approximate the ideal controller in equation (9).

3.2 *Adaptive fuzzy SMC Design*

It has been proved that the fuzzy systems can approximate an arbitrarily continuous function to a given accuracy. The fuzzy systems perform a mapping from the sliding surface vector to the control effort vector. Consider the fuzzy system with m inputs and a single output, the input linguistic variables are $s_i(i=1,\cdots,m)$ and the output linguistic variable is u, where u represents the elements in the control effort vector. The fuzzy system is characterized by a set of if-then rules in the following form:

$$R^{(l)}:\text{If } s_1 \text{ is } F_1^l \text{ and} \cdots \text{and } s_m \text{ is } F_m^l, \text{then } u \text{ is } C^l\ (l=1,2,\cdots r) \tag{12}$$

where F_i^l and C^l denote the linguistic variables of the input and output of the fuzzy sets, r is the number of model rules.

The fuzzy system with singleton fuzzifier, product inference engine, center average defuzzifier is in the following form:

$$u(\boldsymbol{s})=\frac{\sum_{l=1}^{r}\bar{u}^l\left(\prod_{i=1}^{m}\mu_{F_i^l}(s_i)\right)}{\sum_{l=1}^{r}\prod_{i=1}^{m}\mu_{F_i^l}(s_i)} \tag{13}$$

where $\boldsymbol{s}=[s_1,\cdots,s_m]^T\in\Omega_s\subset R^m$, $\mu_{F_i^l}(s_i)$ is the membership function of the linguistic variable s_i, $\bar{u}^l$ represents a crisp value at which the membership function μ_{C^l} for the output fuzzy set reaches its maximum.

By introducing the concept of fuzzy basis function vector, the final output of the fuzzy system can be expressed in the following compact form:

$$u(\boldsymbol{s})=\boldsymbol{\xi}^T(\boldsymbol{s})\boldsymbol{\theta} \tag{14}$$

where $\boldsymbol{\theta}=[\bar{u}^1,\cdots\bar{u}^r]^T$ is the parameter vector and $\boldsymbol{\xi}(\boldsymbol{s})=[\xi_1(\boldsymbol{s}),\cdots,\xi_r(\boldsymbol{s})]^T$ is the fuzzy basis function vector defined as

$$\xi_l(\boldsymbol{s})=\frac{\sum_{l=1}^{r}\left(\prod_{i=1}^{m}\mu_{F_i^l}(s_i)\right)}{\sum_{l=1}^{r}\prod_{i=1}^{m}\mu_{F_i^l}(s_i)} \quad l=1,\cdots,r \tag{15}$$

Based on the fact that a fuzzy system is a universal approximator, we use a fuzzy Takagi-Sugeno system in the form of (12) to approximate each component of the ideal input control vector $\boldsymbol{u}^*=[u_1^*,\cdots,u_m^*]^T$ as follows:

$$u_i^*=\xi_i^T(\boldsymbol{s})\theta_i^*+\varepsilon_i(\boldsymbol{s}) \quad i=1,2,\cdots,m \tag{16}$$

where $\varepsilon_i(\boldsymbol{s})$ is the minimum fuzzy approximation error, $\boldsymbol{\theta}_i^*$ is an unknown optimal fuzzy parameter vector which minimizes $|\varepsilon_i(\boldsymbol{s})|$ over an operating compact set Ω_s and $\boldsymbol{\xi}_i(\boldsymbol{s})$ is a fuzzy basis function vector fixed by the designer.

In this paper, we assume that minimum approximation errors are bounded for all $\boldsymbol{s}\in\Omega_s$, that is, $|\varepsilon_i(\boldsymbol{s})|\le\bar{\varepsilon}_i$, where $\bar{\varepsilon}_i$ is a positive constant. Denote $\boldsymbol{\xi}(\boldsymbol{s})=\mathrm{diag}[\boldsymbol{\xi}_1(\boldsymbol{s}),\cdots,\boldsymbol{\xi}_m(\boldsymbol{s})]$, $\boldsymbol{\theta}^*=[\boldsymbol{\theta}_1^{*T},\cdots,\boldsymbol{\theta}_m^{*T},]^T$, $\boldsymbol{\varepsilon}(\boldsymbol{s})=[\varepsilon_1(\boldsymbol{s}),\cdots,\boldsymbol{\varepsilon}_m(\boldsymbol{s})]^T$, $\boldsymbol{u}^*$ can be expressed as

$$\boldsymbol{u}^*=\xi^T(\boldsymbol{s})\boldsymbol{\theta}^*+\varepsilon(\boldsymbol{s}) \tag{17}$$

Since the ideal parameter vector $\boldsymbol{\theta}^*$ is unknown, it should be estimated by a suitable adaptation law. Define $\hat{\boldsymbol{\theta}}$ is an estimate of the ideal vector $\boldsymbol{\theta}^*$, then we can employ the following fuzzy systems to approximate $\boldsymbol{u}^*$:

$$\boldsymbol{u}_{fz}=\xi^T(\boldsymbol{s})\hat{\boldsymbol{\theta}} \tag{18}$$

Further, to cancel the approximation error between the ideal controller $\boldsymbol{u}^*$ and the fuzzy systems $\boldsymbol{u}_{fz}$, we append a switch control term $\boldsymbol{u}_{vs}=[u_{vs1},\cdots,u_{vsm}]^T$. The switch control term $\boldsymbol{u}_{vs}$ will be designed next. Therefore, the overall control effort can be designed as

$$\boldsymbol{u}(t)=\boldsymbol{u}_{fz}+\boldsymbol{u}_{vs} \tag{19}$$

Define the parameter estimation error $\tilde{\boldsymbol{\theta}}=\hat{\boldsymbol{\theta}}-\boldsymbol{\theta}^*$ and the fuzzy approximation error $\tilde{\boldsymbol{u}}_{fz}=\boldsymbol{u}_{fz}-\boldsymbol{u}^*$, we have

$$\begin{aligned} \tilde{\boldsymbol{u}}_{fz} &= \boldsymbol{u}_{fz}-\boldsymbol{u}^*=\xi^T(\boldsymbol{s})\hat{\boldsymbol{\theta}}-\xi^T(\boldsymbol{s})\boldsymbol{\theta}^*-\varepsilon(\boldsymbol{s}) \\ &= \xi^T(\boldsymbol{s})\tilde{\boldsymbol{\theta}}-\varepsilon(\boldsymbol{s}) \end{aligned} \tag{20}$$

By substituting (5) into (9), we get

$$u^* = -y^{(r)} + u + v \tag{21}$$

Inserting (10) and (19) into equation (21), we obtain

$$u^* = u_{fz} + u_{vs} - \dot{s} \tag{22}$$

After simple manipulation, above equation becomes

$$\dot{s} = \xi^T(s)\tilde{\theta} + u_{vs} - \varepsilon(s) \tag{23}$$

Consider the following switch control term

$$u_{vsi} = -\hat{E}_i \operatorname{sgn}(s_i) \quad i = 1,2,\cdots,m \tag{24}$$

where $\hat{E}_i$ is the estimated switch control gain.

The parameter updated online by the following form

$$\dot{\hat{\theta}}_i = -\eta_i s_i \xi_i(s) \tag{25}$$

$$\dot{\hat{E}}_i = \lambda_i |s_i| \tag{26}$$

where η_i and λ_i are adaptive positive constants.

Theorem 1. For the uncertain nonlinear system(1), if Assumption 1 is satisfied, the controller take the form(18), (19) and (24),the adaptive laws are chosen as (25) and (26), then all signals in the closed-loop system are bounded, the tracking errors and $\tilde{\boldsymbol{\theta}}_i$ converge asymptotically to zero.

Proof: Consider the following Lyapunov function candidate

$$V_1 = \frac{1}{2}s^T s + \sum_{i=1}^{m}\frac{1}{2\eta_i}\tilde{\theta}_i^T\tilde{\theta}_i \tag{27}$$

Differentiating (27) with respect to time and using (23) and (25), we have

$$\begin{aligned} \dot{V}_1 &= s^T\dot{s} + \sum_{i=1}^{m}\frac{1}{\eta_i}\tilde{\theta}_i^T\dot{\tilde{\theta}}_i = s^T\left[\xi^T(s)\tilde{\theta} + u_{vs} - \varepsilon(s)\right] + \sum_{i=1}^{m}\frac{1}{\eta_i}\tilde{\theta}_i^T\dot{\tilde{\theta}}_i \\ &= \sum_{i=1}^{m}\left(s_i\xi_i^T(s)\tilde{\theta}_i + \frac{1}{\eta_i}\tilde{\theta}_i^T\dot{\tilde{\theta}}_i + s_i u_{vsi} - s_i\varepsilon_i\right) \end{aligned} \tag{28}$$

In order to achieve $\dot{V}_1 \leq 0$, u_{vsi} can be designed as

where $\bar{\varepsilon}_i \leq E_i$.
Using (29)

where $\bar{\varepsilon}_i \leq E_i$.

Using (29), (28) becomes

$$\begin{aligned} \dot{V}_1 &= \sum_{i=1}^{m} s_i\left(-E_i \operatorname{sgn}(s_i) - \varepsilon_i\right) \\ &\leq \sum_{i=1}^{m}\left(-E_i|s_i| + |\varepsilon_i||s_i|\right) \\ &= \sum_{i=1}^{m} -\left(E_i - |\varepsilon_i|\right)|s_i| \leq 0 \end{aligned} \tag{30}$$

Since E_i depend on the bound of the fuzzy approximation errors, it is difficult to obtain completely accurate values for these parameters. To satisfy the existence condition, a large approximation bound should be chosen in advance. In this case, the controller results in large implementation cost and leads to chattering efforts. In this paper, the bound of the fuzzy approximation errors are adaptively estimated on-line in order to achieve chattering elimination.

Define the estimate error $\tilde{E}_i = \hat{E}_i - E_i$, choosing a Lyapunov function

$$V = \frac{1}{2}s^T s + \sum_{i=1}^{m}\frac{1}{2\eta_i}\tilde{\theta}_i^T\tilde{\theta}_i + \sum_{i=1}^{m}\frac{1}{2\lambda_i}\tilde{E}_i^T\tilde{E}_i \tag{31}$$

Differentiating (31) with respect to time and using (23),(24)and(25),we obtain

$$\begin{aligned} \dot{V} &= s^T\dot{s} + \sum_{i=1}^{m}\frac{1}{\eta_i}\tilde{\theta}_i^T\dot{\tilde{\theta}}_i + \sum_{i=1}^{m}\frac{1}{\lambda_i}\tilde{E}_i^T\dot{\tilde{E}}_i \\ &= \sum_{i=1}^{m}\left(s_i\xi_i^T(s)\tilde{\theta}_i + \frac{1}{\eta_i}\tilde{\theta}_i^T\dot{\tilde{\theta}}_i - s_i\hat{E}_i\operatorname{sgn}(s_i) - s_i\varepsilon_i + \frac{1}{\lambda_i}\left(\hat{E}_i - E_i\right)\dot{\hat{E}}_i\right) \\ &= \sum_{i=1}^{m}\left(-\hat{E}_i|s_i| - s_i\varepsilon_i + \frac{1}{\lambda_i}\left(\hat{E}_i - E_i\right)\dot{\hat{E}}_i\right) \end{aligned} \tag{32}$$

Selecting the adaptive law (26), then (32) can be rewritten as

$$\begin{aligned} \dot{V} &= \sum_{i=1}^{m}\left(-\hat{E}_i|s_i| - s_i\varepsilon_i + \left(\hat{E}_i - E_i\right)|s_i|\right) \\ &\leq -\sum_{i=1}^{m}\left(E_i - |\varepsilon_i|\right)|s_i| \leq 0 \end{aligned} \tag{33}$$

Therefore, V is reduced gradually and the control system is stable which means that the system trajectories converge to the sliding surfaces $s(t)$ while $\hat{\boldsymbol{\theta}}_i$ and $\hat{E}_i$ remain bounded. Now, according to (33), let

$$\Gamma(t) \equiv \sum_{i=1}^{m}\left(E_i - |\varepsilon_i|\right)|s_i| \leq -\dot{V} \tag{34}$$

Integrating both sides of the above inequality yields

$$\int_0^t \Gamma(\tau)d\tau \leq V\left(s(0),\tilde{\theta},\tilde{E}\right) - V\left(s(t),\tilde{\theta},\tilde{E}\right) \tag{35}$$

Considering $V(\boldsymbol{s}(0), \tilde{\boldsymbol{\theta}}, \tilde{\mathbf{E}})$ is bounded and $V(\boldsymbol{s}(t), \tilde{\boldsymbol{\theta}}, \tilde{\mathbf{E}})$ is non-increasing and bounded, it is concluded that

$$\lim_{t\to\infty}\int_0^t \Gamma(\tau)d\tau \leq \infty \tag{36}$$

Furthermore, $\dot{\Gamma}$ is bounded, so that by Barbalat's lemma, it can be shown that $\lim_{t\to\infty}\int_0^t \Gamma(\tau)d\tau = 0$. That is, $\lim_{t\to\infty} s_i(t) = 0$, which implies that the tracking errors $e_i(t)$ and $\tilde{\boldsymbol{\theta}}_i$ converge asymptotically to zero. As a result, the proposed direct adaptive fuzzy sliding mode controller is asymptotically stable. This completes the proof.

The proposed direct adaptive fuzzy sliding mode controller has two terms: $\boldsymbol{u}_{fz}$ given in (19) with the parameter $\hat{\boldsymbol{\theta}}_i$ adjusted by(25)and the approximation bound $\hat{E}_i$ adjusted by(26). By applying these adaptive laws, the DAFSMC is model free and can be guaranteed to be stable for any nonlinear system with the form of(1).

4 SIMULATION

In this section, an example will be used to test the effectiveness of the proposed controller.

Example. Consider the following nonlinear system.

$$\begin{bmatrix} \ddot{x}_1 \\ \dot{x}_2 \end{bmatrix} = \begin{bmatrix} x_1 + \dot{x}_1^2 + x_2 \\ x_1 + 2\dot{x}_1 + 3x_1x_3 \end{bmatrix} + \begin{bmatrix} 3 & 0.5 \\ 1.5 & 4+\sin(x_1) \end{bmatrix} \begin{bmatrix} u_1 \\ u_2 \end{bmatrix} + \begin{bmatrix} 0.2\cos(3t) \\ 0.5\sin(4t) \end{bmatrix}$$

In this example, we consider that the nonlinear functions in the system are assumed to be completely unknown, that is, the design of the proposed controller does not require the knowledge of the system's mode. Moreover, the external disturbance is also unknown. Here, these functions are only required for simulation purpose.

The reference trajectories for x_1 and x_2 are given as $x_{1d} = 2\sin(0.5t + 0.5)$, $x_{2d} = 2\cos(0.5t)$. The control objective is to control the output $\boldsymbol{y} = [x_1, x_2]^T$ to track the reference trajectory $\boldsymbol{x}_d = [x_{1d}, x_{2d}]^T$.

The integral sliding surfaces (7) are used with $k_{11} = 5$, $k_{21} = 5$ and $k_{21} = 5$. The control law (19), the parameter adaptive law (25) and (26) are used with $\eta_1 = \eta_2 = 50$, $\lambda_1 = \lambda_2 = 0.05$.

For the fuzzy control, two fuzzy systems in the form of (14) are used to generate the fuzzy control signals u_{fz1} and u_{fz2}. Each fuzzy system has $\boldsymbol{s} = [s_1, s_2]^T$ as input, and for each input variable $s_i (i = 1, 2)$, five fuzzy sets are characterized by the following membership functions. Then, we apply $r = 5 \times 5 = 25$ fuzzy rules for constructing u_{fz1} and u_{fz2} respectively.

$$\begin{cases} \mu_{F_i^1} = \exp\left[\dfrac{-(s_i+1.5)^2}{2}\right] \\ \mu_{F_i^2} = \exp\left[\dfrac{-(s_i+0.75)^2}{2}\right] \\ \mu_{F_i^3} = \exp\left(\dfrac{-s_i^2}{2}\right) \\ \mu_{F_i^4} = \exp\left[\dfrac{-(s_i-0.75)^2}{2}\right] \\ \mu_{F_i^5} = \exp\left[\dfrac{-(s_i-1.5)^2}{2}\right] \end{cases}$$

The initial states of the system are chosen as $x_1(0) = -2$, $\dot{x}_1(0) = 0.25$ and $x_2(0) = 0.5$. The initial parameter values are selected as $\hat{\boldsymbol{\theta}}_1(0) = \hat{\boldsymbol{\theta}}_2(0) = [0.1, \cdots, 0.1]_{1\times 25}$ and $\hat{E}_1(0) = \hat{E}_2(0) = 0.01$. The simulation results are shown in Figures 1–6. The response

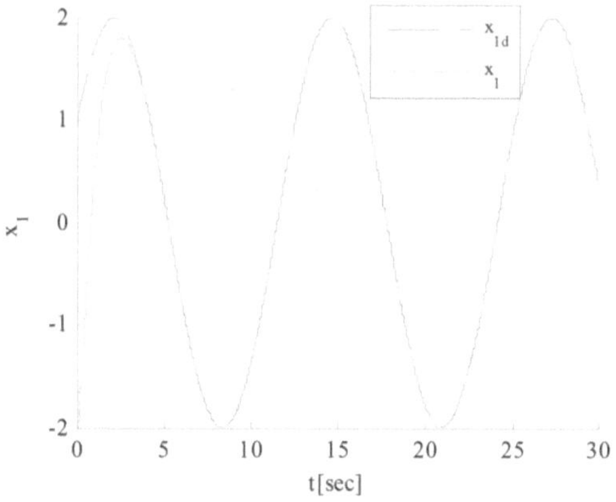

Figure 1. The time response of $x_1(t)$.

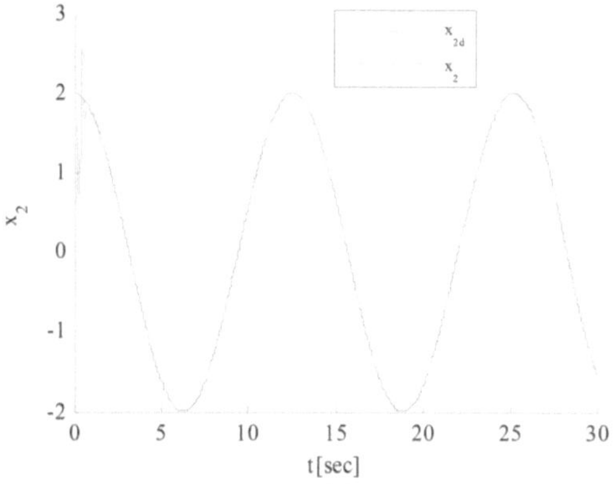

Figure 2. The time response of $x_2(t)$.

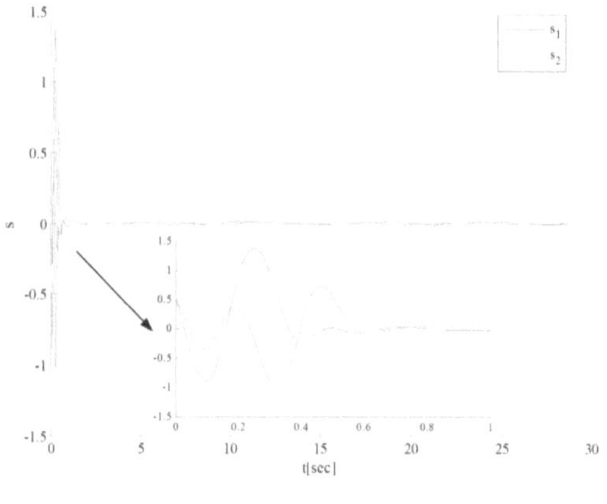

Figure 3. Evolution curves of sliding mode.

curves of the state tracking trajectories $x_1(t)$ and $x_2(t)$ are illustrated in Figure 1 and Figure 2, respectively, from which it can be seen that the actual state trajectories (solid-lines) asymptotically track the reference trajectories (dotted-lines) as desired. Figure 3 depicts the convergent processes of the sliding mode surfaces; we can note that the sliding mode surfaces are driven

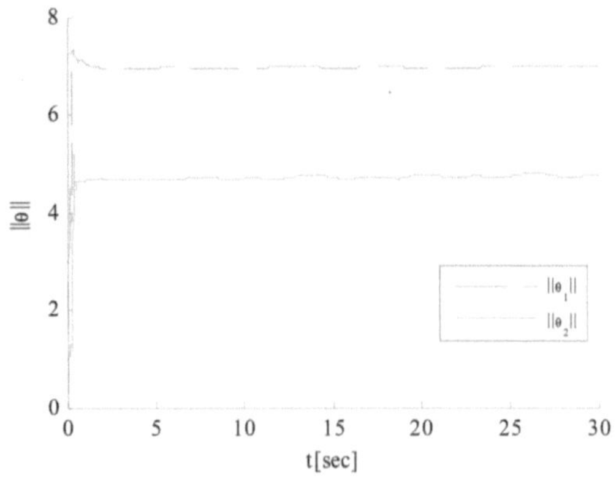

Figure 4. Evolution curves of the fuzzy parameters Euclidian norm $\|\hat{\boldsymbol{\theta}}_i\|$.

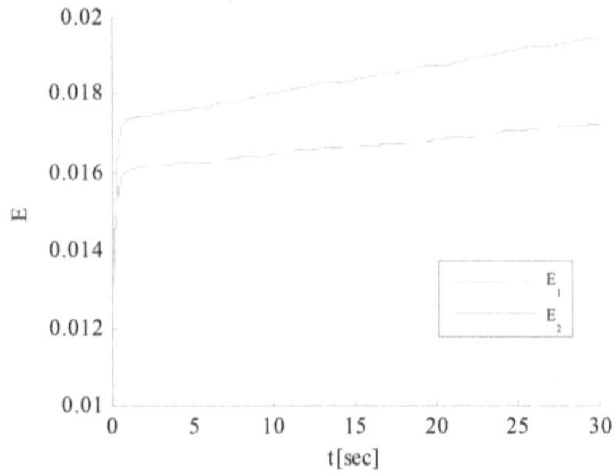

Figure 5. Evolution curves of the estimated switch gains.

to zeros by the proposed direct adaptive fuzzy control. The Euclidian norm of the estimated fuzzy parameter vector $\hat{\boldsymbol{\theta}}_1$ and $\hat{\boldsymbol{\theta}}_1$ are shown in Figure 4 and the estimated switch gain $\hat{E}_1$ and $\hat{E}_2$ are shown in Figure 5, which illustrates that the estimated parameters are bounded. From Figure 6, we can see that the control signal is bounded and smooth. Therefore, these simulation results demonstrate the tracking capability of the proposed controller and its effectiveness for control tracking of MIMO uncertain nonlinear systems.

5 CONCLUSION

In this paper, a direct adaptive fuzzy sliding mode controller is proposed for a class of uncertain MIMO nonlinear systems with unknown control gain matrix and external disturbances. In this study, the control gain matrix is just required to be nonsingular but not to be symmetric positive definite, which relaxed the constraint on the control gain matrix.

The integral sliding surfaces are introduced to construct an unknown ideal control law. Within the direct control scheme, fuzzy adaptive systems are used to approximate the ideal control law. Moreover, an adaptive switch control term is appended to eliminate the fuzzy approximation error. The fuzzy parameters and switch control parameters are adjusted automatically online, yielding relatively rapid adaptation. All adaptive learning laws in the proposed control system are derived from the Lyapunov stability theorem, ensuring the convergence and stability of the control system. A simulation example have illustrated that the actual trajectories converge to the desired trajectories, the control signals are almost smooth and the parameters estimates are bounded. These simulation results have demonstrated that the tracking capability of the proposed controller for uncertain MIMO nonlinear systems with unknown nonsingular control gain matrix and external disturbances.

The major contributions of this paper are that it presents a simple and economic direct adaptive fuzzy sliding mode controller for uncertain MIMO nonlinear systems with unknown nonsingular control gain matrix, and provides mathematical proof of the stability and convergence of the control system using the Lyapunov stability theorem.

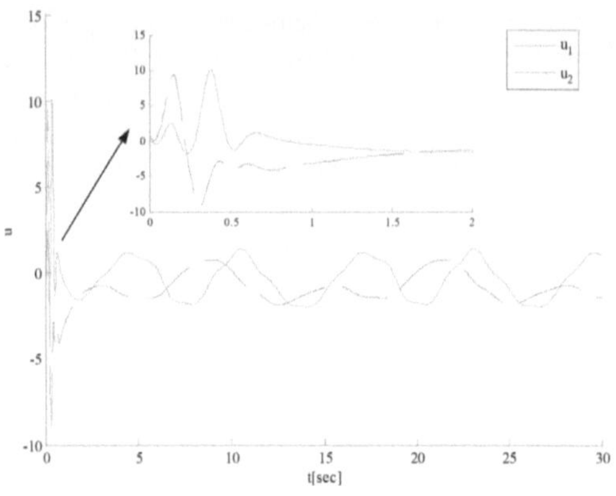

Figure 6. Evolution curves of the control inputs.

REFERENCES

Bing, Chen., Xiaoping, Liu., & Kefu, Liu. (2009). Direct adaptive fuzzy control of nonlinear strict-feedback systems. *Automatica*, 45(6), 1530–1535.

Chiu-Hsiung, Chen., Chih-Min, Lin., & Te-Yu, Chen. (2008). Intelligent adaptive control for MIMO uncertain nonlinear systems. *Expert Systems with Applications,* 35(3), 865–877.

Chun-Sheng, Chen. (2012). Direct Adaptive Output Feedback CMAC Control with Unknown Control Gain for a Class of Nonlinear Systems. *Procedia Engineering*, 29, 3966–3971.

Essounbouli, N., & Hamzaoui, A. (2006). Direct and indirect robust adaptive fuzzy controllers for a class of nonlinear systems. *International Journal of Control Automation and Systems*, 4(2), 146–154.

Hai-Jun, Rong., & Guang-She, Zhao.(2013). Direct adaptive neural control of nonlinear systems with extreme learning machine. *Extreme Learning Machine's Theory and Application*, 22(3), 577–586.

Hanxiong, Li., & Shaocheng, Tong. (2003). A hybrid adaptive fuzzy control for a class of nonlinear MIMO systems. *IEEE Transactions on Fuzzy Systems*, 11(1), 24–34.

Labiod, S., & Guerra, T. M. (2010). Direct and Indirect Adaptive Fuzzy Control for a Class of MIMO Nonlinear Systems. *Advances in Robot Manipulators*, InTech.

Labiod, S., & Guerra,T. M. (2007). Direct adaptive fuzzy control for a class of MIMO nonlinear systems. *International Journal of Systems Science*, 38(8), 665–675.

Nekoukar, V., & Erfanian, A.(2011). Adaptive fuzzy terminal sliding mode control for a class of MIMO uncertain nonlinear systems. *Fuzzy Sets and Systems*,179(1), 34–49.

Phan, P.A., & Gale, T. J. (2008). Direct adaptive fuzzy control with a self-structuring algorithm. *Fuzzy Sets and Systems*, 159(8), 871–899.

Shaocheng, Tong., & Han-Xiong, Li. (2003). Fuzzy adaptive sliding-mode control for MIMO nonlinear systems. *IEEE Transactions on Fuzzy Systems*,11(3), 354–360.

Shuanghe, Yu., Xinghuo, Yu., & Zhihong, Man. Sydney,Australia, (2000). Robust global terminal sliding mode control of SISO nonlinear unertain systems. *Proceedings of the 39th IEEE Conference on Decision and Control*, 2198–2203.

Sofiane, D., & Khaber, F. (2012). Direct adaptive fuzzy control of a class of MIMO non-affine nonlinear systems. *International Journal of Systems Science*,43(6), 1029–1038.

Wuxi, Shi. (2013). Adaptive fuzzy control for MIMO nonlinear systems with nonsymmetric control gain matrix and unknown control direction. *IEEE Transactions on Fuzzy Systems.*

Future Mechatronics and Automation – Yang (Ed.)
© 2015 Taylor & Francis Group, London, ISBN 978-1-138-02648-3

Adjacent vertex distinguishing total coloring of Cartesian product graphs

Zheng-Qing Chu & Jia-Bao Liu
Course Department, Common Course Department, Anhui Xinhua University, Hefei, China

ABSTRACT: A proper total coloring of G is an adjacent vertex distinguishing total coloring if for any two adjacent vertices, the sets of colors appearing on the vertex and incident edges are different. The smallest number of colors of which such a coloring of G exists is called the adjacent vertex distinguishing total chromatic number and is denoted by $\chi_{at}(G)$. The adjacent vertex-distinguishing total chromatic number of Cartesian Product Graphs $P_l \times P_m \times P_n$ were proposed in this article, furthermore, adjacent vertex distinguishing total chromatic number were also obtained.

1 INTRODUCTION

Graph coloring which has important practical significance and theoretical significance is the main research content. With the wide use of graph coloring in reality, it gradually becomes the important study field for many scholars. Vertex-distinguishing edge coloring or strong edge coloring which is proposed from network problem, biology, information science and computer science is a very difficult problem. In the literature[1–2], vertex-distinguishing edge coloring and adjacent vertex-distinguishing coloring can be studied further. New coloring problems have been proposed constantly, definition of adjacent vertex-distinguishingcoloring, djacent vertex-distinguishing strong and total coloring is proposed in the literature[3–6] related to problems, as well as several kinds of simple graphs related to the coloring chromatic number and some assumptions. This article considers limited simple undirected graph, $V(G)$ and $E(G)$ refers to the graph vertex set and edge set respectively, $\Delta(G)$ refers to the maximum degree of graph G. Unspecified graph theory terms and marks used in this article can be referred to in the literature[9].

2 RELATED DEFINITIONS

Definition 1.1[3]. Assume that Graph G is a connected graph with order at least 2, k is the positive integer, f is the map from $V(G) \cup E(G)$ to $\{1, 2, \ldots, k\} \forall u, v \in V(G)$, so $C(u) = \{f(u)\} \cup \{f(uv) | u, v \in V(G), uv \in E(G)\}$. If (1) for $\forall uv, vw \in E(G), u \neq w$, so $f(uv) \neq f(vw)$;

(2) For any $uv \in E(G)$, so $f(u) \neq f(v), f(u) \neq f(uv)$, $f(v) \neq f(uv)$;

So f is k-normal total coloring of Graph G. If f meets the requirements:

(3) For any $uv \in E(G)$, so $C(u) \neq C(v)$,

So f is called k-adjacent vertex-distinguishing total coloring of Graph G (simply marked as k-*AVDTC*). $\min\{k|_G$ has k-adjacent vertex-distinguishing total coloring$\}$ is adjacent vertex distinguishing total chromatic number of Graph G, it is referred to as $\chi_{at}(G)$, among which, $C(u)$ is called the color set of point u under f.

Definition 2. Assume P_l, P_m, P_n is the simple connection with a length of l, m and n. Graph $P_l \times P_m \times P_n$ refers to the graphs grained by Cartesian product operation of P_l, P_m, and P_n. Assume that the vertex subscript of Graph $P_l \times P_m \times P_n$ starts from the top left corner, from top to down, from left to right, mark on its vertex in a form of S. Among which:

$$V(P_l \times P_m \times P_n) = \{, v_{i1}, v_{i2}, v_{i3}, v_{i4}, \cdots, v_{i,(m+1)(l+1)}, i = 1, 2, \cdots, n+1\},$$

$$E(P_l \times P_m \times P_n) = \{v_{11}v_{12}, v_{12}v_{13}, \cdots, v_{n+1,(m+1)(l+1)-1}v_{n+1,(m+1)(l+1)}\}.$$

3 LEMMAS AND PROOF

Lemma 2.1 If Graph G has two adjacent vertexes with maximum degree, so $\chi_{at}(G) \geq \Delta(G) + 2$.

Proof. Assume u, v are two adjacent vertexes with maximum degree in Graph G, so fro any k-adjacent vertex-distinguishing total coloring f, both $C(u)$ and $C(v)$ have $\Delta(G) + 1$ kinds of colors, while $C(u) \neq C(v)$, so if k-adjacent vertex-distinguishing total coloring is required for Graph G, then $k \geq \Delta(G) + 2$.

4 MAIN THEOREM AND PROOF

Theorem 1. When $l = m = n = 1$, adjacent vertex distinguishing total chromatic number of Graph $P_l \times P_m \times P_n$ is: $\chi_{at}(P_l \times P_m \times P_n) = 5$.

Proof. When $l=m=n=1$, mapping f from $V(P_l \times P_m \times P_n) \cup E(P_l \times P_m \times P_n)$ to $\{1,2,3,4,5\}$ is defined as follows:

$v_{11}=v_{22}=2$; $v_{121}=v_{21}=1$; $v_{13}=v_{24}=2$; $v_{14}=v_{23}=1$;

$v_{11}v_{12}=v_{13}v_{14}=v_{21}v_{22}=v_{23}v_{24}=3$;

$v_{12}v_{13}=v_{11}v_{14}=v_{21}v_{24}=v_{22}v_{23}=4$;

$v_{11}v_{21}=v_{12}v_{22}=v_{13}v_{23}=v_{14}v_{24}=5$.

Obviously, f is 5-normal total coloring of $P_l \times P_m \times P_n$, because $\forall uv \in E(P_l \times P_m \times P_n)$ has $C(u) \neq C(v)$, so, f is 5-*AVDTC* of $P_l \times P_m \times P_n$. *Theorem 2.* When only one of l, m, n equals to 1, adjacent vertex distinguishing total chromatic number of Graph $P_l \times P_m \times P_n$ is: $\chi_{at}(P_l \times P_m \times P_n)=6$.

Proof. Sub-case1.

When $m=n=1, l \neq 1$, mapping f from $V(P_l \times P_m \times P_n) \cup E(P_l \times P_m \times P_n)$ to $\{1,2,3,4,5,6\}$ is defined as follows:

$v_{ij}=2, v_{ij+3}=1, i \equiv 1 (\bmod 2)$,

$j \equiv 1 (\bmod 2), i=1,2,\cdots,l+1$

$v_{ij}=2$, $v_{ij+3}=1$, $i \equiv 0 (\bmod 2)$, $j \equiv 0 (\bmod 2)$,

$i=1,2,\cdots,l+1$;

$v_{ij}=1$, $v_{ij+3}=2$, $i \equiv 1 (\bmod 2)$, $j \equiv 0 (\bmod 2)$,

$i=1,2,\cdots,l+1$;

$v_{ij}=1$, $v_{ij+3}=2$, $i \equiv 0 (\bmod 2)$, $j \equiv 1 (\bmod 2)$,

$i=1,2,\cdots,l+1$;

$v_{i1}v_{i4}=v_{i2}v_{i3}=3$, $i=1,2,\cdots,l+1$; $v_{i1}v_{i2}=v_{i3}v_{i4}=4$,

$i=1,2,\cdots,l+1$;

$v_{i1}v_{ii+1}=v_{i2}v_{i+1,2}=v_{i3}v_{i+1,3}=v_{i4}v_{i+1,4}=5$, $i \equiv 1 (\bmod 2)$,

$i=1,2,\cdots,l+1$;

$v_{i1}v_{i+1,1}=v_{i2}v_{i+i,2}=v_{i3}v_{i+1,3}=v_{i4}v_{i+1,4}=6$,

$i \equiv 0 (\bmod 2)$, $i=1,2,\cdots,l+1$.

Obviously, f is 6-normal total coloring of $P_l \times P_m \times P_n$, because $\forall uv \in E(P_l \times P_m \times P_n)$ has $C(u) \neq C(v)$, so f is 6-*AVDTC* of $P_l \times P_m \times P_n$.

Sub-case 2. When $l=n=1, m \neq 1$, mapping f from $V(P_l \times P_m \times P_n) \cup E(P_l \times P_m \times P_n)$ to $\{1,2,3,4,5,6\}$ is defined as follows: $v_{ij}=2, v_{ij+3}=1$, $i \equiv 1(\bmod 2), j \equiv 1(\bmod 2), i=1,2,\ldots,m+1, v_{ij}=2, v_{ij+3}=1, i \equiv 0(\bmod 2), j \equiv 0(\bmod 2), i=1,2,\ldots,m+1; v_{ij}=1, v_{ij+3}=2, i \equiv 1(\bmod 2), j \equiv 0(\bmod 2), i=1,2,\cdots,m+1$; $v_{ij}=1$, $v_{ij+3}=2$, $i \equiv 0(\bmod 2)$, $j \equiv 1(\bmod 2), i=1,2,\cdots,m+1$;

$v_{i1}v_{i4}=v_{i2}v_{i3}=3, i=1,2,\cdots,m+1$;

$v_{i1}v_{i2}=v_{i3}v_{i4}=4, i=1,2,\cdots,m+1$;

$v_{i1}v_{ii+1}=v_{i2}v_{i+1,2}=v_{i3}v_{i+1,3}=v_{i4}v_{i+1,4}=5$,

$i \equiv 1 (\bmod 2), i=1,2,\cdots,m+1$;

$v_{i1}v_{i+1,1}=v_{i2}v_{i+i,2}=v_{i3}v_{i+1,3}=v_{i4}v_{i+1,4}=6$,

$i \equiv 0 (\bmod 2), i=1,2,\cdots,m+1$.

Obviously, f is 6-normal total coloring of $P_l \times P_m \times P_n$, because $\forall uv \in E(P_l \times P_m \times P_n)$ has $C(u) \neq C(v)$, so f is the 6-*AVDTC* of $P_l \times P_m \times P_n$.

Sub-case3. When $m=l=1, n \neq 1$, mapping f from $V(P_l \times P_m \times P_n) \cup E(P_l \times P_m \times P_n)$ to $\{1,2,3,4,5,6\}$ is defined as follows: $v_{ij}=2, v_{ij+3}=1$, $i \equiv 1(\bmod 2)$,, $j \equiv 1(\bmod 2)$, $i=1,2,\ldots,n+1. v_{ij}=2$, $v_{ij+3}=1$, $i \equiv 0(\bmod 2)$, $j \equiv 0(\bmod 2)$, $i=1,2,\ldots,n+1$; $v_{ij}=1$, $v_{ij+3}=2$, $i \equiv 1(\bmod 2), j \equiv 0(\bmod 2), i=1,2,\ldots,n+1$; $v_{ij}=1$, $v_{ij+3}=2$, $i \equiv 0(\bmod 2), j \equiv 1(\bmod 2), i=1,2,\ldots,n+1$; $v_{i1}v_{i4}=v_{i2}v_{i3}=3$, $i=1,2,\ldots,n+1$; $v_{i1}v_{i2}=v_{i3}v_{i4}=4$, $i=1,2,\ldots,n+1$; $v_{i1}v_{ii+1}=v_{i2}v_{i+1,2}=v_{i3}v_{i+1,3}=v_{i4}v_{i+1,4}=5$, $i \equiv 1(\bmod 2)$, $i=1,2,\ldots,n+1$; $v_{i1}v_{i+1,1}=v_{i2}v_{i+i,2}=v_{i3}v_{i+1,3}=v_{i4}v_{i+1,4}=6$, $i \equiv 0(\bmod 2)$, $i=1,2,\ldots,n+1$.

Obviously, f is 6-normal total coloring of $P_l \times P_m \times P_n$, because $\forall uv \in E(P_l \times P_m \times P_n)$ has $C(u) \neq C(v)$, so f is 6-*AVDTC* of $P_l \times P_m \times P_n$.

Theorem 3. When only one of l, m, n does not equal to 1, adjacent vertex distinguishing total chromatic number of Graph $P_l \times P_m \times P_n$ is: $\chi_{at}(P_l \times P_m \times P_n)=7$.

Proof. Sub-case1. When $l=1, m, n \neq 1$, map f from $V(P_l \times P_m \times P_n) \cup E(P_l \times P_m \times P_n)$ to $\{1,2,\cdots,7\}$ is defined as follows:

$v_{ij}=2, i \equiv 1(\bmod 2), j \equiv 1(\bmod 2)$, $i=1,2,\cdots,n+1$;

$j=1,2,\cdots,2m+2$;

$v_{ij}=1, i \equiv 0(\bmod 2), j \equiv 1(\bmod 2)$,

$i=1,2,\cdots,n+1;\ j=1,2,\cdots,2m+2;$

$v_{ij}=1, i\equiv 1(\bmod 2), j\equiv 0(\bmod 2),$

$i=1,2,\cdots,n+1;\ j=1,2,\cdots,2m+2;$

$v_{ij}=2, i\equiv 0(\bmod 2), j\equiv 0(\bmod 2),\ i=1,2,\cdots,n+1;$

$j=1,2,\cdots,2m+2;$

$v_{ij}v_{i+1j}=6, i\equiv 1(\bmod 2),\ i=1,2,\cdots,n+1;$

$j=1,2,\cdots,2m+2;\ v_{ij}v_{i+1j}=7, i\equiv 0(\bmod 2),$

$i=1,2,\cdots,n+1;\ j=1,2,\cdots,2m+2;$

$v_{ij}v_{ij+1}=3, j\equiv 1(\bmod 2),\ i=1,2,\cdots,n+1;$

$j=1,2,\cdots,2m+1;\ v_{ij}v_{ij+1}=4, j\equiv 0(\bmod 2),$

$i=1,2,\cdots,n+1;\ j=1,2,\cdots,2m+1;$

$v_{ij}v_{ij+3}=5, j\equiv 0,1(\bmod 3), i=1,2,\ldots,n+1; j=1,2,\ldots,2m-1$. Obviously, f is 7-normal total coloring of $P_l\times P_m\times P_n$, because $\forall uv\in E(P_l\times P_m\times P_n)$ has $C(u)\neq C(v)$ so f is 7-*AVDTC* of $P_l\times P_m\times P_n$.

Sub-case2. When $m=1, l, n\neq 1$, mapping f from $V(P_l\times P_m\times P_n)\cup E(P_l\times P_m\times P_n)$ to $\{1,2,\cdots,7\}$ is defined as follows:

$v_{ij}=2, i\equiv 1(\bmod 2), j\equiv 1(\bmod 2),\ i=1,2,\cdots,n+1;$

$j=1,2,\cdots,2l+2;$

$v_{ij}=1, i\equiv 0(\bmod 2), j\equiv 1(\bmod 2),$

$i=1,2,\cdots,n+1;\ j=1,2,\cdots,2l+2;$

$v_{ij}=1, i\equiv 1(\bmod 2), j\equiv 0(\bmod 2),$

$i=1,2,\cdots,n+1;\ j=1,2,\cdots,2l+2;$

$v_{ij}=2, i\equiv 0(\bmod 2), j\equiv 0(\bmod 2),$

$i=1,2,\cdots,n+1;\ j=1,2,\cdots,2l+2;$

$v_{ij}v_{ij+1}=6, j\equiv 1(\bmod 2),$

$i=1,2,\cdots,n+1;\ j=1,2,\cdots,2l+2;$

$v_{ij}v_{ij+1}=7, j\equiv 0(\bmod 2),$

$i=1,2,\cdots,n+1;\ j=1,2,\cdots,2l+2;$

$v_{ij}v_{i+1j}=4, i\equiv 1(\bmod 2),$

$i=1,2,\cdots,n+1;\ j=1,2,\cdots,2l+2;$

$v_{ij}v_{i+1j}=5, i\equiv 0(\bmod 2),$

$i=1,2,\cdots,n+1;\ j=1,2,\cdots,2l+2;$

$v_{i1}v_{i,2l+2}=v_{i,2}v_{i,2l+1}=\cdots=v_{i,l}v_{i,l+3}=3$

$i=1,2,\ldots,n+1$. Obviously, f is 7-normal total coloring of $P_l\times P_m\times P_n$, because $\forall uv\in E(P_l\times P_m\times P_n)$ has $C(u)\neq C(v)$, so f is 7-*AVDTC* of $P_l\times P_m\times P_n$.

The following can be obtained using the similar method of Sub-case1, 2 When $n=1, l, m\neq 1$, Graph $P_l\times P_m\times P_n$ is 7-*AVDTC*.

Theorem 4. When l, m, n do not equal to 1, adjacent vertex distinguishing total chromatic number of Graph $P_l\times P_m\times P_n$ is: $\chi_{at}(P_l\times P_m\times P_n)=8$.

Proof: When $l, m, n\neq 1$, mapping f: $V(P_l\times P_m\times P_n)\cup E(P_l\times P_m\times P_n)$ to $\{1,2,\cdots,8\}$ is $v_{ij}=1, i\equiv 1(\bmod 2), j\equiv 1(\bmod 2),$

$v_{ij}=2, i\equiv 1(\bmod 2), j\equiv 0(\bmod 2),$

$v_{ij}=2, i\equiv 0(\bmod 2), j\equiv 1(\bmod 2),$

$v_{ij}=1, i\equiv 0(\bmod 2), j\equiv 0(\bmod 2),$

$v_{ij}v_{ij+1}=3, i\equiv 1(\bmod 2),$ where

$i=1,2,\cdots,n+1;\ j=1,2,\cdots,(m+1)(l+1).$

$v_{ij}v_{ij+1}=4, i\equiv 0(\bmod 2),\ i=1,2,\cdots,n+1;$

$j=1,2,\cdots,(m+1)(l+1),$

$v_{i,j(l+1)-l}v_{i,(j+1)(l+i)}=v_{i,j(l+1)-l+1}v_{i,(j+1)(l+i)-1}$

$=\cdots=v_{i,j(l+1)-1}v_{i,(j+1)(l+i)-l+1}=4$

$j \equiv 1(\mathrm{mod}\, 2)$, $j = 1,2,\cdots,m$;

$$v_{i,j(l+1)-l}v_{i,(j+1)(l+i)} = v_{i,j(l+1)-l+1}v_{i,(j+1)(l+i)-1}$$
$$= \cdots = v_{i,j(l+1)-1}v_{i,(j+1)(l+i)-l+1} = 5$$

$j \equiv 1(\mathrm{mod}\, 2)$, $j = 1,2,\cdots,m$;

$v_{ij}v_{i+1j} = 7, i \equiv 1(\mathrm{mod}\, 2), i = 1,2,\cdots,n+1$;

$j = 1,2,\cdots,(m+1)(l+1)$; $v_{ij}v_{i+1j} = 7, i \equiv 1(\mathrm{mod}\, 2)$,

$i = 1,2,\cdots,n+1$;

$j = 1,2,\cdots,(m+1)(l+1)$; $v_{ij}v_{i+1j} = 8, i \equiv 1(\mathrm{mod}\, 2)$,

$i = 1,2,\ldots,n+1; j = 1,2,\ldots,(m+1)(l+1)$ Obviously, f is 8-normal total coloring of $P_l \times P_m \times P_n$, because $\forall uv \in E(P_l \times P_m \times P_n)$ has $C(u) \neq C(v)$, so f is 8-*AVDTC* of $P_l \times P_m \times P_n$.

ACKNOWLEDGMENT

The work of J.B. Liu is partly supported by the Natural Science Foundation of Anhui Province of China under Grant

No.KJ2013B105.No.2012tskcx04.

No.2012tskcx05.

REFERENCES

[1] Balister P N, Bollob B, Shelp R H .Vertex distinguishing coloring of graphs with $\Delta(G) = 2$, discrete mathematics, 2002, 252:17–29.

[2] Zhang Zhongfu. On the adjacent vertex distinguish total coloring of graphs, Science in China Ser, A2004, 10:574–583.

[3] Zhang Zhongfu, Chen Xiang'en, Li Jingwen. On the adjacent vertex-distinguishing total coloring of graph, *Science China,* Volume A: Mathematics, 2004, 34 (5): 574–583.

[4] Zhang Zhongfu, Li Jingwen, Chen Xiang'en. Vertex-distinguishing edge coloring of graph whose distance does not exceed β, *Mathematics Journal* (Chinese Version), 2006, 9(3):703–708.

[5] Zhang Zhongfu, Li Jingwen, Chen Xiang'en. Vertex-distinguishing total coloring of graph whose distance does not exceed β, *Science China,* Volume A: Mathematics, 2006, 36(10): 1119–1130.

[6] Zhang Zhongfu, Chen Hui. Yao Bing. On adjacent vertex strong distinguishing total coloring of graph, *Science China,* Volume A: Mathematics, 2007, 1073–1082.

Future Mechatronics and Automation – Yang (Ed.)
 ISBN 978-1-138-02648-3

Design of embedded graphical user interface of a graphics driver library based on STemWin

YanMing Zhou, WeiShan Liang & Lin Qiu
Lushan College of Guangxi University of Science and Technology, Liuzhou, Guangxi, China

ABSTRACT: In order to shorten the development period of creating an embedded graphic interface, STMicroelectronics (ST) developed a software protocol with a SEGGER microcontroller. ST offers a free STemWin graphics driver library and graphic development tool STemWin for Micro Controller Unit (abbreviate as MCU) which are based on the STM32 series of microcontroller integrated circuits. In this paper, firstly we introduce the characteristics of graphics driver databases and the use of STemWin development tools. Then, based on the STemWin graphics driver library, the embedded graphical interface general design method (on which MCU in STM32 series using the graphical development tool STemWin) is followed.

Keywords: Embedded graphical user interface; STM32; STemWin; development tools

1 INTRODUCTION

Embedded system is developing forward to humanization at present. A friendly, simple, easy graphical user interface has become an important part of many embedded systems. In the past, the design of graphical interfaces cost a lot of time and energy, and that conditioned the product as soon as possible to the market. To solve this problem, many semiconductor manufacturers now not only sell a new processor, but also launch processor supporting software at the same time. ST launched the MCU for the STM32 series which is based on an ARM Cortex-M3 core. It comes with the peripheral base, motor driver library, standard protocol stack, and STemWin graphics driver. All developers need to do, is to modify the basic hardware driver interface program. Then the STemWin graphics driver library can run on the MCU of the STM32 series.

2 GRAPHICS DRIVER LIBRARY ARCHITECTURE

The STemWin graphics driver library includes an application layer, control layer, and a driver layer. The most important part in this library is the control layer, which can direct call with the STemWin development tools.

2.1 *Application layer*

The application layer is the most top-level of the STemWin graphics driver library. we can develop Graphical User Interface (abbreviate as GUI) directly with the STemWin development tools. It is an important layer of human-computer interaction because it is user oriented. This application layer consists of some object code files that can be compiled and linked with the STemWin development tools which can then even be downloaded directly onto the STM32 for operation.

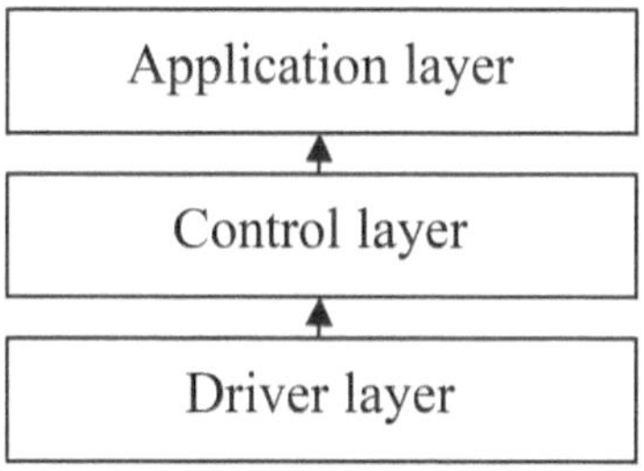

Figure 1. Graphics driver library architecture.

2.2 *Control layer*

The control layer, whose function it is to achieve human-computer interaction, is the basis for building the application layer. The STemWin driver control mode can be called with the STemWin graphics development tools. That makes the design of the human-computer interface more human, simple, and easy. This can directly implement that visual programming of UI design.

2.3 *Driver layer*

The driver layer contains some low drivers of multiple graphics driver chips. If we want to modify a programme, it is only necessary to modify the macro

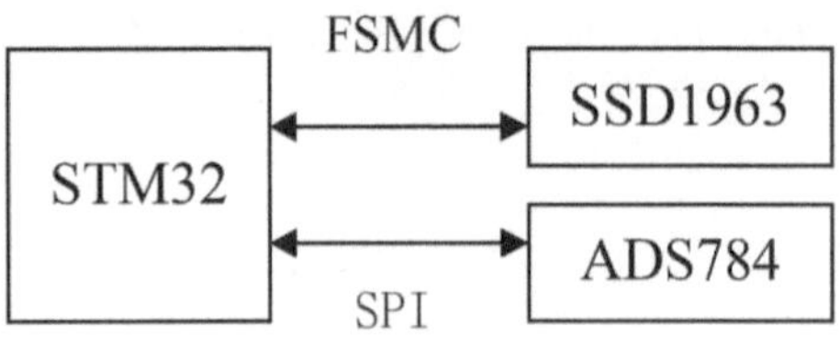

Figure 2. Hardware circuit.

definition of the corresponding driver chip in order for the interface function of the related hardware to work. There are several low drivers of graphics driver chips which can be supported, up to more than ten manufacturers and about more than 30 types of driver chips.

3 DESIGN OF EMBEDDED GRAPHICAL USER INTERFACE

3.1 *Realization of the hardware circuit*

The STM32F103ZET6 chip, whose operating frequency can reach 72 MHZ and can be extended to 8 M RAM and 128 M ROM, is selected as the MCU. The chosen external driving chip is of type SSD1963, which is an Liquid Crystal Display (abbreviate as LCD) controller with a self-contained 1215 KB frame buffer. It can display an 864*480 graphics content of 24 bpp. In this project, we drive a 10.4 inch LCD display with an SSD1963 chip. It is then externally connected to a resistance screen of four-line 10.4 inch and a touch chip of type ADS7843 is used. The chip ADS7843 is a 12 bit sampling analog digital converter that can analyse screen coordinates. These connections between the MCU, LCD controller and the touch chip are as shown in Figure 2.

3.2 *Transplantation of driver layer*

The driver layer of the STemWin graphics driver library integrates the driving programs of the graphics chip SSD1963. On a single chip microcomputer STM32, when running those driving programs of chip SSD1963 inside the graphics driving library, what needs to be done is to modify the macro definition, increase the initialization and reading- writing programs of the STM32 FSMC bus. ST Company provides the driver code for external equipment of the STM32 series. The actual use can be called directly without modification. Then the focus of of the driver layer transplantation is how to configure the STemWin graphics driver as far as possible to use those internal integrated graphics driver programs. The codes of the STemWin graphics driver are as follows:

```
#define DISPLAY_DRIVER GUIDRV_
FLEXCOLOR
#define COLOR_CONVERSION GUICC_M565
GUI_DEVICE * pDevice;
CONFIG_FLEXCOLOR Config = {0};
GUI_PORT_API PortAPI = {0};
//Set the type of displaying driver and the color
conversion
pDevice=GUI_DEVICE_CreateAndLink(DISPLAY_
DRIVER, COLOR_CONVERSION, 0, 0);
//Configure the display resolution
LCD_SetSizeEx (0, XSIZE_PHYS , YSIZE_PHYS);
LCD_SetVSizeEx(0,VXSIZE_PHYS, VYSIZE_
PHYS);
//Configuration display direction
Config.Orientation=0;
GUIDRV_FlexColor_Config(pDevice, &Config);
//Set button layer API of screen display
PortAPI.pfWrite16_A0 = LcdWriteReg;
PortAPI.pfWrite16_A1 = LcdWriteData;
PortAPI.pfWriteM16_A1=LcdWriteDataMultipl;
PortAPI.pfReadM16_A1 = LcdReadDataMultiple;
/*Set the display controller*/
GUIDRV_FlexColor_SetFunc(pDevice,&PortAPI,
GUIDRV_FLEXCOLOR_F66720,GUIDRV_FLEX
COLOR_M16C0B16);
```

Where SSD1963 driver in GUIDRV_FLEXCOLOR _F66720.

Then interface form is GUIDRV_FLEXCOLOR_M16C0B16. The driving layer of the STemWin graphics driver library is thus transplanted, and the control layer can be used.

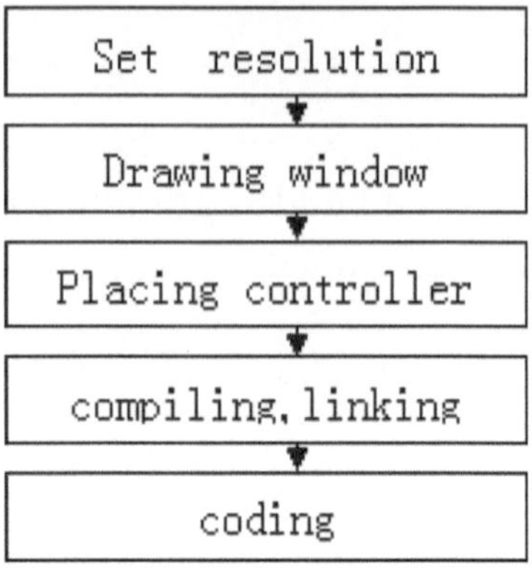

Figure 3. Design process of an interface

3.3 *Graphics interface of STemWin development tools*

STemWin graphical tools is actually a graphic design software, known as STemwin_GUIBuilder. Graphical interface design can begin with any layer of the STemWin graphics driver. Design starting from the control layer, can receive more reliability and stronger structural interface. At the beginning of the design of the interface, open the STemwin_GUIBuilder software then set the screen resolution, background colour and LED's window. The performance process of a graphic design program is shown in Figure 3:

It can be seen from this process that with the STemWin graphics driver library, the design of an embedded graphical user interface becomes so simple instead of that dull code design interface.

4 CONCLUSION

The graphics driver library, developed with ST and based on an STM32 platform, is powerful and easy to use. It can help designers to rapidly develop some beautiful and user-friendly graphical interfaces for products. It can simplify the design process, reduce the cost and has very strong use value.

ACKNOWLEDGEMENT

Innovation training project: This paper is a part of research results of project "Automatic warehouse IOT system" (project number: 2013CXJH14), which is Lushan college of Guangxi University of Science and Technology 2013 autonomous region College Students' Innovation Training Project.

REFERENCES

ADS7843 low voltage i/o touch screen controller [Z].Texas Instruments

SSD1963 Product Preview [Z].S OLOMON S YS TECH SE MICONDU CTOR

ST.Application note. Getting started with STemWin Library [S].ST life augmented

ST.STM32 Microcontroller DATA SHEET [S]. ST life augmented

Future Mechatronics and Automation – Yang (Ed.)

Research and design of the controller for vision-based multi-rotor MAV

Yong-jun Wang, Zhi Li, Shu-bao Pan & Xiang Li
Guilin University of Aerospace Technology, Guilin, P.R. China

ABSTRACT: This work presents a method for controlling an autonomous, multi-rotor MAVs based on visual odometry which consists of two major sections. First, a method for developing an accurate dynamic model of a multi-rotor MAVs based on a combination of first priciples and emperical data is presented. This modeling technique is used to develop a control system which enables trajectory tracking. Second, a vision-based state-estimation technique is presented along with a hardware implementation that enables execution of the algorithms in real-time on board a small vehicle with strict payload constraints. The methods described are implemented in flight-ready hardware with a minimal weight, power and computational footprint. The system is then evaluated on board a small, eight-rotor helicopter. The vision-based trajectory tracking in flight on this vehicle is successfully demonstrated.

Keywords: MAVs, visual odometry, vision-based, state-estimation, multi-rotor, trajectory tracking

1 INTRODUCTION

Autonomous air vehicles frequently rely on GPS as a primary source of state feedback. However, relying on GPS disallows operation in enclosed spaces, under heavy vegetation or near large obstacles since GPS does not provide sufficient accuracy in these environments.

When the loss of GPS navigation occurs, dead reckoning via direction and distance is the next most accurate way of maintaining position location. Similar work in GPS-denied navigation uses laser-based odometry, structured-light or places visual markers in the environment. These approaches are not appropriate for operating in unstructured environments, therefore it is necessary to develop a system around more robust vision-based techniques.

In order to allow a vehicle to explore an environment completely, it is attractive to allow the vehicle to operate inside buildings, near large obstacles, or under heavy vegetation. Unfortunately, these environments prevent the vehicle from receiving a strong, reliable GPS signal. Therefore, we need to develop a method for performing state estimation without the availability of GPS. This work presents a vision-based solution to the state estimation problem which can provide the necessary feedback to stabilize the vehicle's flight.

2 ANALYZE COMPARATIVE STUDY ON CONTROL METHOD

There has been a significant amount of work in the area of GPS-denied navigation for multi-rotor vehicles. The state-estimation systems for these vehicles falls into three broad categories: laser-based, structured-light- based and vision-based. The computation for these systems is done on-board the vehicle in some cases, and is done off-board in others. The controllers used to stabilize these systems are typically either PID or LQR type controllers.

Laser-based state-estimation is a good approach to this type of problem because it yields accurate positioning results and is not as computationally demanding as vision-based solutions. Achtelik et al. [2] present a laser-based localization system for use on a small UAV. While this type of system will work well in structured environments with distinctive 3-D structures near the vehicle, it will not work well outside of these assumptions. A critical failure of this system is that it is limited by the range of the LIDAR. If there is no structure within the sensing range of the laser, the vehicle's motion cannot be calculated. The technique presented in [1] implements a LQR type controller to stabilize the flight of the vehicle. This control technique is applicable because they have also implemented an Extended Kalman Filter which provides full state feedback. Without this filter, the latency of the state estimation system disallows an LQR control scheme.

Structured light is an excellent approach for indoor helicopter navigation since the low-cost Kinect sensor is available that produces depth images directly. Huang et al. [2] present a method for using a Kinect for creating a map of an indoor environment, but this technique will not work outdoors. This is because the brightness of an outdoor environment prevents the sensor from working correctly. Also, this type of sensor only works as short range, so the helicopter will not be able to operate away from obstacles. This group implements a PID type position controller, which is very similar to the technique presented in this work, and they are able to achieve satisfactory position tracking.

There are several groups (Minh and Ha [3]; Meier et al. [4]; Wu, Johnson and Proctor [5]) who have presented work using cameras with fiducials in the environment.

This technique reduces the computational complexity of the vision algorithms significantly which enables the algorithms to be executed on-board the vehicle in real-time. However, this technique is not extensible to a large environment since it is impractical to place markers through out a large region.

Other groups handle the computational complexity of vision-based algorithms by performing the computation on heavy, powerful, ground-based computers. The helicopter must send the images to the ground computer which executes the algorithms and sends back control commands. Achtelik et al. [1] and Blösch et al. [6] both present techniques that require an off-board computer. This solution to the computation problem is not reasonable for vehicles that will travel a long distance because of the unreliability of wireless links. These wireless systems have non-deterministic latency which is unacceptable for stabilizing a vehicle in flight. Conte and Doherty [7] solve the computation problem by using a large vehicle. They use a Yamaha R-Max vehicle which has a total weight of close to 100 kg. This vehicle has a large enough payload capacity that it can carry heavy, powerful computers on board. This type of vehicle is good for flying far away from obstacles, but its size prevents it from operating in confined spaces.

The most similar system to the one presented in this work is described by Voigt et al. in [8]. This vehicle uses a vision-based system for localization and performs all of the computation on-board a small UAV. This general approach meets the requirements of an outdoor helicopter that will carry out long missions, and is very similar to the techniques presented in this work. The limitation of the system described in [8] is that all of the on-board computation power is dedicated to the vision system, leaving no available computation power for other algorithms.

In contrast to the existing work in GPS-denied navigation, this work presents a vision-based system that is capable of operating in unstructured environments while performing all of the required computation on-board a MAVs. This system enables operation of a small helicopter in a wide variety of GPS-denied environments while leaving computational resources available for the other tasks required to carry out a meaningful mission.

3 DYNAMIC MODEL OF THE EIGHT-ROTOR MAV AND CONTROL ARCHITECTURE

In order to develop a trajectory tracking controller, it is first necessary to develop a dynamic model of the system to be controlled. The Eight-Rotor is very

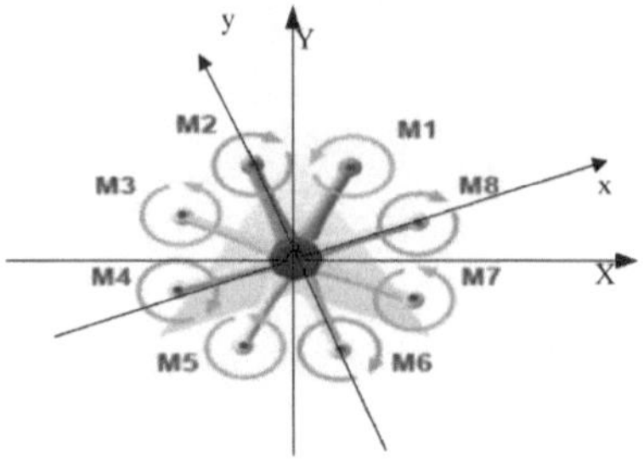

Figure 1. Flight theory of Eight-Rotor MAV.

Table 1. Dynamics Variables

Symbol	Definition
ϕ	roll angle
θ	pitch angle
ψ	yaw angle
Ω	total rotor speed
$I_{x,y,z}$	body inertia
J_r	rotor inertia
U_1	total thrust
U_2	front/back thrust differential
U_3	front/back thrust differential
U_4	cw/ccw torque differential

well modeled, with eight rotors in a cross configuration. The nonlinear model and a linearized model for the use in controller development are described. For the following discussion, the axes of the Octo-rotor vehicle are denoted as (x, y, z) and are defined with respect to the vehicle as shown in Figure 1. Roll, pitch, and yaw are defined as the angles of rotation about the x, y, and = axis, respectively. The global workspace axes are denoted as (X,Y,Z) and are defined with the same orientation as the Octo-rotor sitting upright on the ground. Figure 1 shows the orientation and axes of the Octo-rotor MAVs. The direction of the arrow in diagram indicates the rotation direction of the motor/propeller.

The equations of motion for a multi-rotor helicopter, as described in [9] and [10], are shown in Eq. 1 with the notation shown in Table 1. These equations are derived from first principles and are applicable to a general multi-rotor helicopter. They neglect any aerodynamic effects and do not include any external disturbances.

$$\begin{cases} \ddot{x} &= (cos\phi sin\theta cos\psi + sin\phi sin\psi)\frac{1}{m}U_1 \\ \ddot{y} &= (cos\phi sin\theta sin\psi - sin\phi cos\psi)\frac{1}{m}U_1 \\ \ddot{z} &= -g + (cos\phi cos\theta)\frac{1}{m}U_1 \\ \ddot{\phi} &= \dot{\theta}\dot{\psi}(\frac{I_y - I_z}{I_x}) - \frac{J_r}{I_x}\dot{\theta}\Omega + \frac{l}{I_x}U_2 \\ \ddot{\theta} &= \dot{\phi}\dot{\psi}(\frac{I_z - I_x}{I_y}) + \frac{J_r}{I_y}\dot{\phi}\Omega + \frac{l}{I_y}U_3 \\ \ddot{\psi} &= \dot{\phi}\dot{\theta}(\frac{I_x - I_y}{I_z}) + \frac{1}{I_z}U_4 \end{cases} \quad (1)$$

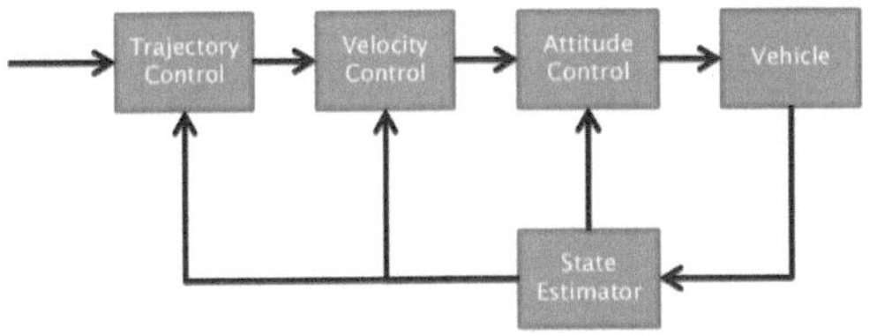

Figure 2. Overall Control System Architectur.

From these equations, it can be seen that the linear dynamics depend upon the vehicle's attitude, the total thrust produced by the vehicle and the vehicle's mass. The rotational dynamics depend on the moments of inertia of the vehicle and rotors as well as the rotor speeds and torques, but do not depend upon the linear state of the vehicle.

Since the rotational dynamics can be considered independently of the linear dynamics, we will first develop a model that describes the state evolution for the rotational subsystem. Since this work is designed to be implemented on an off-the-shelf helicopter, we assume that the vehicle contains an existing attitude controller. This assumption further complicates the rotational dynamics of the system, since we do not know the dynamics of the attitude controller. This means that the attitude dynamics are dependent on several unknown parameters including the physical properties of the vehicle and the unknown dynamics of the low-level control system.

The vehicle is assumed to have an attitude controller embedded in it that will take as commands a desired roll, pitch, yaw rate and total thrust. A cascaded control approach is taken where the inner control loop is a velocity controller that accepts a 3-dimensional velocity command and produces commands applied to the existing low-level attitude controller. The outer loop of the control system achieves trajectory tracking. This loop receives a desired path from the planner and issues commands to the velocity controller to track the path. Figure 2 shows the overall system diagram.

4 STATE ESTIMATION

Section 3 describes a control approach which can be applied to stabilize a helicopter in flight, but it assumes that the vehicle's state can be measured. And in that section the dynamics of a state-estimator is considered, but do not discuss the implementation of such a estimator.

There are several things that must be considered in designing a state-estimation system that will provide feedback for the controllers described in Section 3. The selected method must not require any heavy sensors, the associated algorithms must be able to run in real-time on board the vehicle and it must be robust. Visual odometry is selected as the primary source of state feedback for this vehicle. It is a good choice since the cameras required can be very light and it functions well in an outdoor environment.

Figure 3. Octo-rotor Aircraft of Draganflyer X8.

Vision-based solutions are not ideal since they are often very computationally expensive. This algorithm must execute in real-time, so it must have a significant amount of computational resources allocated to it.

The specific algorithm used for visual odometry is described in [11]. It is a stereobased, frame-to-frame approach. The algorithm extracts corner-like features from two consecutive stereo image pairs. It then matches these features across all four images using gradient block-matching. The matched features are filtered through a spatial bucketing technique which removes some matches in an attempt to distribute the matches evenly over the field of view. Following this, the matches are put through a RANSAC-based outlier rejection process. The resulting inliers are used to calculate the rotation and translation of the vehicle over the time between consecutive frames.

Since visual odometry is a frame-to-frame technique, it has the inherent drawback of allowing the state estimate of the vehicle to drift over time. While this problem is unavoidable in this type of algorithm, the state estimation system will not be useful if this drift rate is too high.

5 TEST RESULTS

The vehicle selected to test the algorithms described above is the Dragan flyer X8, an eight-rotor helicopter shown in Figure 3.

This vehicle has one 16000 mAh 6-cell lithium polymer batteries on board. These batteries provide the vehicle with a 20 minute flight time while carrying a 1.0 kg payload. The cameras are computer-vision grade, global-shutter, high dynamic range CMOS devices. The video device VCP can compress and process video information which can be transmitted to controller via CAN bus.

The INS has an accelerometer, gyroscope, magnetometer, and L1 band GPS receiver. The GPS was not used during any of these experiments. The inertial sensors are all MEMS devices and are relatively low performance. The IMU measurements from this device are available at 50 Hz.

With the control system from Section 3, the state estimation system from Section 4 and the vehicle described in this section, a full-closed loop trajectory tracking system is possible.

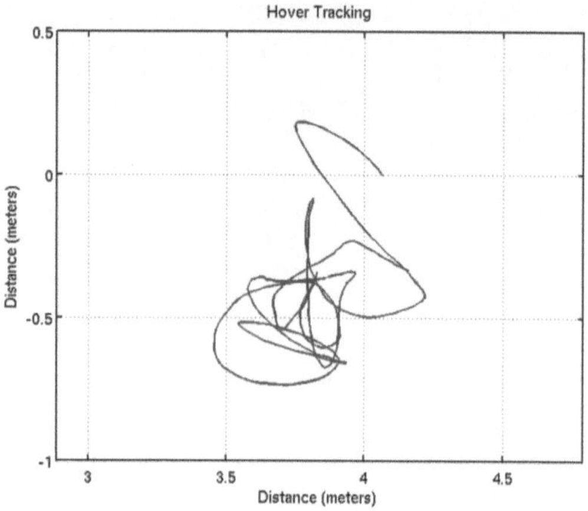

Figure 4. Hover Tracking Performance.

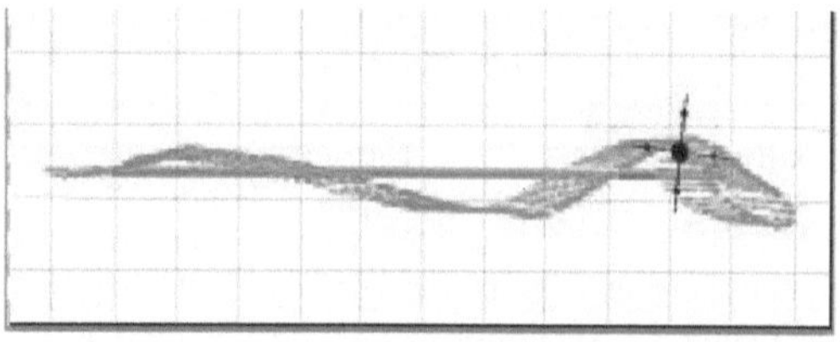

Figure 5. Linear Trajectory Tracking Performance (1 meter grid)

Figure 4 shows the hover performance of the aircraft. This test was conducted by providing the trajectory controller with a path that consists of a single waypoint with a desired velocity of zero. During this 45 second test, the maximum deviation of the vehicle from the set point was less than one meter.

Figure 5 shows the vehicle's performance when commanded to track a straight line trajectory. The desired trajectory is shown in blue, and the vehicles actual rajectory is shown in red with position at the base of the arrow, heading indicated by the direction of the arrow. This trajectory is 10 meters in length and was traversed at 1.5 meters per second. During this test, the cross-track error never exceeded 1 meter.

The total closed-loop performance of a system such as the one described and tested above is affected by the dynamics of all of the constituent systems. The primary limiting factor to the closed-loop performance of the system described here is the latency of the state-estimation algorithm. This latency is about 130 ms, and the controllers must be tuned to allow for this latency without going unstable. In order to prevent instability, the controllers are tuned to be very conservative. While this is necessary to allow for the state-estimation dynamics, it is detrimental to the closed-loop performance of the system.

Since the state-estimation dynamics are included in the vehicle simulation, very similar effects are observed in the simulation environment. If the state-estimation delay is reduced, the vehicle's simulated performance can be improved significantly by tuning the controllers to be more aggressive.

6 CONCLUSION

The work presented in this report shows a method for controlling a small multi-rotor helicopter based on visual odometry state estimation. This techniques described are demonstrated on a small helicopter with a light payload. A novel hardware solution enabling real-time execution of the algorithms necessary for visual odometry is described and demonstrated.

The control approach presented is very general and can be easily applied to any new MAVs of this type. The controllers are deigned with robustness in mind and are able to handle noise and latency from the state estimation system. The controller can maintain a cross-track error of less than 1 meter while traveling at 1.5 meters per second while tolerating 130ms of state estimation latency. The dynamic modeling technique described requires minimal knowledge of the physical parameters of the system, and instead develops a model based upon experimental data collected from the vehicle in flight.

The computing architecture presented allows for the computationally expensive algorithms required for visual odometry to be moved off of the main computer. This architecture enables never before possible computation per gram that enables vision-based state estimation on a class of vehicles that was not previously thought possible.

ACKNOWLEDGEMENT

This work was supported by the Guangxi Key Laboratory of Automatic Detecting Technology and Instruments (YQ14203). Thanks to the project "Research on the construction and error compensated of the strapdown AHRS based on Multi-Sensor Microsystems", which was supported by national natural science foundation of Guilin University of Aerospace Technology (YJI303).

REFERENCES

[1] M. Achtelik, A. Bachrach, R. He, S. Prentice, and N. Roy, Stereo vision and laser odometry for autonomous helicopters in GPS-denied indoor environments, in *Proceedings of the SPIE Unmanned Systems Technology XI*, vol. 7332, (Orlando, F), 2009.

[2] A. S. Huang, A. Bachrach, P. Henry, M. Krainin, D. Maturana, D. Fox, and N. Roy, Visual odometry and mapping for autonomous flight using an rgb-d camera, in Int. *Symposium on Robotics Research (ISRR)*, (Flagstaff, Arizona, USA), Aug. 2011.

[3] L. D. Minh and C. Ha, Modeling and control of quadrotor mav using visionbased measurement, in

Strategic Technology (IFOST), 2010 International Forum on, pp. 70–75, Oct. 2010.

[4] L. Meier, P. Tanskanen, F. Fraundorfer, and M. Pollefeys, Pixhawk: A system for autonomous flight using onboard computer vision, in *Robotics and Automation (ICRA)*, 2011 IEEE International Conference on, pp. 2992–2997, May 2011.

[5] A. Wu, E. Johnson, and A. Proctor, Vision-aided inertial navigation for flight control, in *2005 AIAA* Guidance, Navigation, and Control Conference and Exhibit, pp. 1–13, 2005.

[6] M. Blösch, S. Weiss, D. Scaramuzza, and R. Siegwart, Vision based mav navigation in unknown and unstructured environments, in *Robotics and Automation (ICRA)*, 2010 IEEE International Conference on, pp. 21–28, May 2010.

[7] G. Conte and P. Doherty, An integrated uav navigation system based on aerial image matching, in *Aerospace Conference*, 2008 IEEE, pp. 1–10, March 2008.

[8] R. Voigt, J. Nikolic, C. Hurzeler, S. Weiss, L. Kneip, and R. Siegwart, Robust embedded egomotion estimation, in *Intelligent Robots and Systems (IROS)*, 2011 IEEE/RSJ International Conference on, pp. 2694–2699, sept. 2011.

[9] S. Bouabdallah, Design and control of an indoor micro quadrotor, in *In Proc. of Int.* Conf. on Robotics and Automation, 2004.

[10] P. Pounds, R. Mahony, P. Hynes, and J. Roberts, Design of a four-rotor aerial robot, in *Proc.* 2002 Australasian Conference on Robotics and Automation, vol. 27, p. 29, 2002.

[11] B. Kitt, A. Geiger, and H. Lategahn, Visual odometry based on stereo image sequences with ransac-based outlier rejection scheme, in *Intelligent Vehicles Symposium (IV)*, 2010 IEEE, pp. 486–492, IEEE, 2010.

Future Mechatronics and Automation – Yang (Ed.)
© 2015 Taylor & Francis Group, London, ISBN 978-1-138-02648-3

Tow tension controller for robotic automated fiber placement based on fuzzy parameter self-adjusting PID

Jie Chen & Yugang Duan
State Key Laboratory for Manufacturing Systems Engineering, Xi'an Jiaotong University, Xi'an, Shaanxi Province, P.R. China

ABSTRACT: Robotic automated fiber placement (Robotic AFP) was a cost-effective and highly innovative approach to produce large and complex composite structures. In order to achieve desired qualities, the tow tension of the process required to be controlled accurately. Due to the high nonlinearity of the system, such as the large elastic modulus, flexibility and viscosity of the tow, traditional methods failed to work effectively. Afuzzy parameter self-adjusting PID combining fuzzy logic and PID together was proposed in this work. Fuzzy logic was able to respond quickly to disturbances without the need for an accurate model and PID control could eliminate the steady-state error. With this method, the tow tension could be precisely regulated thus improved the quality of the composite structures.

Keywords: Robotic automated fiber placement, tow tension, fuzzy parameter self-adjusting PID, fuzzy logic.

1 INTRODUCTION

Composites were carefully designed materials suitable for specific applications and have gained intense popularity in products that needed to be strong enough but lightweight in order to bear harsh working conditions. Due to their attractive stiffness-to-weight and strength-to-weight ratios, fiber reinforced composites offered many advantages for aerospace, automotive and marine industries. Traditional approaches of producing fiber reinforced composites included filament winding techniques, manual hand lay-up and tape-laying. But all of these methods were time-consuming, labor intensive and will generate high levels of scrap material, whereas the current industries required more automated and cost-effective processes [1]. Robotic automated fiber placement (Robotic AFP) successfully addressed these industrial requirements and provided high flexibility and output rate. Moreover, it could also reduce the scrap of materials and was quite suitable for manufacturing large and complex composite structures.

Since early 1980s, fiber placement methods have emerged at some companies from ordinary lab machines to fuselage production on aircrafts. Grant et al. [4] provided a general review of automated composite processing methods that were currently being used to fabricate aircraft structures and presented a detailed description of the automated tape layer process and the fiber placement process. Shirinzadeh et al. [5] presented the overall strategy for the establishment of robotic automated fiber placement facilities and described the methodology for developing process planning, programming and simulation. In 2006, Tierney et al. [6] presented a model for predicting through-thickness heat transfer and bond strength development based on intimate contact and healing at the ply interface. In order to figure out optimum process parameters for automated thermoplastic tow-placement systems, Heider et al. [7] demonstrated the use of online optimization algorithms based on artificial neural networks. And the method derived a model for material quality as a function of process parameters.

The control of the tow tension in AFP process was an important issue. Without the appropriate tension, the tow would either wind (when the tow tension was too small) or get damaged due to tensile deformation (when the tow tension was too large). Furthermore, AFP process required fast, accurate and independent tow tension controller in order to increase the production and improve the quality. But due to the large elastic modulus, flexibility and viscosity of the tow, traditional control approaches, such as PID, failed to accomplish this task. In this work, a novel tension controller for AFP process was developed based on fuzzy parameter self-adjusting PID. In section II, the fuzzy parameter self-adjusting PID was discussed in detail. In section III, the experimental setup for the whole tension control system was established based on FANUC six degrees of freedom manipulator and ARM-based microprocessor. The experimental results were discussed and analyzed as well. And in section IV, the conclusion was drawn.

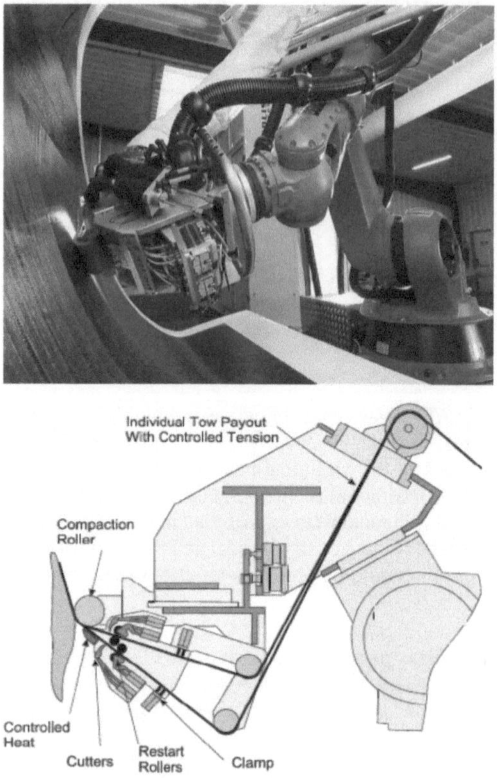

Figure 1. 6-axis robotic platform made by KUKA robots and the fiber placement head [2] [3].

2 FUZZY PARAMETER SELF-ADJUSTING PID

In this section, a novel control method combining fuzzy logic and traditional PID was discussed in detail. Fuzzy control had the advantages of fast response and high robustness without the need for a precise system model. But its steady-state accuracy was relatively low. Traditional PID could compensate fuzzy control in this regard. Therefore, these two methods were combined together to build a fuzzy parameter self-adjusting PID in this paper. In order to improve the system performance, a simplified mathematical model for the robotic AFP tension system was derived firstly. Moreover, a simulation model of the fuzzy parameter self-adjusting PID was established and analyzed in SIMULINK-MATLAB.

2.1 *Simplified mathematical model for the tension system*

A schematic of the mechanical model for the fiber roll in robotic AFP was depicted in Fig. 2.

According to the principle of torque balance, the following equation can be acquired [8]:

$$T(t)R(t) = M(t) + J(t)\dot{\omega}(t) \quad (1)$$

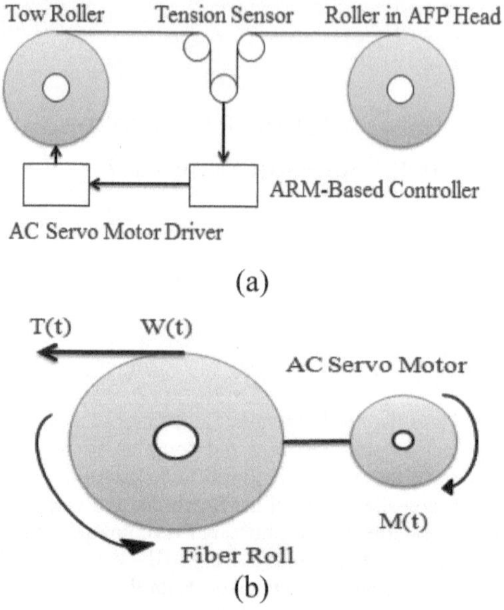

Figure 2. (a) Schematic for the robotic AFP; (b) Mechanical model for the tow roll.

Eq. (2)–(5) can be derived based on AC servo motor characteristics and circuit theories:

$$M(t) = K_1\Phi I \quad (2)$$

$$\varepsilon_1 = -K_2\omega(t) \quad (3)$$

$$\varepsilon_2 = -L\dot{I} \quad (4)$$

$$I = \frac{\varepsilon_1+\varepsilon_2+U}{r} \quad (5)$$

where T was the fiber tension, R was the radius of the roll, M was the resistance moment of the AC servo motor, J was the inertia moment of the roll, ω was the angular velocity of the roll, K_1 was motor structure coefficient K_2 was Back-EMF coefficient of the motor, Φ was the flux, I was the armature current, ε_1 was the Back-EMF, ε_2 was EMF of the motor, U was the input motor control voltage and r was the resistance of the motor.

Combining (2)–(5), Eq. (1) can be simplified to the following expression:

$$T(t) = \frac{K_1\Phi(u-K_2\omega(t)-L\dot{I_d})+J(t)r\omega(t)}{rR(t)} = \frac{K_1\Phi(30u-K_2\pi n(t)-30L\dot{I_d})+J(t)r\pi n(t)}{30rR(t)} \quad (6)$$

2.2 *Simulation model of the fuzzy parameter self-adjusting pid in simulink-matlab*

Due to its simple structure, high stability and reliability, traditional PID controller has been widely used to control industrial processes. But it was mainly applied

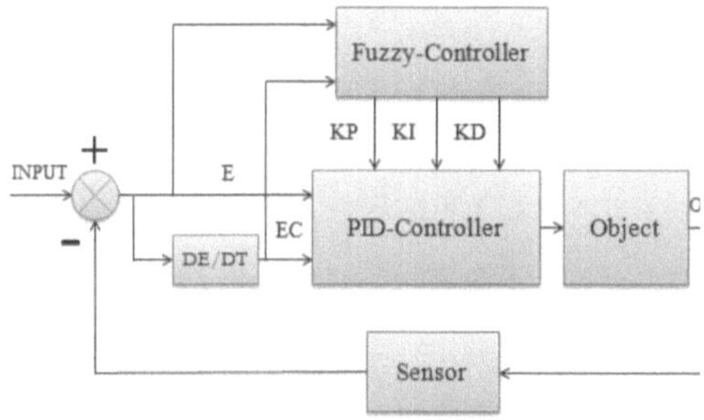

Figure 3. The novel controller combining fuzzy logic and PID.

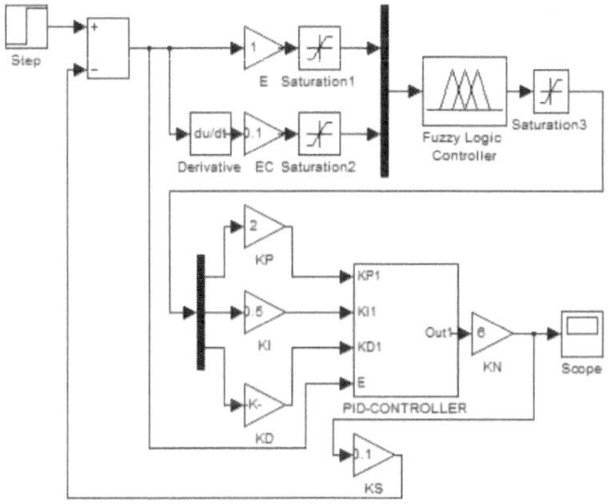

Figure 4. Simulation model for the control system in SIMULINK-MATLAB.

to the linear systems with exact mathematical models, and failed to achieve the desired control goals for the nonlinear, large delay and time-varying uncertain systems. The tension control system in robotic AFP was such a typical nonlinear system that traditional PID failed to accomplish this task. In this section, fuzzy logic was used to adjust the parameters of PID controller online [9], namely scale coefficient KP, integral coefficient KI and differential coefficient KD. Fig.3 showed a schematic of the novel controller which combined fuzzy logic and traditional PID together.

A simulation model for the control system in SIMULINK-MATLAB was depicted in Fig. 4. In next section, the proposed model was used to build an ARM-based controller and the experimental results were shown as well.

3 EXPERIMENTAL SECTION

3.1 *Experimental setup.*

A tension sensor was applied to measure the fiber tension, a Panasonic AC servo motor was used to drive the fiber roll, and in order to execute the algorithm an ARM-based microprocessor was used with its input connected to the sensor and its output connected with

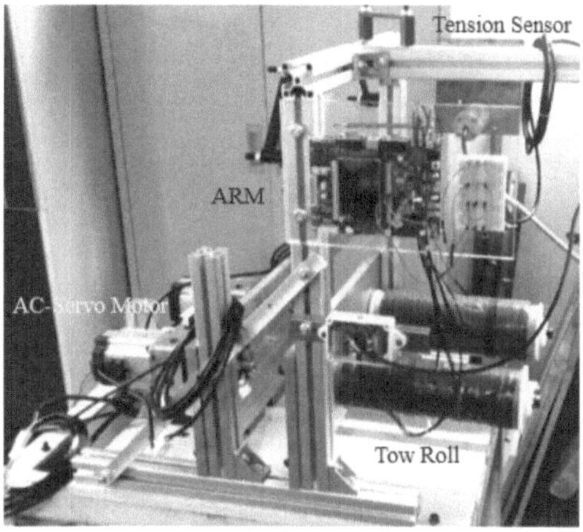

Figure 5. Experimental setup for the tension control system.

the motor driver Moreover, a LCD screen was installed to monitor the process parameters. A photo of the whole experimental setup was shown in Fig. 5.

In Fig. 1(a), when the process began, the tow roller rotated to release the tows and the roller on the AFP head rotated as well to collect the tows and place them on the mold. The 3-wheel tension sensor measured the tow tension with its output connected to the input of the ARM based controller. The output of the controller was used to control the AC servo motor thus regulated the tow tension.

3.2 *Discussion and analysis of the experimental data*

It could be seen from Fig.6(a) that a Sudden change happened around 10ms causing the fiber tension to change from 10 N to 4 N. Due to the fuzzy-adaptive PID controller, the tension was regulated to about 10 N within 80 ms. Fig. 6(b) showed that the tension maintained around 10 N after the sudden change disappeared which proved the stability of the control system. Fig. 7 showed the real time fiber tension during the placement process by PID. It could be clearly seen that traditional PID failed to regulate the tension.

4 SUMMARY

Composite materials have been widely used in many industry sectors due to their superior properties, such as attractive strength-to-weight and stiffness-to weight ratios over the traditional structural materials, thus suitable for many applications in the aerospace and automotive industries. Robotic automated fiber placement was an innovative and highly automatic method in composites manufacturing, in order to achieve desired qualities, the tow tension needed to be controlled properly. Due to the high nonlinearity of the system, such as the large elastic modulus, flexibility and viscosity of the tow, traditional methods failed to

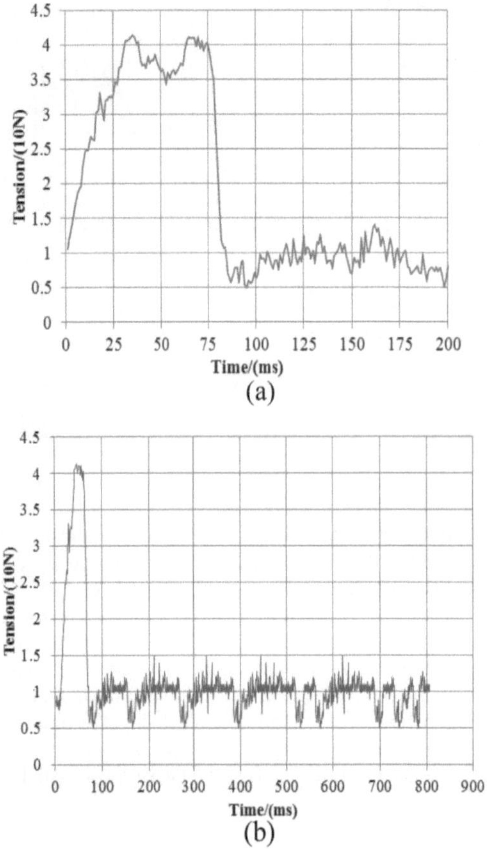

Figure 6. Real time fiber tension during the placement process by fuzzy parameter self-adjusting PID.

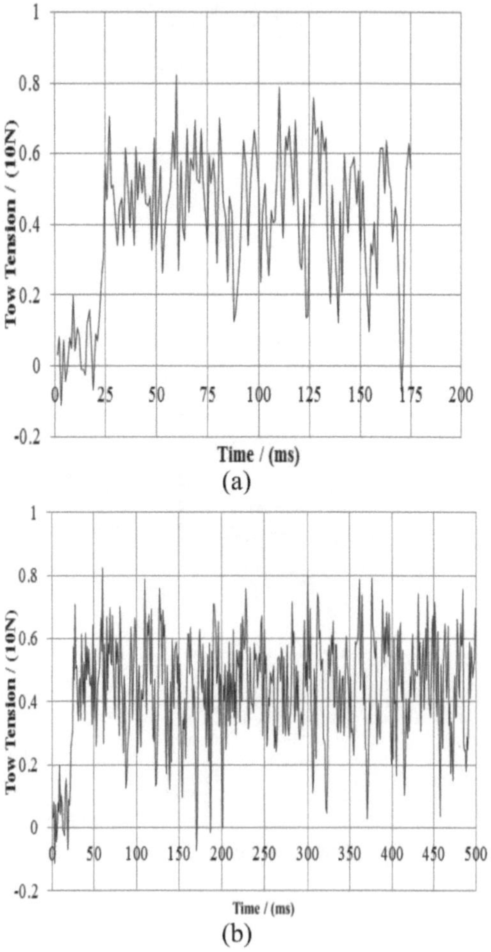

Figure 7. Real time fiber tension during the placement process by PID.

work effectively.In this paper, a fuzzy parameter self-adjusting PID, which combined PID and fuzzy logic together, was proposed to solve the problem. Fuzzy logic had the advantage of fast response without the need for an accurate mathematical model, and PID was able to eliminate the steady-state error. Experimental results showed that this novel control strategy had high stability, accuracy and responded fast to disturbances. Therefore, the conclusion could be safely drawn that, with the proposed method in this work it was able to precisely control the fiber tension and thus improved the quality of the composite products.

ACKNOWLEDGEMENTS

This research was supported by NCET-11-0419, Program for New Century Excellent Talent in University, National High Technology Research and Development Program 863 [2012AA040209], and was also supported by National Major Projects Machine Tool [2014ZX04001091].

CORRESPONDING AUTHOR

Yugang Duan, Email: ygduan@mail.xjtu.edu.cn, Tel: (+86)13609187679.

REFERENCES

[1] Shirinzadeh B, Alici G, Foong CW, Cassidy G, Fabrication process of open surfaces by robotic fibre placement, Robotics and Computer-Integrated Manufacturing, 2004, 20:17–28.

[2] D. Eva, "Automated processing of aerospace composite components," World Wide Web: <http://www.cincinnatilamb. com>, Nov. 2004.

[3] Lorient, "Fiber placement robotic cell," World Wide Web: <http://www.coriolis-composites.com/technologie.php>, 2014.

[4] Grant C, Automated processes for composite aircraft structure, Industrial Robot: An International Journal, 2006, 33(2): 117–121.

[5] Shirinzadeh B, Foong CW, Tan BH, Robotic fibre placement process planning and control, Assembly Automation, 2000, 20(4):313–320.
[6] Tierney J, Gillespie Jr, Modelling of in situ strength development for the thermoplastic composite tow placement process,Journal of Composite Materials, 2006, 40(16): 1487–1506.
[7] Heider D, Piovoso MJ, Gillespie JW Jr, A neural network model-based open-loop optimization for the automated thermoplastic composite tow-placement system, Composites A, 2003, 34(8): 791–799.
[8] Ren Shengle, Wang Yongzhang, Lu Hua, Su Guosen, A precision tension control system based on PIC, Materials Science Forum, 2006, 532~533: 97~100.
[9] Woo Z W, Chung H Y, Lin J J, A PID type fuzzy controller with self-tuning scaling factors, Fuzzy Sets and Systems, 2000, 115(2): 321–326.

Future Mechatronics and Automation – Yang (Ed.)
© 2015 Taylor & Francis Group, London, ISBN 978-1-138-02648-3

The research and design of an internal cooling control system for plastic film production based on Cortex M3

Hua Guo
College of Electronics, Communication and Physics, Shandong University of Science and Technology, Qingdao, Shandong, China

Sheng-Wen Yu
College of Geomatics, Shandong University of Science and Technology, Qingdao, Shandong, China

ABSTRACT: According to the requirements of the internal cooling system in plastic film production, based on the analysis of the internal cooling principle and the deficiencies of existing internal cooling systems, a single chip solution based on STM32F103 has been proposed, which adopts the Proportion, Integration, and Differentiation (PID) algorithm. This realized high precision control of the film bubble size in the whole production process and has been tested in more than 50 greenhouse film producers for a long time; the actual control precision can be up to±1mm, and get the user acceptance.

Keywords: Greenhouse film; internal cooling; PID; ARM; STM32F103

1 INTRODUCTION

Plastic film has plays an important role in agricultural production in China. The internal cooling system is one of the key methods used to improve production efficiency and product quality of the film. In foreign countries, the internal cooling technology has been applied to practical film production in the last century. Similar techniques have been successfully developed in 2003 in China, and rapidly popularized. However, in practice we found that the domestic internal cooling system has the disadvantages of poor control precision, long adjustment period, lack of stabilization, complicated operation and other shortcomings. For stringent specifications, these disadvantages often make it difficult to meet the requirements, which results in large losses. Therefore, the development of a high control accuracy, long-term, stable internal cooling system is necessary.

2 FUNCTIONAL ANALYSIS OF THE INTERNAL COOLING SYSTEM

As shown in Figure 1, the internal cooling system is mainly composed of a microcontroller, two frequency converters, two fans, and the ultrasonic ranging sensors (ultrasonic probe). One of the frequency converters and fans is used to generate wind to cool the film; the other one is used to suck out high temperature waste air from the film bubble, using the ultrasonic probe. The microcontroller obtains the current bubble size and compares it with the production specification, according to a control algorithm, it then changes the running speed of the fans by changing the frequency converters output frequency in order to keep the film bubble size stable at the required specification size. In order to maintain stability in this stage, it is vital to ensure the dynamic balance of the air volume that is pumped in and sucked out. A stable bubble has a significant impact on the film's finish, transparency, thickness and the uniformity of tensile strength, because at high production speeds, the film needs rapid cooling down from the high temperature, if the internal cooling system were not stable, this would lead to instability of the film quality.

3 DESIGN OF INTERNAL COOLING CONTROL SYSTEM

3.1 *Three ultrasonic probe structures*

Bubble size can be obtained through the ultrasonic probe; obtaining the actual size of the bubble is the foundation of the whole control system.

This scheme adopts three probes to detect the size of the bubble, as shown in Figure 2. The probes' positions have been equidistantly distributed on the XY axis with the origin at the centre of the mould. Under ideal condition, the bubble's radius R can be calculated using the following formula:

$$R= |OA|-|MA|;\ R= |OB|-|NB|;\ R= |OC|-|PC|;\ R= (|CA|-|CP|-|MA|)/2; \tag{1}$$

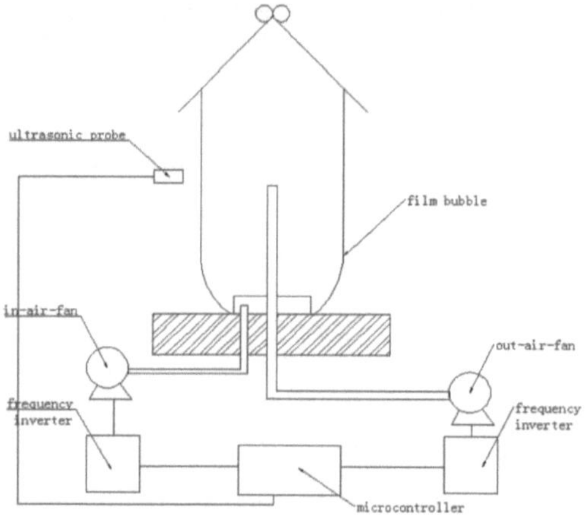

Figure 1. Diagram of the internal cooling control system structure.

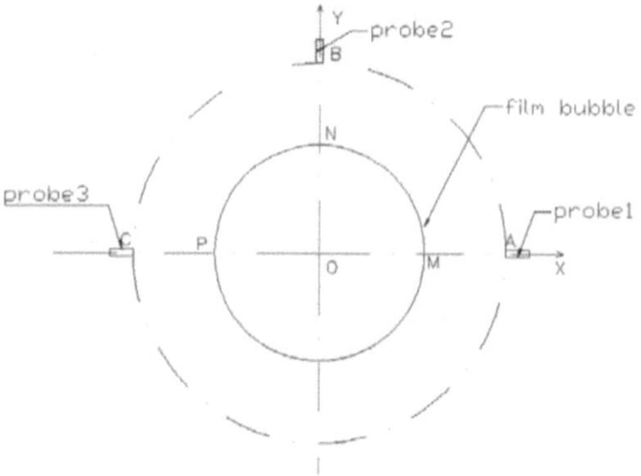

Figure 2. Ideal state in vertical view.

where |MA|, |NB| and |PC| can be measured by the ultrasonic probe, which are known values. |OA| = |OB| = |OC|, which are the distances from ultrasonic probes to the mold centre O, which are determined when the probes are installed; these are known values too.

However, in the actual production process, due to the existence of installation errors, different raw plastic materials, and the effect of natural wind, the film bubble centre will deviate from the mold centre and dynamically change. Figure 3 depicts a common scenario. In this case, formula (1) is no longer suitable; the calculation method for the actual bubble radius R can be deduced as follows:

By the Pythagorean Theorem:

$$a=|N'P'|=\sqrt{|P'O|^2+|N'O|^2}$$
$$b=|N'M|=\sqrt{|N'O|^2+|M'O|^2} \qquad (2)$$
$$c=|P'M'|=|P'O|+|M'O|$$

According to the geometrical relationship:

$$|N'O|=|BO|-|BN'|$$
$$|P'O|=|CO|-|CP'| \qquad (3)$$
$$|M'O|=|AO|-|AM'|$$

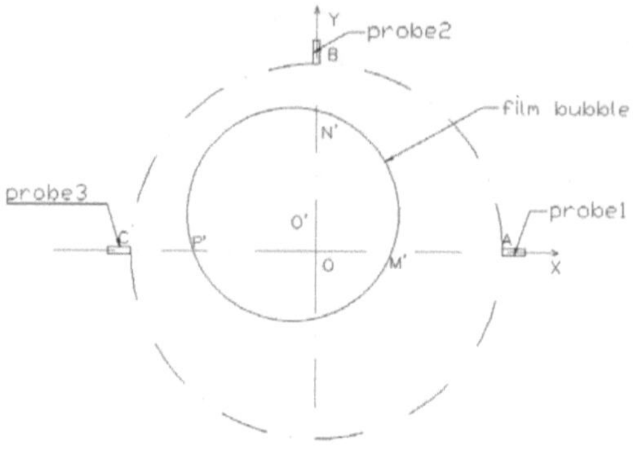

Figure 3. Actual state in vertical view.

where |OA| = |OB| = |OC| is the distance from the ultrasonic probe to the centre of the mould; these are known values and |BN'|, |CP'|, |AM'| can be measured by the ultrasonic probes as known values too. Put all these known values into the three formulas in (2), and a, b, and c can be worked out. From the cosine theorem of $a^2 = b^2 + c^2 - 2bc * cosA$, cosA can be calculated then we can get sinA = 1-cosA.

By the sine theorem:

$$a^2=b^2+c^2-2bc*cosA$$

R is the radius of the circumscribed circle and the bubble's radius is equal to the radius of the circumscribed circle:

$$R=a/2sinA \qquad (4)$$

Through the above derivation, no matter how the bubble's centre changes position, its real radius R can be accurately calculated.

3.2 *The hardware design of the control system*

As shown in Figure 4, the design adopts the 32bit ARM CortexT-M3 CPU STM32F103 as the core processor, which runs at speed up to 72 MHZ, with single-cycle hardware multiply and divide, integrated 16 channel 12 bit 1M/s analogue to digital (AD) converter, dual 12 bit digital-to-analogue (DA) converter and up to 4 Mbit/s of the Universtall Asynchronous Receiver/Transmitter(UART) serial communication interface, These specifications fully meet the design requirements. In the actual operation environment, a large number of high power motors and frequency converters produce strong electromagnetic interference, the system therefore uses a two-way isolation power supply and uses the isolation op-amp to isolate the signal in the input and output circuit, in order to ensure the stability of the whole system. The signal flow is as follows: first, the three-way "0-10V" ultrasonic signal is selected by the multichannel switcher– CD4051, then the signal voltage will have been converted to the range of 0-2.5 V by the isolation op-amp circuits, finally it enters the

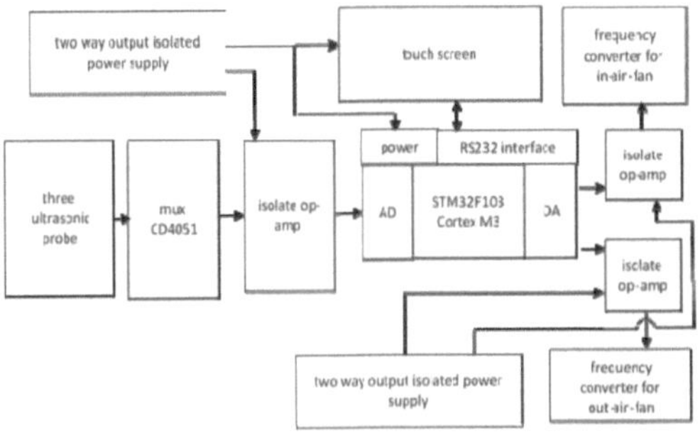

Figure 4. Principle diagram of the hardware design.

STM32F103's AD conversion process. The dual DA conversion, which was also converted from the range "0-2.5V" to standard industrial control signal range in "0-10V" by the isolated op-amp circuits, and respectively connected to the two frequency converters for controlling the two fans' running speed.

4 SOFTWARE DESIGN OF THE INTERNAL COOLING CONTROL SYSTEM

Based on an analysis of the internal cooling system's features, The most important thing is to ensure that the air entering and exiting the bubble always maintains a balance. This homeostasis plays a decisive role to keep the bubble at a specified size. If the bubble is too large or too small, an adjustment in the in and out of the air in a timely manner should make it return to the set size. Meanwhile, the cooling effect must be stable in the whole production process. To achieve this goal, this paper proposes a method that keeps the in air fan's running speed unchanged, only adjusting the out air fan's running speed to keep the bubble stable. Because in this method the pumped in air speed is constant, it ensures the consistency of the cooling effect.

In the field of industrial control, PID (Proportion, Integration, and Differentiation) is one of the most widely used automatic controllers. For the control objects of a typical process – "first-order lag + pure lag" and "second-order lag + pure lag" control object, PID is the optimal controller. PID's regulation method is an effective way to maintain continuous dynamic quality correction. In the film production process, dynamically adjusting the size of the bubble is a typical lag control object. The system therefore uses PID as the core algorithm.

The program flowchart of the whole system is shown in Figure 5. First, the system operating parameters, such as P, I, D, the bubble's setting size, the initial values of the two-way DA converters, are initialized in the initialization parameter section. Second, according to the DA converters' initial values, start the one way DA output, control the output frequency of the frequency conversion, and make the pump in air fan run steady. Third, the microprocessor samples the three ultrasonic probes' current values in order to avoid

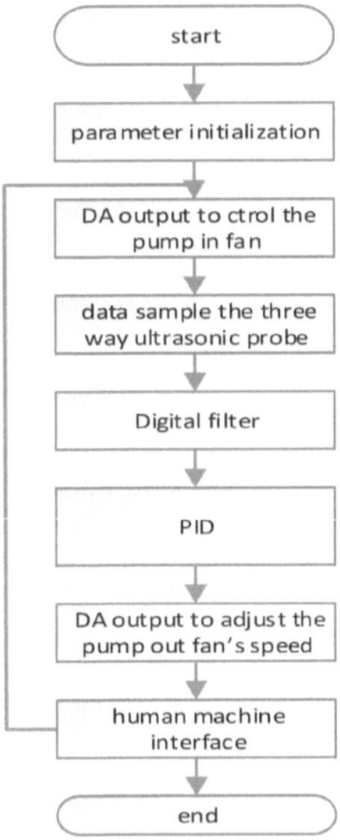

Figure 5. Flowchart of the whole system.

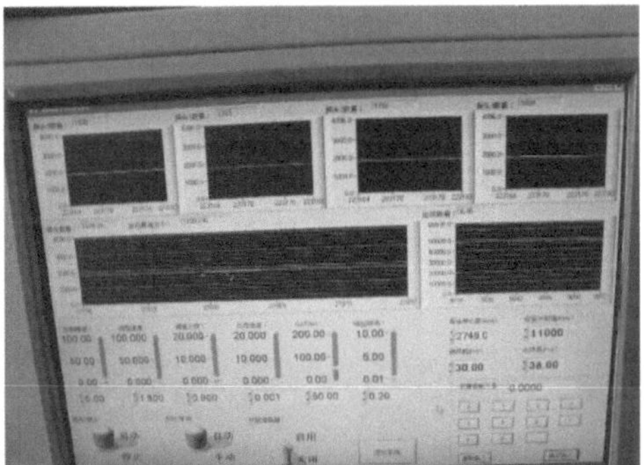

Figure 6. Human operation interface.

interference effects; the sampled data is further digitally filtered. Finally, the microprocessor calculates the output data based on the PID algorithm. The data will have been output through the other way DA converter to control the pump out air fan's running speed, in order to realize the dynamic adjustment of the bubble size.

5 TEST AND SUMMARY

The system has been continuously tested by more than fifty greenhouse film manufacturers in China for two years. Figure 7 shows the actual produced film; from the picture, it can be seen that the film surface is

Figure 7. The actual production effect.

smooth and neat, indicating that the bubble size was stable. According to manufacturers' feedback, the system control precision is 5 to 10 times higher than the traditional system that used Programmable Logic Controller (PLC). The error of the diameter of bubble can be controlled within ±1 mm; the production can be directly controlled to achieve the required standards to export to Europe and the United States. The system has a fast adjustable speed and automatic control, which greatly reduces the waste production. In addition, the system users unanimously agree that the system has a long-term stable operation, an intuitive and friendly human-computer interaction interface, and is simple to operate. It is, therefore, worthy of popularization and application.

CORRESPONDING AUTHOR

Name: Hua Guo; Email: stone_strong@163.com; Mobile phone: +86-13954287475

REFERENCES

Guo Chan-chan, Zhou Nan-qiao, Yong, Peng Xiang-fang. Internal Cooling System for Blown Film Bubble [J]. Plastics, 2003, 32(5): 41–44.

Guo Hua. Internal cooling control system and method for plastic film production [P]. China: ZL201010227802.8, 2013.4.24.

Song Yuan-bin. Design and Implementation of a Cooler of Data Acquisition and Control System based on STM32 [D]. Dalian, China: Dalian University of Technology, 2013.

Wang Lei, Song Wen-zhong. PID Control [J]. Process automation instrumentation, 2004, (4): 1–5.

XIAO Qian-jun. Multi-parameter data acquisition card design based on STM32 and MODBUS protocol [J]. Manufacturing Automation, 2010, 32(12): 205–208

Zhang Jiao-jiao, Cao Sen, Guo jian-yi, Su Guo-hua. The design of device data acquisition system based on STM32F103 [J]. Equipment Manufacturing Technology, 2012, (7): 307–311.

Zhao Xiao-xiao. Research on Fuzzy PID Control Method Combined Fuzzy Theory and Conventional PID Control [J]. Shandong Electric Power, 2009, (6): 54–56.

Future Mechatronics and Automation – Yang (Ed.)
© 2015 Taylor & Francis Group, London, ISBN 978-1-138-02648-3

Author index